# BIOCHEMISTRY AND METABOLISM

# BIOCHEMISTRY AND METABOLISM

**Prof. Harmeet Kour**
Deptt. of Botany, Govt. P.G. College for Women,
Gandhinagar, Jammu

**Prof. Veena Tandon**
Deptt. of Botany, Govt. P.G. College for Women,
Gandhinagar, Jammu

**ANMOL PUBLICATIONS PVT. LTD.**
**NEW DELHI - 110 002 (INDIA)**

**ANMOL PUBLICATIONS PVT. LTD.**
**H.O.:** 4374/4B, Ansari Road, Daryaganj,
New Delhi-110 002 (India)
Ph.: 23278000, 23261597
**B.O.:** No. 1015, Ist Main Road, BSK IIIrd Stage
IIIrd Phase, IIIrd Block,
Bangalore - 560 085 (India)
*Visit us at*: www.anmolpublications.com

*Biochemistry and Metabolism*

First Published, 2009
ISBN 978-81-261-4055-8



PRINTED IN INDIA

Printed at Balaji Offset, Delhi.

# Contents

# Preface

Biochemistry is the chemistry of life. Biochemists study the elements, compounds and chemical reactions that are controlled by enzymes that are taken place in all living organisms. The findings of biochemistry are used in many areas, from genetics to molecular biology and from agriculture to medicine.

It provides a logical and accessible introduction to biochemical principles. Emphasis is placed on the relationship between molecular architecture and the functional properties of biomolecules, and the dynamic, unceasing, and self-regulating nature of living processes.

## Chapter 1

# Water and Its Properties

### WATER CHEMISTRY IN FRESHWATER AQUACULTURE

Water quality determines not only how well fish will grow in an aquaculture operation, but whether or not they survive. Fish influence water quality through processes like nitrogen metabolism and respiration. Knowledge of testing procedures and interpretation of results are important to the fish farmer. Some water quality factors are more likely to be involved with fish losses such as dissolved oxygen, temperature and ammonia. Others, such as pH, alkalinity, hardness and clarity affect fish, but usually are not directly toxic. Each water quality factor interacts with and influences other parameters, sometimes in complex ways. What may be toxic and cause mortalities in one situation, can be harmless in another. The importance of each factor, the determination method and frequency of monitoring depends upon the type and rearing intensity of the production system used. Water Quality Variables

### TEMPERATURE

All biological and chemical processes in an aquaculture operation are influenced by temperature. Fish adjust their body temperature and metabolic rate by moving into cooler or warmer water. Each species has a preferred or optimum temperature range where it grows best. At temperatures above or below optimum, fish growth is reduced. Mortalities may occur at extreme temperatures.

## DISSOLVED OXYGEN

The minimum dissolved oxygen (DO) level that fish can safely tolerate depends upon temperature and to a certain extent the species. Volubility of oxygen increases as temperature decreases. In ponds, DO can change dramatically over a 24 hour period. During the day oxygen is produced by photosynthesis, the process by which green plants convert water and carbon dioxide in the presence of light, to oxygen and carbohydrates. During the night and day oxygen is consumed by respiration, the process by which plants and animals use oxygen to produce carbon dioxide as they burn carbohydrates, but in the day photosynthesis usually produces more oxygen than is used. Typically, oxygen levels are lowest just before dawn and highest in the late afternoon.

DO in a culture system must be maintained above levels considered stressful to fish. Warmwater fish (species that grow best at temperatures above 80°F) can tolerate lower DO concentrations than coldwater fish (species that grow best at temperatures below 60°F). As a rule of thumb, DO should be maintained above 3.0 ppm (parts per million; frequently used interchangeably with milligrams per liter, mg/L) and 5.0 ppm for warm and coldwater fish, respectively. Prolonged exposure to low, nonlethal levels of DO constitutes a chronic stress and will cause fish to stop feeding, reduce their ability to convert ingested food into fish flesh and make them more susceptible to disease. Intensive fish production in ponds, cages, flow-through and recirculating systems requires aeration or oxygenation to maintain DO at safe levels.

## NITROGENOUS WASTES

Most fish and freshwater invertebrates excrete ammonia as their principle nitrogenous waste. Analytical methods are used to determine total ammonia-nitrogen (TAN). The proportion of TAN that exists in ionized and un-ionized form varies with pH and temperature. As pH and temperature increase, the amount of TAN in the toxic un-ionized form increases Fish continuously exposed to more than 0.02 ppm of the un-ionized form may exhibit reduced growth and increased susceptibility to disease. When fish are cultured intensively and fed protein-rich feeds they can produce high concentrations of ammonia in the water. Ammonia and other metabolic wastes are gradually removed by natural processes in ponds or through the use of biological filters in recirculating and reuse systems. Ammonia is removed by bacteria that initially convert it into nitrite and subsequently into nitrate. Nitrite is toxic to fish and causes "brown blood" disease. Concentrations of 0.5 ppm have reduced growth and adversely affected fish. Fish can tolerate nitrate to several hundred ppm. Removal or detoxification of ammonia is facilitated by providing and maintaining an optimal environment for the appropriate bacteria (pH between 7-9; temperature approximately 75-85° F).

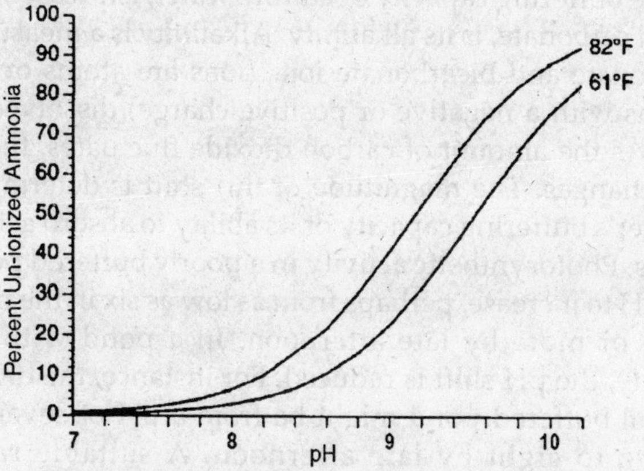


**Fig.** Relationship between pH, Temperature and Unionized Ammonia at 61°F and 82°F

## pH

The concentration of bases and acids in the water determines its pH. A low pH is acidic and a high pH is basic; a pH of 7 is neutral. Fish survive and grow best in waters with a pH between 6-9. If pH readings are outside this range, fish growth is reduced. At values below 4.5 or above 10, mortalities occur. In well-buffered ponds , pH typically fluctuates one or two units daily. In the morning, carbon dioxide levels are high and pH is low as a result of respiration during the night (carbon dioxide forms a mild acid when dissolved in water). After sunrise, algae and other green plants produce carbohydrates and oxygen from carbon dioxide and water by photosynthesis. As carbon dioxide is removed from the water, its pH increases. The lowest pH of the day is typically associated with the lowest level of dissolved oxygen. The highest pH of the day is typically associated with the highest level of dissolved oxygen. In recirculating systems, vitrification and respiration of both fish and biofilter bacteria decrease pH. Frequently, a buffer such as sodium bicarbonate is added to prevent the pH from falling too much.

## ALKALINITY

The buffering capacity of culture water, expressed as ppm calcium carbonate, is its alkalinity. Alkalinity is a measurement of carbonate and bicarbonate ions (ions are atoms or groups of atoms with a negative or positive charge) dissolved in the water. As the amount of carbon dioxide fluctuates, the pH of water changes. The magnitude of this shift is determined by the water's buffering capacity or its ability to absorb acids and/or bases. Photosynthetic activity in a poorly buffered pond can cause pH to increase, perhaps from as low as six in the morning to nine or more by late afternoon. In a pond with higher alkalinity, the pH shift is reduced. For instance, the daily shift in a well buffered pond might be from a pH of seven in the morning to eight by late afternoon. A suitable range of alkalinity is 20 to 300 ppm. Alkalinity in excess of 300 ppm does not adversely affect fish, but it does interfere with action of certain commonly used chemicals (e.g., copper sulfate). Alkalinity remains relatively constant in ponds, but decreases steadily in nonsupplemented, recirculating systems. Alkalinity can be increased by adding agricultural limestone to ponds or sodium bicarbonate to recirculating systems.

## HARDNESS

Calcium and magnesium ions comprise hardness. Test procedures usually determine both ions as "total hardness," expressed as ppm calcium carbonate. In most waters the concentrations of alkalinity and hardness are similar, but they can differ vastly as alkalinity measures negative ions (carbonate, bicabonate) and hardness measures positive ions (calcium, magnesium). Hardness is important, especially in the culture of several commercial species such as striped bass and catfish. If hardness is deficient, these species do not grow well. Hardness should be above 50 ppm; low hardness can be adjusted by the addition of lime or calcium chloride.

## CARBON DIOXIDE

Only when using groundwater, transporting fish at high densities, or in recirculating systems are carbon dioxide problems likely to develop. At high concentrations, carbon dioxide causes fish to lose equilibrium, become disoriented and possibly die. Testing groundwater before use and aerating it, if necessary, will reduce carbon dioxide to acceptable levels. Careful planning, aeration or oxygenation and buffering of water will keep carbon dioxide at acceptable levels when large numbers of fish are hauled extended distances or cultured in recirculating systems.

## SALINITY

The total concentration of all ions in the water is its salinity. Freshwater fish exhibit a range in salinity tolerance. Many commercially important species (e.g., channel catfish, *Ictalurus punctatus;* largemouth bass, *Micropterus salmoides;* tilapia, *Tilapia* sp.) survive and grow well in slightly salty water. After they smelt, salmon and trout can tolerate salt water. Salinity not only affects osmoregulation it also influences the concentration of un-ionized ammonia. During the planning stage of an aquaculture operation, salinity should be measured and the water's appropriateness determined.

## IRON

Many groundwaters contain elevated levels of dissolved iron. When exposed to the air, this iron interacts with oxygen, becomes insoluble and forms a red deposit. Small clumps of iron are produced that can settle on fish gills, causing irritation and stress. Problems can be avoided if the iron-bearing water is exposed to air and the resultant clumps of iron removed by settling or filtration before the water enters the culture system.

## CHLORINE

To control bacteria, municipal water supplies are typically treated with chlorine at 1.0 ppm. If municipal waters are used to culture fish, residual chlorine must be removed by aeration, with chemicals such as sodium thiosulfate, or filtration through activated charcoal. Chlorine levels as low as 0.02 ppm can stress fish.

## HYDROGEN SULFIDE

Ponds with oxygen-poor bottoms and accumulated organic material can release hydrogen sulfide when seined or disturbed. Substratum beneath heavily fed cages/net pens can accumulate wastes (e.g., uneaten food, feces) and produce hydrogen sulfide gas if oxygen becomes deficient. Hydrogen sulfide gas has a rotten egg odor and is extremely toxic to fish. Any detectable odors or levels should be avoided and extreme care should be taken when handling fish in an afflicted pond. Ponds can be drained, exposed to air and/or excavated to correct the problem.

## WATER CLARITY

In pond and cage culture, water clarity can affect fish. If fish that prefer turbid waters (e.g., bullhead, catfish, walleye) are cultured in relatively clear water they will experience stress; survival and growth will be adversely affected. Accumulation of suspended solids and discoloration of culture water occur in recirculating systems which can irritate fish and precipitate disease. Some suspended and dissolved materials can cause off-flavour in fish. Filtration and flocculent can be used to remove solids and reduce discoloration.

## MONITORING METHODS

A variety of methods are available to monitor water

quality In pond, cage and low intensity culture, the high precision of sophisticated analytical methods is not needed to make informed management decisions However, intensive culture in recirculating and reuse systems requires frequent and sophisticated monitoring.

If fish are maintained at high densities, then temperature, dissolved oxygen, ammonia, nitrite and pH should be monitored daily or more frequently (e.g., continuous monitoring of dissolved oxygen in recirculating systems). Water clarity, alkalinity and hardness can be measured less frequently, perhaps one or two times per week, as they do not fluctuate as rapidly.

Salinity, iron and chlorine should be determined when a potential water source is first examined so corrective measures may be incorporated into the production system during the design or planning stage. Carbon dioxide should be measured when first using a new groundwater source and routinely in recirculating systems.

When hydrogen sulfide and carbon dioxide problems are likely, systems should be monitored closely and the means to correct problems should be readily available. At lower stocking densities, water quality parameters can be monitored less frequently or not at all.

Regardless of the frequency, monitoring should be conducted at a standard time and depth where fish are located. Time of measurement and observed values should be recorded; good record keeping is essential to successful aquaculture. In pond and cage culture it is preferable to monitor dissolved oxygen early in the morning, when conditions stressful to fish are most likely to occur (e.g., low oxygen). Conversely, temperature and pH in ponds are best measured during the late afternoon.

## THE HYDROGEN BOND

The hydrogen bond gives water its unique properties. To understand the hydrogen bond, we first take a close look at the water molecule itself. The water molecule is composed of two hydrogen (H) atoms and one oxygen (O) atom. The two

H—atoms in their neutral state each have only one electron. The neutral oxygen atom has six electrons in its outer shell and needs two further electrons to fill all eight places available in this shell (to satisfy the so called *octet rule*). The H—O bond is a normal covalent bond; each atom contributes one electron to that of its neighbour.

The hydrogen atoms each contribute their one and only electron, while the oxygen contributes its two unpaired electrons. Two of these bonds in the structure H—O—H fill up the two vacancies in the oxygen electron structure to complete the octet. Each shared electron pair is concentrated between the two atoms along the bond axis between H and O.

The H—O—H structure has a V—shape. The angle at the oxygen between the two hydrogens is degrees.

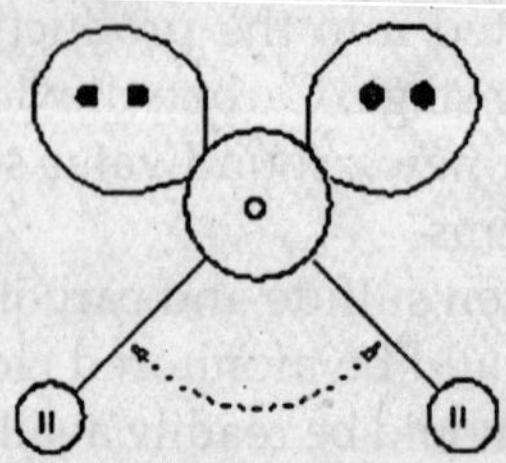

**Fig.** A Water Molecule. The Two "Bubbles" Represent the Two Unshared Electron Pairs

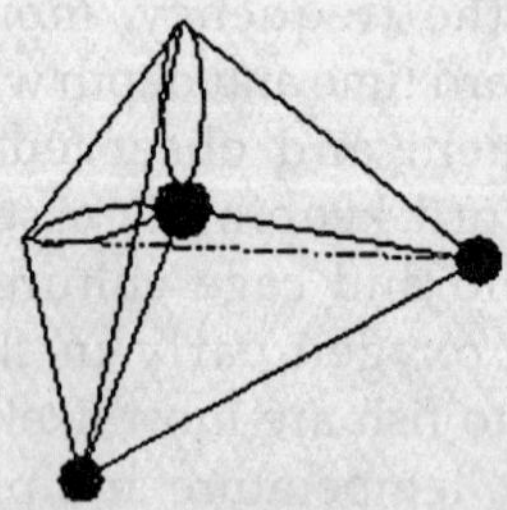

**Fig.** The Tetrahedral Structure of Water. The Cenre Ball Represents the Oxygen Atom, the Two Outer Black Balls the Hydrogen Atoms and the Two Clouds the Unpaired Electrons

Now we come to the hydrogen bond. It is what we call a *shared electron bond* that is a special case of the *coordinative covalent bond*. The oxygen has two unshared electron pairs that do not contribute to the H—O bonds of the water molecule. These electron pairs are oriented outwards, away from the H—atoms. An unshared electron pair can interact with a *nearly naked* hydrogen nucleus (proton) to form what is called a *coordinative covalent bond*, or *hydrogen bond* The hydrogen nucleus appears *nearly naked* from outside the water molecule, because the hydrogen electron is (mostly) on the inside of the molecule, located along the H—O bond.

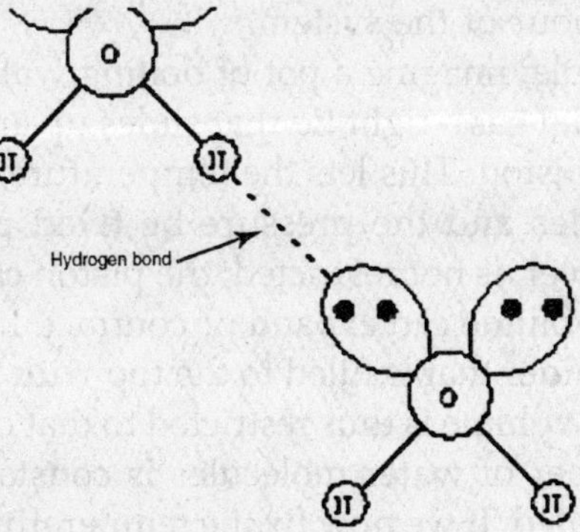


**Fig.** A Hydrogen Bond between Two Water Molecules

The hydrogen bond is approximately 30 times weaker than a normal covalent bond, because only one of the contributing atoms is supplying electrons to it; the two electrons stay mainly concentrated near the oxygen. Because the hydrogen bond is so weak, it is easily broken. At room temperature, thermal energy is enough to break hydrogen bonds. In liquid water the whole network of hydrogen bonds "flickers," each bond making and breaking again in a millionth of a microsecond. It is this network of flickering hydrogen bonds that gives liquid water its unique properties.

The coordinative covalent bond is not necessarily weak. The nitrogen atom in the ammonia molecule $NH_3$ has one lone electron pair. When a hydrogen ion $H^+$ is added to make the ion, $NH^+$, the four hydrogen atoms share equally in the resulting bond structure. As a results, the bond energy between N and each of the four H—atoms is far stronger than the hydrogen bond between water molecules and is not broken by thermal agitation at room temperature.

## ENSEMBLE

Typically we use temperature, volume and pressure to a large system of particles. We call such a system and all its thermodynamic properties an *ensemble*. The *Wasser* programme uses four thermodynamic properties: pressure P, volume V, temperature T and number of particles N. The values of three of these four properties can be fixed at one time to the water *ensemble*; the value of the fourth property results from the natural behaviour of the system.

For example, imagine a pot of boiling water. Put a lid on the pot, a special gas—tight lid that slides up and down inside the pot like a piston. This lets the temperature, the number of water molecules and the pressure be fixed properties. The volume, however, is not restricted; the piston can slide up and down, so the volume can expand or contract. In contrast, now picture a container that is filled to the top with water and then sealed off. The volume is thus restricted to that of the container, the total number of water molecules is constant because the container is sealed. If we now fix the temperature, the pressure has to adjust accordingly.

In the *Wasser* programme, the user can choose between two different ensemble setups: the NVT ensemble or the NPT ensemble. In the NVT ensemble, the number of particles N, the volume V and the temperature T can be set to desired values; the pressure will be a result of the system behaviour based on your choices. In the NPT ensemble, the number of particles N, the pressure P and the temperature T can be fixed and the volume will result from the system behaviour.

## PERIODIC BOUNDARY CONDITIONS

We run computer simulation programs to predict and study the properties of a system in bulk. For the present we are not interested in surface effects. But our simulations track only a small number of particles in order not to slow down the computation. As a result, most molecules are near the edge of the sample, that is near its "Surface." Therefore, it looks like we cannot avoid surface effects in our computations. The system size would have to be extremely large to ensure that the surface has only a small influence on the bulk properties, but this system would be too large to simulate. A glass of water, for example, has a very large system size in comparison to its surface. In contrast, water in a drinking straw has a much smaller system size compared with its surface.

To eliminate surface effect from the computation we use a trick. Surface effects can be ignored for all system sizes if we use periodic boundary conditions. In periodic boundary conditions, the cubical simulation box is replicated throughout space to form an infinite lattice. In the course of the simulation, when a molecule moves in the central box, its periodic image in every one of the other boxes moves with exactly the same orientation in exactly the same way. Thus, as a molecule leaves the central box, one of its images will enter through the opposite face. There are no walls at the boundary of the central box and the system has no surface. The central box simply forms a convenient coordinate system for measuring locations of the N molecules.

As a particle moves through a boundary, all its corresponding images move across their corresponding boundaries. The number of particles in the central box (and hence in the entire system) is conserved.

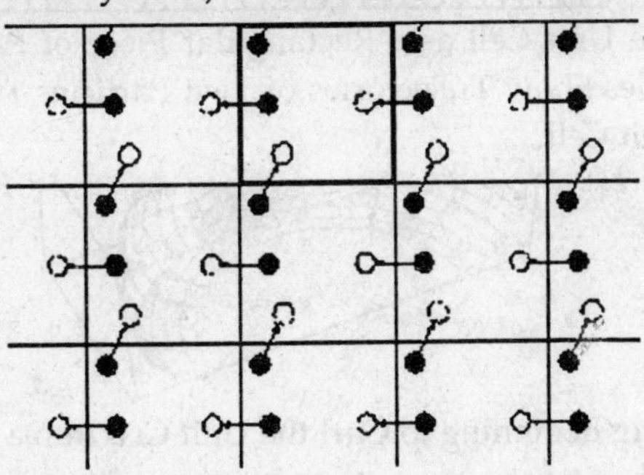

**Fig.** Periodic Boundary Conditions. The Central Box is Outlined by a Thicker Line

It is not necessary to store the coordinates of all images in a simulation (this would be an infinite number), just those of the molecules in the central box. When a molecule leaves the box by crossing a boundary, attention may be switched to the *identical* molecule just entering from the opposite side. It is sometimes useful to picture the single *central* simulation box in our two dimensional example as being rolled up to form the surface of a three dimensional torus or donut Think of drawing two perpendicular dashed lines across the middle of a piece of typing paper parallel to its edges. Now roll up the paper so that two sides meet to form a hose with circular ends. Next bend the hose until the two circular ends meet. The results is a torus or *donut* Each original dashed line now closes on itself. A particle following either one of these lines comes back to the same point on the donut, just as a particle in the central box of the infinite lattice in comes back to the same location if it follows a straight line.

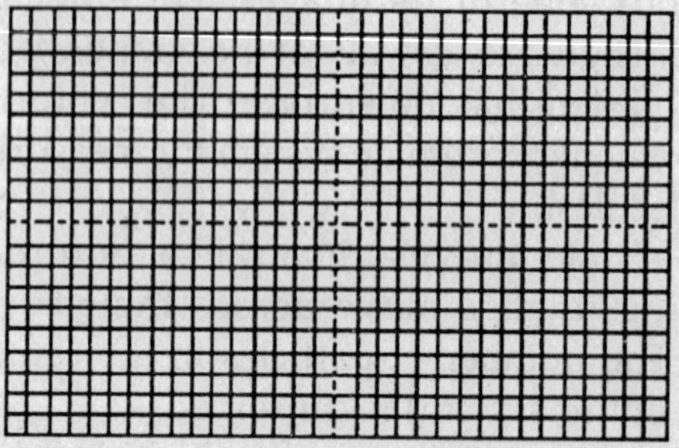

**Fig.** The Unit Cell as a Rectangular Piece of Paper. Dashed Lines Show Trajectories of Two Particles Traversing the Unit Cell

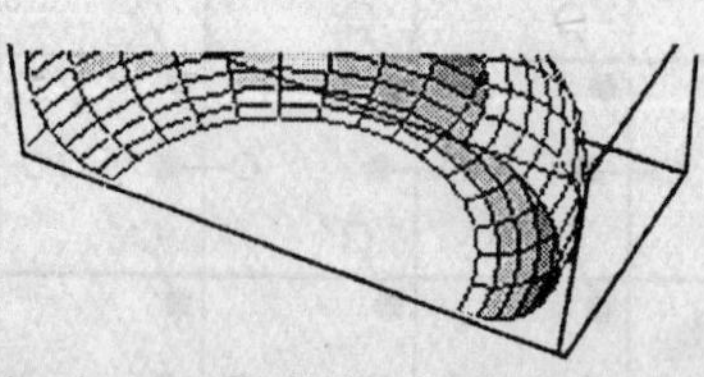

**Fig** Beginning to Curl the Unit Cell Into a Tube

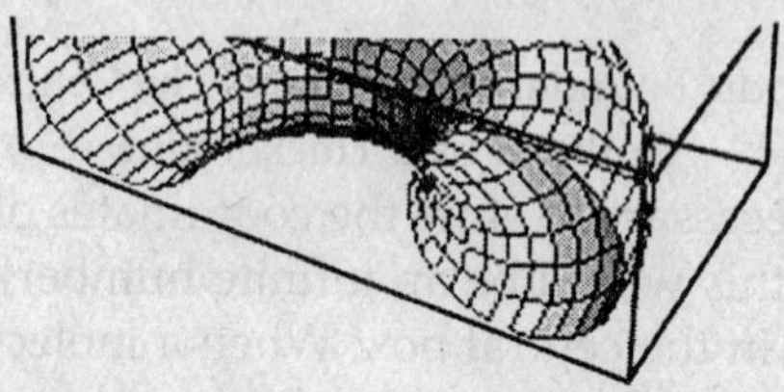

**Fig.** The Unit Cell as a Tube

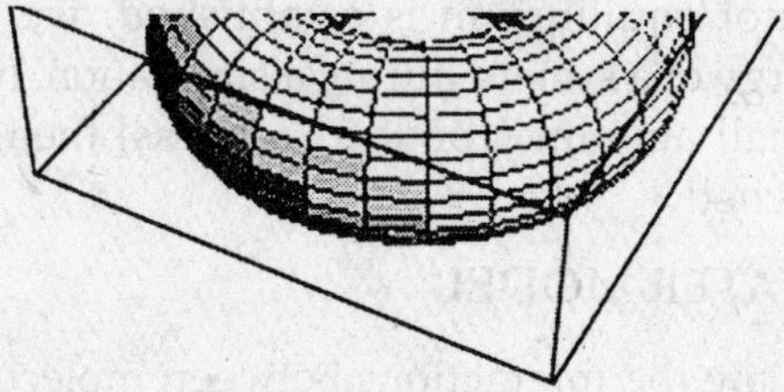

**Fig.** Connecting the Ends of the Unit Cell as a Tube to form a Donut. Particles Traversing the Dashed Lines in the Original Unit Cell will Return to the Same Location on the Donut.

With the *donut* configuration there is no need to consider an infinite number of replicas of the system, nor any image particles. The *donut* correctly represents the topology of the two–dimensional system, if not the geometry. A similar analogy exists for a three–dimensional periodic system, but this is more difficult to visualize.

## EQUILIBRIUM

When investigating the microscopic properties of a system, we often refer to the fact that the system should be *in equilibrium* to ensure reproduceable and realistic results. *Equilibrium* means that the thermodynamic properties of the system, such as temperature and pressure, will not change when the system is left alone. Technically, equilibrium means that the system has found a potential energy minimum for a given set of thermodynamic properties. The system moves naturally toward a potential energy minimum:

When the *Wasser* programme first starts, it randomly distributes the molecules in the simulation box on a cubic lattice and gives them mixed velocities and directions of motion. As the programme runs, the molecules move about, interact and reorient themselves, automatically tending to a state of minimum potential energy. Once this process no longer causes changes in the energy or thermodynamic properties (such as temperature and pressure), we say that the system is *in equilibrium*. The molecules still move about, but in such a way that the potential energy stays near a minimum. To check whether or not equilibrium is established, we monitor the potential energy of a system during a simulation. If it fluctuates by only a small amount (about 3% or less) then equilibrium has been reached.

## THE SPC WATER MODEL

To calculate the interactions between molecules in liquid water, we need a model of the individual water molecule that tells us where the charges reside. In the *Wasser* simulation programme we use a very simple model to represent the water molecule called the SPC model. In the SPC, or Simple Point Charge model, the water molecule has three centers of concentrated charge: a predominance of positive charge on the H—atoms and excess negative charge on the O—atom. Our assumption that there are point charges is an approximation that leads to an incorrect value for the permanent dipole moment of the water molecule. To correct this, the H—O—H bond angle is changed to degrees in the model (compared with the experimentally—found H—O—H bond angle of degrees.).

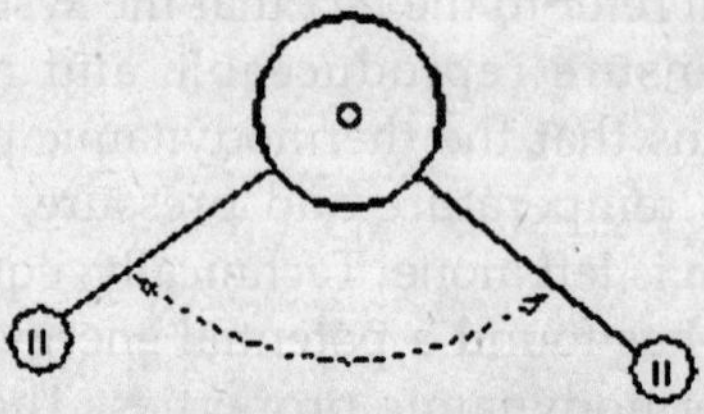

**Fig.** The SPC—water Molecule. To Yield Overall Neutrality, the Negative Partial Charge on the Oxygen Atom has Twice the Magnitude of the Partial Positive Charge on Each of the Two Hydrogen Atoms

The results of the charge concentration and the *widened* V—shaped bond angle is that the permanent dipole moment of the SPC—model water molecule has a value close to that measured in experiment. It also has an effect on the mobility of the molecules. The molecule is able to move faster than it would in "real" water due to the missing two lone electron pairs in the SPC—model. This effect however decreases when the temperature is increased. In the *Wasser programme* the oxygen atoms are represented by a red colour, the hydrogen atoms by a white colour. The hydrogen bonds are represented by a turquoise colored "stick" between an oxygen and a hydrogen.

## SIMULATE THE LIQUID—TO—GAS PHASE TRANSITION OF WATER

We are confronted daily with the phase transition from liquid to gas — for example every time we boil water. Now we use the simulation to help our understanding of the microscopic properties of water that change when a phase transition occurs. At the boiling point of any liquid, a large fraction of molecules possess kinetic energy high enough to break bonds with the bulk liquid. That is, the pressure within the liquid is greater than the "outside" pressure and the liquid "evaporates," resulting in a departure of molecules from the bulk fluid. The PV phase diagram of water shows that as we lower the pressure, the volume increases. However this curve does not continue linearly; its decline comes to a halt and for a period the volume can increase without a change in pressure. The curve has a plateau in this region. Then the curve continues its downward slope. This interruption marks the phase transition from liquid to gas. When the liquid evaporates, its pressure does not rise, because the system expands at the same time (provided the the NPT ensemble has been chosen.)

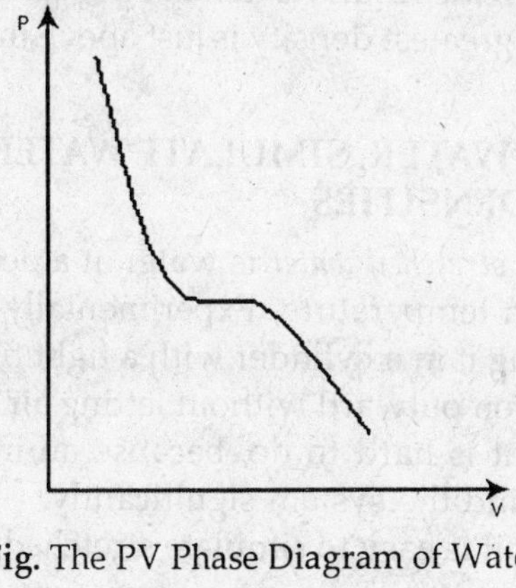


**Fig.** The PV Phase Diagram of Water

## SIMULATE THE SOLID—TO—LIQUID PHASE TRANSITION OF WATER

The solid to liquid phase transition is the melting process of ice. Any phase transition occurs due to a change in kinetic energy of the participating particles. If the system is in the solid phase and the kinetic energy is sufficiently increased, the system changes from solid to liquid. In the solid phase, the network strives for the lowest energy conformation. In ice, a water molecule has four nearest neighbors to which it is bonded via hydrogen bonds (two from its hydrogen atoms and two from the lone electron pairs on the oxygen.) In the SPC—model, this structure is a result of the dipole moment and the partial charges on the atoms. The geometry leads to a rather open hexagonal structure, each of the four bonds representing a lowered overall energy. When the average kinetic energy is raised, the additional jostling begins to destroy the open hexagonal structure. Paradoxically, this allows the molecules to move closer to each other, making and breaking hydrogen bonds much more rapidly. On average, there can now be more than four nearest neighbors at a time, lower energy and a higher density in the just—melted liquid system. When rising temperature increases average kinetic energy yet further, molecules move through the liquid so fast that fewer bonds are formed at any one time, shortening the average time each bond exists. The result is higher total energy and thus lower density. Overall, the results is that at atmospheric pressure the temperature of greatest density is just above the melting point, at Celsius.

## STRETCHED WATER, SIMULATE WATER FOR DIFFERENT DENSITIES

So—called *stretched water* is water at a density lower that at normal room temperature. Experimentally one can stretch water by placing it in a cylinder with a tight fitting piston and pulling the piston outward without letting air into the system. This experiment is hard to do, because an immense force is needed to expand the system significantly.

In contrast it is easy to simulate stretched water using the

*Wasser* programme. Simply specify a density less than $1\frac{\theta}{cm^3}$ using the NVT ensemble or set the pressure to a value smaller than 62 Mpa in the NPT ensemble.

*Background:* When looking at molecules of water at density $1.0\frac{\theta}{cm^3}$ we see a randomly structured hydrogen bond network. When we start decreasing the density, the pressure of the system starts to decrease and gradually takes negative values.

To understand negative pressure we can picture a rubber band that we hold between our fingers. When we stretch the rubber band, it tries to snap back to its relaxed position. We can say that the "outward force of the rubber band on our fingers is negative." The same thing happens when we put water into a closed container without bubbles then stretch the container. Both the container and the water try to snap back to their relaxed conformation. "Pressure" ordinarily means "outward force per unit area." In this sense, the pressure in the water goes negative.

Stretching the water makes a greater volume available to the molecules. The stretched network strives for the energetically most favoured conformation. This conformation is closely related to that of ice, since a water molecule now, due to the tension, has fewer nearest neighbors than it had at density. Under stretching, the system will gradually go from a random network to the geometrically favoured open hexagonal structure as the density is decreased. If the density is lowered too far, however, the system reaches its mechanical stability point; any further reduction in density will "tear the system apart." This occurs at a density of approximately. The system will no longer be in the liquid phase, it will be like a droplet in vacuum.

## The Hydration Shell of an Ion in Water

When an ion is inserted into a water configuration, it changes the structure of the hydrogen bond network. A water molecule tends to rotate (reorient) so that its polarized charge concentration faces the opposite charge of the ion. As the water molecules orient themselves towards the ion, they break the hydrogen bonds to their nearest neighbors. The group of water molecules oriented around an ion is called a *hydration shell*. The orientation of the molecules in the hydration shell results in a net charge on the outside of this shell, a charge of the same sign as that of the ion in the cenre. The charge on the outside of the hydration shell tends to orient water molecules in the vicinity, leading to a *second hydration shell*. The result of forming hydration shells is to weaken the structure of the hydrogen bond network. This explains the observation that salt water has a lower freezing point than pure water: Each ion in the liquid has a hydration shell of oriented water molecules around it, which prevents the water molecules from forming the hexagonal structure of ice.

## SIMULATE THE HYDRATION OF AN ION IN STRETCHED WATER

When simulating an ion in water, one would assume that a smaller ion is always easier to insert than a large one. This is the case for water at the density. $1.0\frac{\theta}{cm^3}$ However, when we stretch water so that its density is lowered, we observe an opening of the water structure; the space available to an ion grows. Very small ions tend to counteract this opening of the structure because they try to reorient the surrounding water molecules with the water molecule's charge concentrations towards the charge of the ion and to pull these reoriented molecules inward, tending to collapse the structure. A larger ion (the appropriate size depending on the density of the system) can however stabilize the open structure. This is due to the fact that a larger ion can "fill" the hole much easier without pulling water molecules towards itself. The hydration shell of the larger ion now has some similarity with the hydration shell of a noble gas. The water molecules (depending on the density) do not reorient towards the ion in the same way for stretched water as they do for regular water with a density of $1\frac{\theta}{cm^3}$. Because of this, the water molecules can still make a small contribution to the hydrogen bond network.

## VARY THE NUMBER OF MOLECULES IN THE SYSTEM

A simulation is only a model. For this reason, whenever looking at the microscopic properties of water, one should question whether or not the investigated system size is large enough. In statistics we learn that statistical accuracy increases with the number of particles in the sample. For example, if we investigate a 96 particle system, we are looking at $2.8695 \times 10^{-24}$ kg of water. In contrast, even a milliliter of water at room temperature has a much greater mass: $1 \times 10^{-3}$ kg and a corresponding larger number of particles. A good method of determining the accuracy of the simulation is to calculate the deviations of fluctuating quantities. For example, given any equilibrated system size, the potential energy should fluctuate by a maximum of about. 3% Other properties that can be studied, depending on the simulation setup (NVT or NPT ensemble), are the pressure $p$ and the density $p$. You will observe that the pressure deviation is very large in comparison to the smaller deviations of the density. This arises from the fact that water has a very small compressibility factor ($R \ll 1$). The fluctuation of the volume (and thus of the density) and of the pressure are dependent on the compressibility, but in different ways. The result is that the deviations of the volume are diminished by small compressibility whereas the deviations of the pressure are enhanced.

Think of a ball bearing bouncing on a block of steel. The steel has a very small compressibility factor, so the ball bearing experiences large forces ("*pressure*") at every bounce, while the density of the steel block remains nearly constant. This image corresponds to the large fluctuations in pressure and low fluctuations in density in a small sample of water.

A side effect has to be kept in mind when simulating an ion in water. The theoretical assumption is that the ion is in infinite dilution (meaning an extremely low concentration.)

This assumption implies that inter—ionic interactions (interactions between one ion and a neighboring ion) do not have to be considered. However we use periodic boundary conditions in our simulation. Therefore in a system that is too small, the ion will interact *with its own image.* As far as the computer simulation is concerned, all the other images of ions are "really there", with consequences for interaction energy that cannot be ignored. If the central box size is large enough (system size sufficiently great), the interaction energy will be close to zero and its contribution can be neglected. To test this, one should compare the simulated box size with a normal plot of the Coulomb potential. If the potential is not close to zero at a distance equal to half the simulated box size, then the box size should be increased to embrace a larger system (half the box size because of the periodic boundary conditions - this is the maximum distance between two particles.

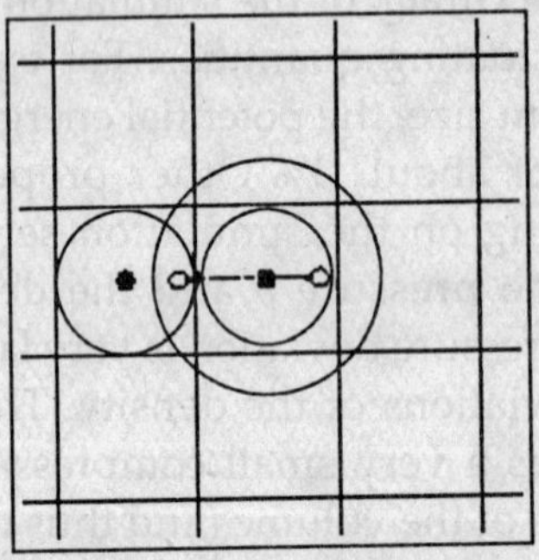

**Fig.** The Maximum Distance between two Particles Must be less than half the box length

A similar difficulty arises for noble gases. The potential used to calculate the interaction between water molecules and a noble gas is the Lennard—Jones potential.

We decrease the "false" interaction of a particle with its own image by increasing the size of the simulation box. To do this we keep the density of particles the same but increase the *number* of particles in the box, thus pushing out its boundaries. (Density of particles is defined as the number of particles per unit volume.) More particles means a greater computational load on the computer, which is why the *Wasser* programme requires fast computers! The box length $L_{Box}$ can be determined by calculation from the density and the number of particles,

$$LBox = \left[\frac{N}{\left(\frac{\rho N_A}{M}\right)}\right]^{\frac{1}{3}}$$

where $N_A = 6.022 \times 10^{23}$ mol$^{-1}$ is the Avogadro constant, ρ is the density in $\frac{g}{cm^3}$, N is the number of particles in the system and M is the molar mass of water. Please keep the units in mind - the box length is usually given in units of nm ($10^{-9}$ m).

## PHYSICO-CHEMICAL PROPERTIES OF WATER

Any experimental procedure is available for detecting the physicochemical changes that homeopathic succussion or other similar techniques can produce in water. Recently, a thermodynamic procedure was developed and proved effective in characterizing homeopathic remedies. It consists in the comparison of the heat-of-mixing reaction of acids or bases in various concentrations with extremely diluted solutions and with bi-distilled water as the control.

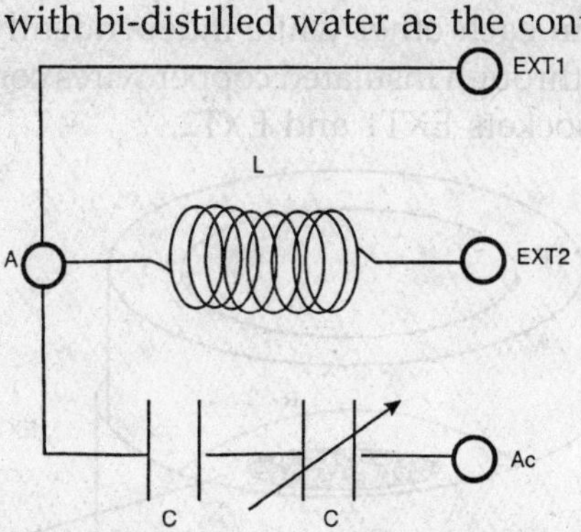


**Fig.** Modified Callegari Radionic Clrcuit

In 100% of the cases, the homeopathically treated solutions showed a heat excess with respect to the control water. We based our early observations on the effects of ultra-low electromagnetic fields mainly on the Sensitive Crystallization

Method and on biological tests. Incidentally, these hinted at the existence of a possible relationship between homeopathy and ultra-low electromagnetic fields. The biological effects of weak electromagnetic fields have been investigated by C.W. Smith since 1974. Recently, he established equivalence between specific electromagnetic frequencies and homeopathic remedies and acupuncture meridians.

## Material & Methods

*The Apparatus:* We standardized on a common experimental procedure and used the same exposure device in all the tests we performed. Three different resonant circuits were used with this device for exposing the water samples. *The Exposure Device* This couples the electromagnetic field from the resonator to the water sample so that the water only comes into physical contact with a glass beaker and never with the device itself. The device is composed of two elements, i.e., an external brass ring (diameterf = 24 mm, 10 mm high, 1 mm thick) and an internal concentric copper dish (diameterf = 12 mm) with curvature corresponding to one third of a 10 mm radius sphere. An insulating plastic ring is interposed between the two above elements. Thus, they are kept electrically separated from each other while independently connected to the resonator through insulated copper wires connected in turn to the plugs/sockets EXT1 and EXT2.

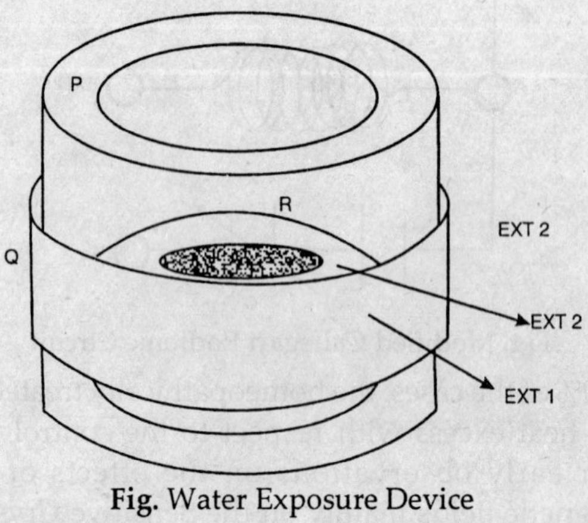


**Fig.** Water Exposure Device

## *The Passive RLC Circuits*

The first resonator circuit we used to irradiate water samples was a modified Callegari Radionic Circuit and the essential features of the Water Exposure Device (*R* indicates the internal copper disk, *Q* the external brass ring and *P* the plastic ring). We later developed and used two other circuits that were quite similar to each other.

Although the component elements and the electrical parameters are the same as those in the previous diagram, a ferrite bar has been added. The set-up and the electrical parameters in the previous references but for the sake of clarity, we briefly repeat them here. The aerial, A, is a brass sphere with diamter = 5.5 mm having a capacitance C = 900 pF; C1 is a variable 30–300 pF capacitor; the inductance of L = 0.02 H. EXT1 and EXT2 refer to the sockets on the resonator posure device. Finally, Ac is an unused socket serving as an elementary aerial compensating device.

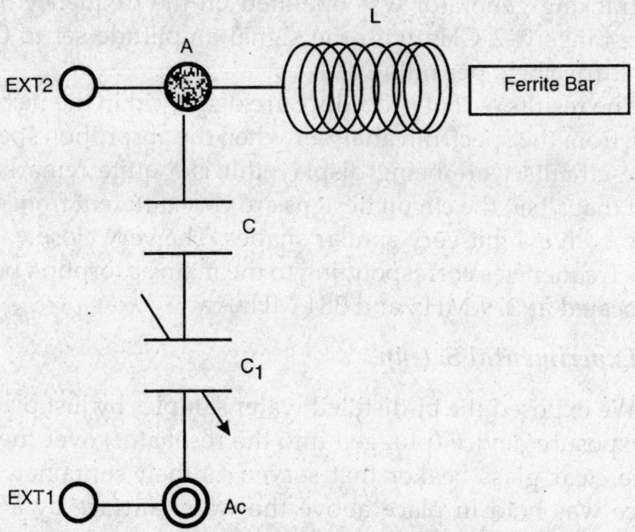


**Fig.** The Ferrite Circuit

## *Methods for Using the Apparatus*

We exposed the bi-distilled water samples contained in glass vials or beakers by placing them under the exposure device. In all the circuits, one of the plugs is always located at the end opposite to the inductance (EXT2) and in line with the aerial (terminal Ac). It is important to note that the whole apparatus has no connection to any external or internal electric supply, or to a generator, or to any other source of energy or oscillations. It is completely passive. The inductance axis in the operating condition must lie parallel to the local geomagnetic/ geographical E-W direction.

### *Electromagnetic Characterization of the Circuits*

For quite a long time, it seemed almost impossible to have an electromagnetic fingerprint of the circuits. This proved to be a much more difficult task than having clear evidence of effects on the bi-distilled water samples. After three years of fruitless trials, finally Dr. Clarbruno Vedruccio provided a completely satisfactory characterization of the circuits by means of a spectrum analyzer linked to a tracking generator The tracking generator was operated on the frequency band in the range 0–2 GHz with the signal amplitude set to 0 dB mW- 1 (decibels per milliwatt).

The results of that procedure are displayed in the pictures taken from the spectrum analyser when the absorption spectra of the circuits were being displayed.It is a quite remarkable event that albeit the circuit designs are very different from each other, both exhibit very similar shapes. The very close values of the frequencies corresponding to the main absorption peaks are located at 379 MHz and 384 MHz.

### *The Experimental Set-up*

We exposed the bi-distilled water samples by just placing the exposure device (plugged into the resonator) over the top of the clear glass beaker that served as their container. The device was held in place above the water surface by a thin cardboard disk acting also as the beaker's cover.

### *Sample Preparation*

We always used fresh bi-distilled water (from Carlo Erba, Italy) which was kept in polyethylene until used. Exposure was always performed over Pyrex glass beakers. Several water volumes were used in the experiments: 80 ml, 100 ml, 250 ml or 400 ml. Two different irradiation times were tested in general: 6 hours or 24 hours.

Once the exposure time had elapsed, the samples were always transferred into dark glass bottles. In all experiments, a control comprising the same volume of the same non-exposed bi-distilled water was kept in the same kind of dark glass bottles.

The bottles containing the samples were kept far apart from the controls until measured. Both calorimetric and pH-metric measurements were performed. These were carried out at various times after the sample production ranging from a few hours to 14 days in order to determine the stability of the effects over time.

### *Microcalorimetric Apparatus*

We measured the heats of mixing by means of a Thermal Activity Monitor (TAM) from Thermometric with 0.1 ¼W sensitivity. It is equipped with flow-mixing and a batch titration vessel. In the case of the flow-mixing vessel, two peristaltic pumps convey solutions into the calorimeter through Teflon tubes. The flow rates through the inlet pipes are equal and constant, so that the solution coming out of the calorimeter has a concentration one half of the initial concentration. Our mass flow rate was constant to within 1% and its value

$$P_w = 3.10^{-3}\ gs^{-1}$$

remained the same in all the experiments. The following relation defines the enthalpies of mixing $\Delta H_{\text{mix}}$:

$$\Delta H_{\text{mix}}\ (mix \rightarrow m_x^f) = \frac{dQ}{dt} Pw$$

where $\frac{dQ}{dt}$ is the heat flux $s^{-1}$ (Watt), *PW* is the total mass flow rate of the solvent (kg $s^{-1}$) and $m^i_x$ and $m^f_x$ are the initial and final molalities, respectively, of the solution. $\Delta H_{\text{mix}}$ is expressed in J $kg^{-1}$ of solvent.

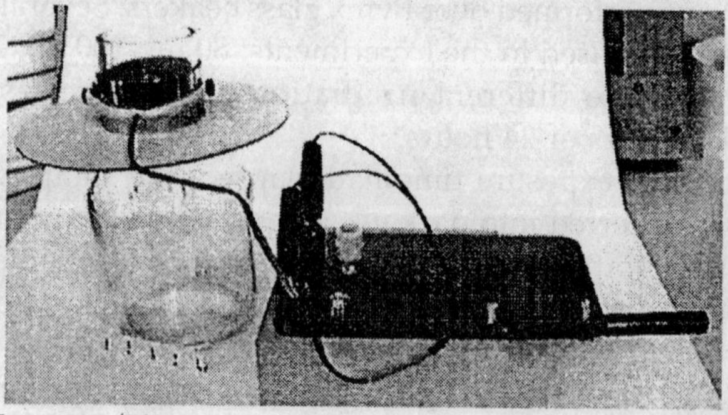

**Fig.** Exposure Device Placed Over a Beaker of Bi-distilled Water Showing Connections to the Resonator Circuit

### *pH Measurements*

We performed pH measurements on 29 out of the 60 samples. The apparatus we used was equipped with an electrode for low molecular force solutions. The values that we measured on irradiated water samples were compared to their corresponding controls and the results were expressed as a ΔpH increase. The generally observed pattern in these measurements was an increase toward basic ΔpH values as shown by the ΔpH changes.

The pH values ranged between 0.05 and 2.60 ΔpH units; in five cases, the ΔpH change shifted toward acid values as negative values. All the negative pH values corresponded to negative or zero excess heats, but there were positive ΔpH changes with negative excess heats.

## THERMODYNAMICS OF WATER-OCTANOL AND WATER-CYCLOHEXANE

The partition coefficient (P) of a compound has been shown over the years to be the single most important property controlling the transport in living organisms of xenobiotics such as drugs and environmental pollutants. Thousands of quantitative structure-activity relationships (QSARs) have been derived that utilise log P as a descriptor. The 1-octanol-water system is generally used for the measurement or calculation of log P, although many other solvent pairs have also been used, especially the cyclohexane-water system. P is an equilibrium constant and thus, from the van't Hoff isotherm, is a free energy term:

$$\Delta G = 2.303 \log P$$

where ΔG is the Gibbs free energy of transfer from water to pinguid solvent. However, it should be remembered that the Gibbs free energy is the difference between the enthalpy (ΔH) and the entropy (ΔS) of transfer:

$$\Delta G = \Delta H - T\Delta S$$

where T is the temperature (K). It is thus possible for two compounds with very similar log P values to have widely differing enthalpies and entropies of partitioning, indicating different partitioning mechanisms. Indeed, it is sometimes said that the log P term hides more than it reveals, at least so far as partitioning mechanisms are concerned.

Because of this it is important, in considering partitioning mechanisms, to examine not only the free energy of partitioning but also the enthalpy and entropy of partitioning. In an investigation of the effects of hydrogen bonding and steric effects on partitioning behaviour, we have determined the thermodynamics of partitioning of 45 compounds, comprising 34 phenols, 5 benzoic acids and 6 acetanilides, in both the water-octanol and the water-cyclohexane systems. We have used the van't Hoff approach, which involves measuring the partition coefficient of a chemical at a number of different temperatures and obtaining from the van't Hoff isochore the enthalpy of partitioning.

$$\text{Log } P = -2.303\ \Delta H/RT + \text{constant}$$

The method relies on the enthalpy of partitioning being constant over the temperature range used, which may not be the case when there is considerable mutual solubility of the two solvents, as in the case of octanol and water (Water-saturated octanol contains about 27 mole % water at 25oC).

## RESULTS AND DISCUSSION

The Gibbs free energy, enthalpy and entropy of partitioning of each of the 45 chemicals in the water-octanol and water-cyclohexane systems. All of the van't Hoff plots of log P *vs.* 1/

T for water-cyclohexane and most of those for water-octanol, were rectilinear. For six compounds with non-rectilinear plots, tangents were taken at 25°C. It is accepted that for these compounds the thermodynamic parameters probably contain greater error. We believe, however, that the consistency of the results means that they are reasonably accurate.

| Compound | $\Delta G_{w\to o}$[a] | $\Delta H_{w\to o}$[a] | $\Delta S_{w\to o}$[b] | $\Delta G_{w\to c}$[a] | $\Delta H_{w\to c}$[a] | $\Delta S_{w\to c}$[b] |
|---|---|---|---|---|---|---|
| Phenol | -13.7 | -8.4 | 17.7 | -0.3 | 15.3 | 52.5 |
| 2-Chlorophenol | -17.0 | -8.4 | 28.9 | -9.1 | 8.6 | 59.4 |
| 3-Chlorophenol | -18.6 | -10.9 | 26.4 | -4.2 | 10.2 | 48.3 |
| 4-Chlorophenol | -18.6 | -10.5 | 26.9 | -2.8 | 14.0 | 56.4 |
| 2-Nitrophenol | -15.4 | -2.5 | 43.3 | -12.5 | 5.7 | 61.0 |
| 3-Nitrophenol | -16.2 | -15.3 | 3.0 | 4.5 | 11.5 | 23.6 |
| 4-Nitrophenol | -16.1 | -19.2 | -10.3 | 6.4 | 23.0 | 55.5 |
| 2-Nitroresorcinol | -19.8 | -7.3 | 41.9 | -10.2 | -3.3 | 23.4 |
| 2-Hydroxybenzaldehyde | -14.7 | -3.7 | 37.1 | -12.3 | -3.5 | 29.6 |
| 3-Hydroxybenzaldehyde | -13.2 | -8.9 | 14.4 | 6.7 | 7.9 | 3.8 |
| 4-Hydroxybenzaldehyde | -13.3 | -9.6 | 12.4 | 8.5 | 15.3 | 22.8 |
| Benzoic acid | -15.6 | -8.4 | 24.1 | 0.4 | 6.6 | 20.8 |
| 2-Hydroxybenzoic acid | -18.0 | -17.9 | 0.6 | 3.6 | 11.5 | 26.5 |
| 3-Hydroxybenzoic acid | -14.1 | -19.2 | -16.9 | [c] | [c] | [c] |
| 4-Hydroxybenzoic acid | -13.9 | -23.0 | -30.4 | [c] | [c] | [c] |
| 2,6-Dihydroxybenzoic acid | -14.8 | -17.9 | -10.3 | 14.6 | 12.5 | -7.2 |
| 3,5-Dihydroxybenzoic acid | -11.3 | -23.0 | -39.1 | [c] | [c] | * |
| 3-Methyl-2-nitrophenol | -17.6 | -12.8 | 16.3 | -12.0 | -5.7 | 20.9 |
| 4-Methyl-2-nitrophenol | -18.1 | -2.2 | 53.4 | -15.9 | -9.9 | 20.2 |
| 5-Methyl-2-nitrophenol | -18.6 | -3.4 | 51.2 | -14.3 | -5.7 | 28.6 |
| 6-Methyl-2-nitrophenol | -20.0 | 6.8 | 89.8 | -15.6 | -1.0 | 48.8 |
| 2-Methylphenol | -15.9 | -7.7 | 28.3 | -5.3 | 15.6 | 70.1 |
| 3-Methylphenol | -16.2 | -8.1 | 27.5 | -3.6 | 19.2 | 76.2 |
| 4-Methylphenol | -16.1 | -8.4 | 26.7 | -3.6 | 15.3 | 63.4 |
| 2,3-Dimethylphenl | -18.3 | -9.1 | 30.9 | -7.3 | 15.3 | 75.9 |
| 2,4-Dimethylphenol | -18.8 | -9.6 | 31.4 | -8.2 | 12.8 | 70.2 |
| 2,5-Dimethylphenol | -18.5 | -9.9 | 29.1 | -8.6 | 14.3 | 76.7 |
| 2,6-Dimethylphenol | -18.1 | -3.6 | 48.6 | -10.0 | 9.2 | 64.4 |
| 3,4-Dimethylphenol | -18.0 | -10.1 | 27.2 | -5.5 | 19.2 | 82.6 |
| 3,5-Dimethylphenol | -18.6 | -10.5 | 26.9 | -6.0 | 16.4 | 75.0 |
| 2,3,5-Trimethylphenol | -20.6 | -6.7 | 47.6 | -10.6 | 15.3 | 86.9 |
| 2,3,6-Trimethylphenol | -19.8 | -5.7 | 47.9 | -14.3 | 17.9 | 107.9 |
| 2,4,6-Trimethylphenol | -20.1 | -3.8 | 55.7 | -14.0 | 19.7 | 113.0 |

- Units are kJ mol-1
- Units are J mol-1 K-1
- Compound too insoluble in cyclohexane for log P to be measured

Table. Thermodynamic parameters (mole fraction scale) of partitioning from water to octanol (w'→o) and from water to cyclohexane (w'→c). It has been postulated that DH and DS should be rectilinearly correlated for a system or process operating by a single mechanism, although it has been pointed out that since DH and DS are derived from the same set of data, such enthalpy-entropy compensation could be a statistical artefact. The use of $\Delta G - \Delta H$ is preferred for this reason. In fact, for our data, there is good enthalpy-entropy correlation in the water-octanol system (n = 45, r2 = 0.852) but not in the water-cyclohexane system (n = 42, r2 = 0.529. There is virtually no $\Delta G - \Delta H$ correlation in either system; however, when sub-sets of compounds are examined separately, there are good $\Delta G$-$\Delta H$ correlations in both systems.

For example, for the four methyl-2-nitrophenols in the water-octanol system the coefficient of variation (r2) is 0.865. We interpret this to mean that our data are valid and that there are multiple mechanisms of partitioning within our data-set, with distinctions being made even between, for example, methylphenols and dimethylphenols.

## INTRAMOLECULAR HYDROGEN BONDING

2-Chlorophenol, 2-nitrophenol and the four methyl-2-nitrophenols, 2-nitroresorcinol, 2-hydroxybenzaldehyde, 2-hydroxybenzoic acid (salicylic acid) and 2,6-dihydroxybenzoic acid are all capable of intramolecular hydrogen bonding.

This reduces polarity and intermolecular hydrogen bonding ability and might therefore be expected to increase partition coefficient.

However, there is no consistent effect of intramolecular hydrogen bonding on octanol-water partition coefficients; 2-hydroxybenzaldehyde is more hydrophobic ($\Delta G$ more negative) than are its 3- and 4-isomers, whereas 2-nitrophenol is less hydrophobic than are its isomers.

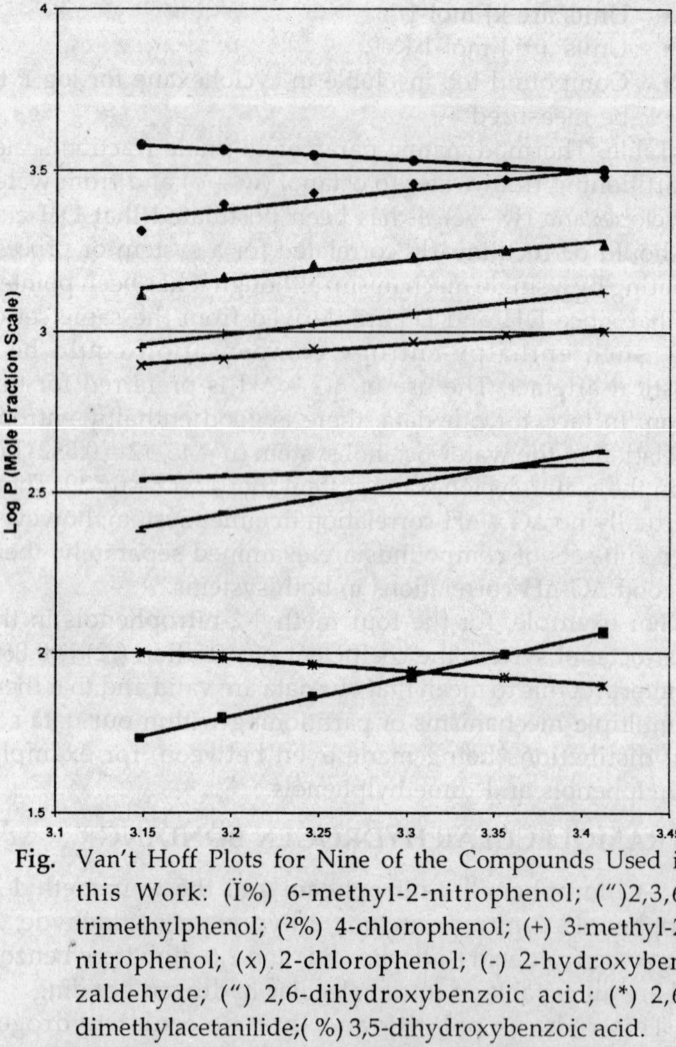


**Fig.** Van't Hoff Plots for Nine of the Compounds Used in this Work: (Ï%) 6-methyl-2-nitrophenol; (")2,3,6-trimethylphenol; (²%) 4-chlorophenol; (+) 3-methyl-2-nitrophenol; (x) 2-chlorophenol; (-) 2-hydroxybenzaldehyde; (") 2,6-dihydroxybenzoic acid; (*) 2,6-dimethylacetanilide;( %) 3,5-dihydroxybenzoic acid.

The hydrogen bond donor ability of water is much greater than is that of octanol, whilst their hydrogen bond acceptor abilities are about the same. The involvement of the carbonyl oxygen of 2-hydroxybenzaldehyde in intramolecular hydrogen bonding thus reduces its interaction with water more than it reduces its interaction with octanol, thereby increasing hydrophobicity. The involvement of the nitro group of 2-

nitrophenol in intramolecular hydrogen bonding still leaves one nitro group oxygen atom available for intermolecular hydrogen bonding with water, thus preventing an increase in hydrophobicity. The situation with salicylic acid is more complex; intramolecular hydrogen bonding leaves a carbonyl oxygen and a hydroxyl group still available for intermolecular hydrogen bonding and it is difficult to predict the overall effect on hydrophobicity. In fact, salicylic acid is more hydrophobic (in the water-octanol system) than are its 3- and 4-isomers.

This is confirmed by its low aqueous solubility (0.18% w/v at 20oC) relative to that of its 3-isomer (0.92% w/v at 20°C). Theoretical studies have also shown that salicylic acid is intramolecularly hydrogen bonded in aqueous solution. The rather lower hydrophobicity of 2,6-dihydroxybenzoic acid suggests that this compound may possess only one intramolecular hydrogen bond in the water-octanol system.

The similarity of the hydrophobicities and aqueous solubilities of all the chlorophenols suggests that 2-chlorophenol may not be intramolecularly hydrogen bonded in water or octanol. It has also been shown that the intramolecular hydrogen bond in 2-chlorophenol is broken in the presence of the hydrogen bond acceptor ethyl propionate.

In the water-octanol system, the enthalpy and entropy of transfer of 2-chlorophenol, 2-nitrophenol, 2-hydroxybenzaldehyde and 2-hydroxybenzoic acid are in each case less negative than are those of the corresponding 3- and 4-isomers. It is probably better to compare values between the 2- and 4-isomers, since the 3-isomers are not capable of classical resonance interaction between substituents.

**Table:** Difference between 2 and 4 substituted phenols of enthalpies (ΔΔH) and entropies (ΔΔS) of transfer from water to octanol and from Water to cyclohexane

| Compound | $\Delta\Delta H_{w-o}$ [a] | $\Delta\Delta S_{w-o}$ [b] | $\Delta\Delta H_{w-c}$ [a] | $\Delta\Delta S_{w-c}$ [b] |
|---|---|---|---|---|
| Chlorophenol | 1.9 | 2.0 | -5.4 | 3.0 |
| Nitrophenol | 16.7 | 33.0 | -17.3 | 5.5 |
| Hydroxybenzaldehyde | 5.9 | 24.7 | -18.8 | 6.8 |
| Hydroxybenzoic acid | 5.1 | 3.0 | [c] | [c] |

- Units are kJ $mol^{-1}$
- Units are J $mol^{-1}$ $K^{-1}$
- Compond too insoluble in cyclophexane for log P to be measured

The positive ΔΔH values for transfer from water to octanol are a consequence of the weak hydrogen bond donor ability of octanol relative to that of water; since an intramolecularly hydrogen bonded –OH group is unavailable for intermolecular hydrogen bond donation to solvent, there is more solute-solvent interaction in water than in octanol, for intramolecularly hydrogen bonded compounds. Hence transfer from water to octanol is enthalpically less favourable. The positive ΔΔS values for transfer from water to octanol confirm, since less solute-solvent interaction in octanol than in water gives rise to greater disorder of solvent molecules consequent upon solute transfer.

The negative ΔΔH values for transfer from water to cyclohexane indicate greater interaction of intramolecularly hydrogen bonded compounds with the non-polar solvent, probably reflecting the less polar nature of these compounds relative to their 4-isomers. This greater interaction accounts also for their low ΔΔS values relative to those for transfer from water to octanol.

We interpret this to indicate that the intramolecular hydrogen bond in 2-chlorophenol is very weak or non-existent in water and octanol, as already suggested by consideration of hydrophobicities. However, the negative ΔΔH value for transfer from water to cyclohexane indicates at least partial intramolecular hydrogen bonding of 2-chlorophenol in cyclohexane, leading to greater interaction with the non-polar solvent. In support of this, there is infrared spectroscopic evidence of such hydrogen bonding of 2-chlorophenol in tetrachloromethane.

Entropically, the increased interaction with cyclohexane is more than offset by the release of structured water upon transfer from the aqueous phase. The strongly positive $\Delta\Delta H_{w'\rightarrow o}$ and $\Delta\Delta S_{w'\rightarrow o}$ values for transfer of 2-nitrophenol indicate that the intramolecular hydrogen bond is intact in this compound in both solvents. The significantly negative $\Delta\Delta H_{w'\to c}$ value, indicating strong solute-cyclohexane interaction, also confirms the existence of the intramolecular hydrogen bond of 2-nitrophenol in cyclohexane solution.

As with 2-chlorophenol, the effect of such increased interaction is more than offset by the release of structured water upon transfer from the aqueous phase, leading to a slightly positive $\Delta\Delta S_{w'\to c}$ value. Similar interpretations can be placed on the thermodynamic parameters for phase transfer of 2- hydroxybenzaldehyde. The low $\Delta\Delta H_{w'\to o}$ value for this compound may indicate that its intramolecular hydrogen bond is not wholly intact in polar solvents.

The $\Delta\Delta H_{w'\to c}$ and $\Delta\Delta S_{w'\to c}$ values, however, confirm that it is intact in cyclohexane solution. The $\Delta\Delta Hw'\to o$ and $\Delta\Delta Sw'\to o$ values for transfer of 2-hydroxybenzoic acid are, like those of 2-chlorophenol, quite close to zero. Aqueous solubility values indicate that salicylic acid is intramolecularly hydrogen-bonded in water and therefore presumably also in octanol. The low $\Delta\Delta H_{w'\to o}$ and $\Delta\Delta S_{w'\to o}$ values for 2-hydroxybenzoic acid are probably due to the fact that the compound still has an –OH group able to form intermolecular hydrogen bonds with both water and octanol. Both 2-nitroresorcinol and 2, 6-dihydroxybenzoic acid are theoretically capable of forming two intramolecular hydrogen bonds.

The increase in hydrophobicity in the water-octanol system of 2-nitroresorcinol relative to that of 2-nitrophenol implies that the compound does possess two intramolecular hydrogen bonds. However, the lower $\Delta Hw'\to o$ value of the former suggests that only one intramolecular hydrogen bond exists in 2-nitroresorcinol in water and octanol solutions, as does the close similarity of the $\Delta S_{w'\to o}$ values of 2-nitrophenol and 2-nitroresorcinol. Berthelot *et al.* have shown that compounds with the potential to form two intramolecular hydrogen bonds actually form only one in the presence of a polar hydrogen bond donor.

The situation appears different for 2-nitroresorcinol in cyclohexane solution. Its $\Delta H_{w'\to c}$ value is more negative than is that of 2-nitrophenol, which suggests that the compound possesses two intramolecular hydrogen bonds in cyclohexane solution. Its relatively low $\Delta S_{w'\rightarrow c}$ value confirms this, since it suggests that there is relatively little structured water around 2-nitroresorcinol molecules in aqueous solution, leading to a relatively small entropy increase upon transfer to cyclohexane. Boykin has observed both intramolecular hydrogen bonds of 2-nitroresorcinol to be intact in toluene solution.

Ultraviolet spectroscopic data support the above hypothesis. In contrast to the behaviour of 2- nitrophenol, we found a pronounced hypsochromic shift of the spectrum of 2-nitroresorcinol on going from cyclohexane to ethanolic or aqueous solution, consistent with breaking of an intramolecular hydrogen bond and consequent loss of planarity of the nitro group. The lower hydrophobicity of 2,6-dihydroxybenzoic acid relative to that of salicylic acid suggests that this compound also has only one intramolecular hydrogen bond in water and octanol. Its entropy of transfer from water to octanol is considerably lower than that of salicylic acid, suggesting hydrogen bond donation to octanol.

The enthalpies of transfer of the two compounds are the same, arising presumably from the very similar hydrogen bond acceptor abilities of water and octanol. That there is certainly one intramolecular hydrogen bond present in 2,6-dihydroxybenzoic acid is shown by the compound's greater hydrophobicity and less negative $\Delta Hw'\rightarrow o$ and $\Delta Sw'\rightarrow o$ values relative to 3,5-dihydroxybenzoic acid.

The water-cyclohexane data also indicate that, unlike the situation in 2- nitroresorcinol, there is only one intramolecular hydrogen bond present in 2,6-dihydroxybenzoic acid in cyclohexane solution. This follows because the changes in $\Delta G_{w'\rightarrow c}$, $\Delta H_{w'\rightarrow c}$ and $\Delta S_{w'\rightarrow c}$ relative to those for salicylic acid all mirror those for transfer from water to octanol. *Steric effects* In the compounds that we have examined, two main steric effects may be postulated: shielding of a polar group and twisting of a polar group out of the plane of the aromatic ring, by adjacent alkyl substitution.

The $\Delta G_{w'\rightarrow o}$, $\Delta H_{w'\rightarrow o}$ and $\Delta S_{w'\rightarrow o}$ values of the three cresols (methylphenols) are all very similar, suggesting that there is no steric shielding of the –OH group in 2-methylphenol. For transfer from water to cyclohexane a slight effect could be apparent, although, bearing in mind the experimental errors in the determination of thermodynamic parameters by the van't Hoff method, this could be an artefact. If the effect is real, our interpretation is that shielding of the –OH group renders 2-methyl-phenol less polar, thereby increasing the partition coefficient and lowering $\Delta G_{w'\rightarrow c}$. However, this is not confirmed by the $\Delta H_{w'\rightarrow c}$ and $\Delta S_{w'\rightarrow c}$ values, so our conclusion must be that the data show no consistent indication of steric shielding in 2-methylphenol. 2,6-Xylenol (2,6-dimethylphenol) does, however, show a significant shielding effect. Transfer from water to octanol involves a less negative $\Delta H_{w'\rightarrow o}$ value and a more positive $\Delta S_{w'\rightarrow o}$ value than those of the other isomers; its $\Delta G_{w'\rightarrow o}$ value, however, is very close to those of its isomers.

The data are interpreted as indicating more structured water around the molecule in aqueous solution, leading to breaking of hydrogen bonds between water molecules upon transfer to octanol and a consequent increase in disorder. Transfer from water to cyclohexane involves a more negative $\Delta G_{w'\rightarrow c}$ value and less positive $\Delta H_{w'\rightarrow c}$ and $\Delta S_{w'\rightarrow c}$ values than those of the other isomers. These all indicate a significant solute-cyclohexane interaction, consistent with appreciable shielding of the –OH group and lowering of effective polarity. Additional methyl substitution does not appear to give rise to additional shielding, even in 2,3,6- trimethylphenol and 2,3,5,6-tetramethylphenol, where a buttressing effect of two adjacent methyl groups might have been expected.

The methyl-substituted benzoic acids show a somewhat similar pattern to the methyl-substituted phenols. Despite the greater size of the carboxyl group compared with the hydroxyl group, there is no evidence of any steric shielding or twisting in 2-methylbenzoic acid in the water-octanol system, since the thermodynamic parameters for all three isomers are very similar. There is some evidence of steric shielding from the water-cyclohexane data, with ΔHo w'→c being less positive and $\Delta G_{w'\rightarrow c}$ being slightly more negative, for 2-methylbenzoic acid compared with the 4- isomer, as is the case for the xylenols. However, the low $\Delta S_{w' \to c}$ value for 2-methylbenzoic acid is not in line with dimethylphenol behaviour and could indicate some twisting of the carboxyl group, thereby reducing the extent of water-structuring in the aqueous phase. Cisarova *et al.* have shown by X-ray diffraction that the carboxyl group of 2,3-dimethylbenzoic acid is non-planar. The data for 3-methylbenzoic acid appear to be anomalous and we have no explanation for this, other than that the data are erroneous. 2,6-Dimethylbenzoic acid, however, shows clear evidence of steric twisting of the carboxyl group, leading to loss of conjugation of the group with the aromatic ring. This renders the group more hydrophilic and hence both $\Delta G_{w' \to o}$ and $\Delta G_{w' \to c}$ are less negative than the corresponding values for 3,5-dimethylbenzoic acid.

Similarly, $\Delta H_{w' \to o}$ and $\Delta H_{w' \to c}$ are both less negative than those of 3,5-dimethylbenzoic acid, which may be attributed to strong interaction of the hydrophilic non-planar carboxyl group of 2,6-dimethylbenzoic acid with water. There is little effect of non-planarity of the carboxyl group on $\Delta S_{w' \to o}$. The $\Delta S_{w' \to c}$ value of 3,5-dimethylbenzoic acid is anomalously low, which may be attributable to self-association of this compound in cyclohexane; benzoic acids are known to undergo self-association at very low concentrations. 2,6-Dimethylbenzoic acid would not be expected to self-associate because its carboxyl group is shielded. Dearden and O'Hara drew attention to the unusual partitioning behaviour of 2-methylacetanilide in the water-octanol system and attributed it to loss of planarity of the acetamido group.

The thermodynamic data confirm this, as $\Delta G_{w' \to o}$ and $\Delta G_{w' \to c}$ are less negative than the corresponding values for both the 3- and 4-isomers, indicating the greater hydrophilicity of the non-planar acetamido group. $\Delta Hw' \to o$ is less negative (more positive), which probably indicates a reduced ability of octanol to interact with a shielded acetamido group; the high $\Delta S_{w' \to o}$ value confirms this, being indicative of greater disorder in octanol than in aqueous solution. $\Delta H_{w' \to c}$, however, is less positive, suggesting a partial screening of polarity and thus greater interaction with cyclohexane. Again, the entropy data confirm this, with $\Delta S_{w'\rightarrow c}$ indicating a lower increase in disorder on water-cyclohexane transfer than for the 3- and 4-isomers. The data for 2,6-dimethylacetanilide, relative to its 3,5-isomer, are entirely consistent with the above explanations. A very interesting aspect of steric effects is shown by the methyl-2-nitrophenols. Dearden and Forbes used ultraviolet absorption spectroscopy to show that in 3-methyl-2-nitrophenol steric hindrance by the methyl group puts a strain upon the intramolecular hydrogen bond. In cyclohexane solution this strain is insufficient fully to rupture the hydrogen bond, as was observed by Boykin for the compound in toluene solution. However, in hydrogen bonding solvents such as water and ethanol, competing solute-solvent hydrogen bonding is sufficient to cause rupture of the intramolecular hydrogen bond.

In 6-methyl-2-nitrophenol, on the other hand, the effect of the methyl group is to push the hydroxyl group closer to the nitro group, thus strengthening the intramolecular hydrogen bond. The partitioning data largely support the above. $\Delta H_{w'\rightarrow o}$ is much more negative for 3-methyl-2-nitrophenol than for the other isomers, suggesting less structured water present and therefore fewer water-water hydrogen bonds to be broken on transfer to octanol; this is confirmed by the very low $\Delta S_{w'\rightarrow o}$ value for 3-methyl-2-nitrophenol. Conversely, for 6-methyl-2-nitrophenol, with a very strong intramolecular hydrogen bond, $\Delta Hw'\rightarrow o$ is actually positive and $\Delta S_{w'\rightarrow o}$ is much more positive than for the other isomers.

The water-cyclohexane transfer data indicate that the intramolecular hydrogen bond in 3-methyl-2-nitrophenol is intact in cyclohexane solution, as $\Delta Hw'\rightarrow c$ values are similar in 3-, 4- and 5-methyl-2-nitrophenols. The high strength of the intramolecular hydrogen bond in 6-methyl-2- nitrophenol is shown by its less negative $\Delta H_{w'\rightarrow c}$ value and its highly positive $\Delta S_{w'\rightarrow c}$ value, both of which indicate a high level of structured water in aqueous solution, leading to rupture of water-water hydrogen bonds and concomitant increase in disorder upon transfer to cyclohexane. Analysis of the thermodynamics of partitioning between water and octanol and between water and cyclohexane has shown that the existence of intramolecular hydrogen bonding in compounds such as 2-nitrophenol, 2-hydroxybenzaldehyde and salicylic acid can be clearly detected.

However, in 2-nitroresorcinol and 2,6-dihydroxybenzoic acid a second intramolecular hydrogen bond does not appear to be formed in polar solvents. Our results also show that no intramolecular hydrogen bond exists in 2-chlorophenol in polar solvents, although it appears to exist in cyclohexane solution. Steric effects such as shielding of polar groups and twisting of conjugated groups out of the plane of the aromatic ring can be seen from the thermodynamics of partitioning of compounds such as 2,6-dimethylphenol, 2,6-dimethylbenzoic acid and 2-methylacetanilide.

## EXPERIMENTAL

Compounds were obtained from a variety of commercial sources and were purified by crystallisation or distillation. Cyclohexane and 1-octanol were of spectroscopic grade quality. Partition coefficients in the octanol-water and cyclohexane-water systems were measured over the temperature range 20-45°C using the filter-probe method modified by Kinkel *et al.* from the original method developed by Cantwell and Mohammed.

It comprised a thermostatted mixing chamber whose contents could be vigorously stirred using a magnetic stirring bar; the output from the chamber was passed through either a hydrophilic or a hydrophobic filter paper and the filtered solution passed into a thermostatted Perkin Elmer 550 ultraviolet absorption spectrophotometer and then returned to the mixing chamber.

A full description of the apparatus is given in Kinkel *et al.*. Solvents were pre-equilibrated for 24 hours in order to achieve mutual saturation. All spectrophotometric calibration plots were rectilinear over the concentration ranges used and all determinations were carried out in duplicate. Partition coefficients were obtained at six temperatures from 20 to 45°C and were calculated on a mole fraction basis and converted to ΔG values using eq.

- ΔH values were obtained from the slope of the log P-1/T plots and ΔS values were obtained from eq.
- Most of the log P-1/T plots were rectilinear, including all those for the cyclohexane-water system. A few showed a slight convex curvature in the octanol-water system and for those compounds the slope was taken as the tangent at 25°C. The compounds with non-rectilinear plots were 3-chlorophenol, 4-chloro-phenol, 2,3,5-trimethylphenol, 2,3,6-trimethylphenol, 2,4,6-trimethylphenol and 3,5-dimethylacetanilide.

## DISSOCIATION AND ASSOCIATION CONSTANT

In chemistry and biochemistry, a dissociation constant is a specific type of equilibrium constant that measures the propensity of a larger object to separate reversibly into smaller components, as when a complex falls apart into its component molecules, or when a salt splits up into its component ions. The dissociation constant is usually denoted $K_d$ and is the inverse of the association constant. In the special case of salts, the dissociation constant can also be called an ionization constant.

For a general reaction

$$A_x B_y \rightleftharpoons xA + yB$$

in which a complex $A_xB_y$ breaks down into $x$ $A$ subunits and $y$ $B$ subunits, the dissociation constant is defined

$$K_d = \frac{[A]^x \times [B]^y}{[A_xB_y]}$$

where $[A]$, $[B]$, and $[A_xB_y]$ are the concentrations of $A$, $B$, and the complex $A_xB_y$, respectively.

### Protein-Ligand binding

The dissociation constant is commonly used to describe the affinity between a ligand (L) (such as a drug) and a protein (P) i.e. how tightly a ligand binds to a particular protein.

Ligand-protein affinities are influenced by non-covalent intermolecular interactions between the two molecules such as hydrogen bonding, electrostatic interactions , hydrophobic and Van der Waals forces. They can also be affected by high concentrations of other macromolecules, which causes macromolecular crowding.

The formation of a ligand-protein complex (C) can be described by a two-state process

$$C \rightleftharpoons P + L$$

the corresponding dissociation constant is defined

$$K_d = \frac{[P][L]}{[C]}$$

where $[P]$, $[L]$ and $[C]$ represent the concentrations of the protein, ligand and complex, respectively.

The dissociation constant has molar units ($M$), which correspond to the concentration of ligand $[L]$ at which the binding site on a particular protein is half occupied, i.e. the concentration of ligand, at which the concentration of protein with ligand bound $[C]$, equals the concentration of protein with no ligand bound $[P]$.

The smaller the dissociation constant, the more tightly bound the ligand is, or the higher the affinity between ligand and protein. For example, a ligand with a nanomolar ($nM$) dissociation constant binds more tightly to a particular protein than a ligand with a micromolar ($\mu M$) dissociation constant.

Sub-nanomolar dissociation constants as a result of non-covalent binding interactions between two molecules are rare. Nevertheless, there are some important exceptions. Biotin and avidin bind with a dissociation constant of roughly $10^{-15}\,M = 1\,fM = 0.000001\,nM$.

Ribonuclease inhibitor proteins may also bind to ribonuclease with a similar $10^{-15}\,M$ affinity. The dissociation constant for a particular ligand-protein interaction can change significantly with solution conditions (e.g. temperature, $pH$ and salt concentration). The effect of different solution conditions is to effectively modify the strength of any intermolecular interactions holding a particular ligand-protein complex together.

Drugs can produce harmful side effects through interactions with proteins for which they were not meant to or designed to interact. Therefore much pharmaceutical research is aimed at designing drugs that bind to only their target proteins with high affinity (typically 0.1-10 *nM*) or at improving the affinity between a particular drug and its *in-vivo* protein target.

### Notation

A dissociation constant $K_a$ is sometimes expressed by its $pK_a$, which is defined as:

$$pK_a = -\log_{10} K_a$$

These $pK_a$'s are mainly used for covalent dissociations (i.e., reactions in which chemical bonds are made or broken) since such dissociation constants can vary greatly.

## DISSOCIATION CONSTANT OF WATER

As a frequently used special case, the dissociation constant of water is often expressed as $K_w$:

$$K_w = [H^+][OH^-]$$

The concentration of water $[H_2O]$ is not included in the definition of $k_w$, for reasons described in the article equilibrium constant. This variation must be taken into account when making precise measurements of quantities such as *pH*.

| *Water temperature* | $K_w$*$10^{-14}$ | $pK_w$ |
|---|---|---|
| 0°C | 0.1 | 14.92 |
| 10°C | 0.3 | 14.52 |
| 18°C | 0.7 | 14.16 |
| 25°C | 1.2 | 13.92 |
| 30°C | 1.8 | 13.75 |
| 50°C | 8.0 | 13.10 |
| 60°C | 12.6 | 12.90 |
| 70°C | 21.2 | 12.67 |
| 80°C | 35 | 12.46 |
| 90°C | 53 | 12.28 |
| 100°C | 73 | 12.14 |

## ACID BASE REACTIONS

For the deprotonation of acids, $K$ is known as $K_a$, the acid dissociation constant. Stronger acids, for example sulfuric or phosphoric acid, have larger dissociation constants; weaker acids, like acetic acid, have smaller dissociation constants.

A molecule can have several acid dissociation constants. In this regard, that is depending on the number of the protons they can give up, we define *monoprotic, diprotic* and *triprotic* acids. The first (e.g. acetic acid or ammonium) have only one dissociable group, the second (carbonic acid, bicarbonate, glycine) have two dissociable groups and the third (e.g. phosphoric acid) have three dissociable groups. In the case of multiple $pK$ values they are designated by indices: $pK_1$, $pK_2$, $pK_3$ and so on. For amino acids, the $pK_1$ constant refers to its carboxyl (–COOH) group, $pK_2$ refers to its amino (–$NH_3$) group and the $pK_3$ is the $pK$ value of its side chain.

$$H_3B \rightleftharpoons H^+ + H_2B^- \quad K_1 = \frac{[H^+].[H_2B^-]}{[H_3B]}$$

$$pK_1 = -\log K_1$$

$$H_2B^- \rightleftharpoons H^+ + HB^2 \quad K_2 = \frac{[H^+].[HB^{-2}]}{[H_2B^-]}$$

$$pK_2 = -\log K_2$$

$$HB^{-2} \rightleftharpoons H^+ + B^{-3} \quad K_3 = \frac{[H^+].[B^{-3}]}{[HB^{-2}]}$$

$$pK_3 = -\log K_3$$

### *pH*

*pH* is a measure of the acidity or basicity of a solution. It is defined as the cologarithm of the activity of dissolved hydrogen ions ($H^+$). Hydrogen ion activity coefficients cannot be measured experimentally, so they are based on theoretical calculations. The pH scale is not an absolute scale; it is relative to a set of standard solutions whose *pH* is established by international agreement.

The concept of *pH* was first introduced by Danish chemist

Søren Peder Lauritz Sørensen at the Carlsberg Laboratory in 1909. Sørensen suggested the notation "*pH*" for convenience, standing for "power of hydrogen", using the cologarithm of the concentration of hydrogen ions in solution, *p* Although this definition has been superseded *p* can be measured if an electrode is calibrated with solution of known hydrogen ion concentration.

Pure water is said to be neutral. The *pH* for pure water at 25°C is close to 7.0. Solutions with a *pH* less than 7 are said to be acidic and solutions with a pH greater than 7 are said to be basic or alkaline. *pH* measurements are important for medicine, biology, chemistry, food science, environmental science, oceanography and many other applications.

## Applications

Pure water has a *pH* around 7; the exact values depends on the temperature. When an acid is dissolved in water the *pH* will be less than 7 and when a base, or alkali is dissolved in water the *pH* will be greater than 7. A solution of a strong acid, such as hydrochloric acid, at concentration 1 mol $dm^{-3}$ has a *pH* of 0.

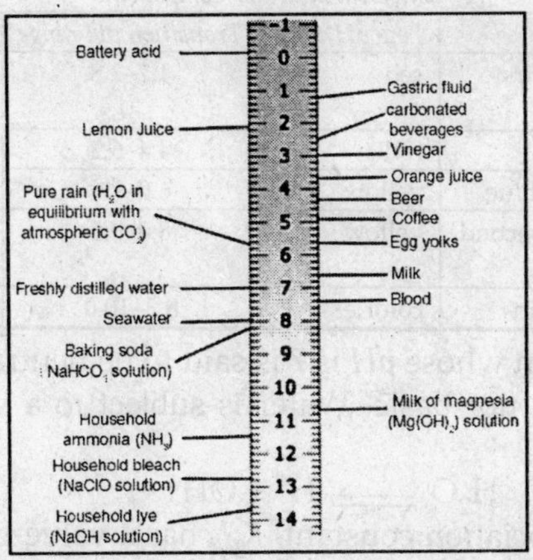


**Fig.** Some Typical *pH* Values

A solution of a strong alkali, such as sodium hydroxide, at concentration 1 mol $dm^{-3}$ has a *pH* of 14. Thus, measured *pH* values will mostly lie in the range 0 to 14. Since *pH* is a logarithmic scale a difference of one *pH* unit is equivalent to a ten-fold difference in hydrogen ion concentration.

Because the glass electrode (and other ion selective electrodes) reponds to activity, the electrode should be calibrated in a medium similar to the one being investigated. For instance, if one wishes to measure the pH of a seawater sample, the electrode should be calibrated in a solution resembling seawater in its chemical composition.

An approximate measure of *pH* may be obtained by using a pH indicator. A *pH* indicator is a substance that changes colour around a particular *pH* value. It is a weak acid or weak base and the colour change occurs around 1 *pH* unit either side of its acid dissociation constant, or $pK_a$, value.

For example, the naturally occuring indicator litmus is red in acidic solutions ($pH < 7$) and blue in alkaline ($pH > 7$) solutions.

Universal indicator consists of a mixture of indicators such that there is a continuous colour change from about *pH* 2 to *pH* 10. Universal indicator paper is simple paper that has been impregnated with universal indicator.

*Universal Indicator Components*

| *Indicator* | *Low pH color* | *Transition pH range* | *High pH color* |
|---|---|---|---|
| Thymol blue (first transition) | red | 1.2–2.8 | orange |
| Methyl red | red | 4.4–6.2 | yellow |
| Bromothymol blue | yellow | 6.0–7.6 | blue |
| Thymol blue (second transition) | yellow | 8.0–9.6 | blue |
| Phenolphthalein | colorless | 8.3–10.0 | purple |

A solution whose *pH* is 7 is said to be neutral, that is, it is neither acidic nor basic. Water is subject to a self-ionisation process.

$$H_2O \rightleftharpoons H^+ + OH^-$$

The dissociation constant, $K_W$, has a value of about $10^{-14}$, so in neutral solution of a salt both the hydrogen ion concentration and hydroxide ion concentration are about $10^{-7}$ mol dm$^{-3}$. The *pH* of pure water decreases with increasing temperatures. For example, the *pH* of pure water at 50°C is

6·55. Note, however, that water that has been exposed to air is mildly acidic. This is because water absorbs carbon dioxide from the air, and carbon dioxide is an acid. After absorption it is slowly converted into the weak acid, carbonic acid, which then dissociates to liberate hydrogen ions.

$$CO_2 + H_2O \rightleftharpoons H_2CO_3 \quad HCO_3^- + H^+$$

The *pH* of distilled water at room temperature is about 5.7.

## CALCULATION OF PH FOR WEAK AND STRONG ACIDS

In the case of a strong acid, there is complete dissociation, so the pH is simply equal to minus the logarithm of the acid concentration. For example, a 0.01 molar solution of hydrochloric acid has a pH of –log(0.01), that is, $pH = 2$.

The *pH* of a solution of a weak acid may be calculated by means of an ICE table. For acids with a $pK_a$ value greater than about 2,

$$pH = \frac{1}{2}(pK_a - \log c_0),$$

where $c_0$ is the concentration of the acid. This is equivalent to Burrows' weak acid *pH* equation

$$pH = -\log_{10}\left(\sqrt{K_a \times c_0}\right)$$

A more general method is as follows. Consider the case of dissolving a weak acid, HA, in water. First write down the equilibrium expression.

$$HA \rightleftharpoons A^- + H^+$$

The equilibrium constant for this reaction is specified by

$$K_a = \frac{[A^-][H^+]}{[HA]}$$

where [] indicates a concentration. The analytical concentration of the two reagents, $C_A$ for $[A^-]$ and $C_H$ for $[H^+]$ must be equal to the sum of concentrations of those species that contain the reagents. $C_H$ is the concentration of added mineral acid.

$$C_A = [A^-] + K_a[A^-][H^+]$$
$$C_H = [H^+] + K_a[A^-][H^+]$$

From the first equation

$$[A^-]=\frac{C_A}{1+K_a[H^+]}$$

Substitution of this expression into the second equation gives

$$C_H=[H^+]+\frac{K_aC_A[H^+]}{1+K_a[H^+]}$$

This simplifies to a quadratic equation in the hydrogen ion concentration

$$K_a[H^+]^2+(1+(C_A-C_H)K_a)[H^+]-C_H=0$$

Solution of this equation gives [H$^+$] and hence *pH*.

This method can also be used for polyprotic acids. For example, for the diprotic acid oxalic acid, writing $A^{2-}$ for the oxalate ion,

$$C_A = [A^{2-}] + \beta_1[A^{2-}][H^+] + \beta_2[A^{2-}][H^+]^2$$
$$C_H = [H^+] + \beta_1[A^{2-}][H^+] + 2\beta_2[A^{2-}][H^+]^2$$

where $\beta_1$ and $\beta_2$ are cumulative protonation constants. Following the same procedure of substituting from the first equation into the second, a cubic equation in [H$^+$] results. In general, the degree of the equation is one more than the number of ionisable protons. The solution of these equations can be obtained relatively easily with the aid of a spreadsheet such as EXCEL or Origin.

## *pH* IN NATURE

*pH*-dependent plant pigments that can be used as *pH* indicators occur in many plants, including hibiscus, marigold, red cabbage (anthocyanin), and red wine.

### Seawater

The *pH* of seawater is very important and there is evidence for ocean acidification. Distinct *pH* scales exist depending on the method used to calibrate the electrode.

1. *Using standard buffers:* The ionic strength of standard buffer solutions is much lower, at about 0.1 M, than that of seawater, which is about 0.7 M. Consequently

they are not recommended for use in measuring the *pH* of seawater.

2. A set of buffers based on artificial seawater was developed. This *pH* scale is referred to as the total scale, denoted by $pH_T$. The total scale was defined using a medium containing sulfate ions, which are subject to the proton absorbing equilibrium $H^+ + SO_4^{2-} \rightleftharpoons HSO_4^-$.
3. The free scale, denoted by $pH_F$, omits the effect of sulfate ions and focuses solely on $[H^+]_F$, in principle making it a simpler representation of hydrogen ion concentration. Analytically, only $[H^+]_T$ can be determined, therefore, $[H^+]_F$ must be estimated using the $[SO_4^{2-}]$ and the dissociation constant constant of $HSO_4^-$. The utility of this scale is limited by the complexity of the calculations. *pH* values measured on the the free scale differ by up to 0.12 *pH* units from both the total and seawater scales.
4. The seawater scale, denoted by $pH_{SWS}$, takes account of the fact that hydrogen fluoride is a weak acid, $H^+ + F^- \rightleftharpoons HF$. However, the concentration of sulfate ions is about 400 times larger than the concentration of fluoride, so the difference between the total and seawater scales is very small.

## BUFFER SOLUTION

A buffer solution is an aqueous solution consisting of a mixture of a weak acid and its conjugate base or a weak base and its conjugate acid. It has the property that the pH of the solution changes very little when a small amount of acid or base is added to it. Buffer solutions are used as a means of keeping *pH* at a nearly constant value in a wide variety of chemical applications.

In a simple buffer solution there is an equilibrium between a weak acid, HA, and its conjugate base, $A^-$.

$$HA + H_2O \rightleftharpoons H_3O^+ + A^-$$

When hydrogen ions are added to the solution the equilibrium moves to the left, in accordance with Le Chatelier's principle, as there are hydrogen ions on the right-hand side of the equilibrium expression. When hydroxide ions are added the equilibrium moves to the right as hydrogen ions are removed in the reaction $H^+ + OH^- \rightarrow H_2O$. Thus, some of the added reagent is consumed in shifting the equilibrium and the *pH* changes by less than it would do if the solution were not buffered.

The acid dissociation constant for a weak acid, HA, is defined as

$$K_a = \frac{[H^+][A^-]}{[HA]}$$

Simple manipulation with logarithms gives the Henderson-Hasselbalch equation, which describes *pH* in terms of $pK_a$

$$pH = pK_a + \log_{10}\frac{[A^-]}{[HA]}.$$

In this equation $[A^-]$ is the concentration of the conjugate base and [HA] is the concentration of the acid.

It follows that when the concentrations of acid and conjugate base are equal, often described as half-neutralization, $pH = pK_a$. In general, the *pH* of a buffer solution may be easily calculated, knowing the composition of the mixture, by means of an ICE table.

The same considerations apply to a mixture of a weak base, B and its conjugate acid $BH^+$.

$$B + H_2O \rightleftharpoons BH^+ + OH^-.$$

The $pK_a$ value to be used is that of the acid conjugate to the base.

In general a buffer solution may be made up of more than one weak acid and its conjugate base; if the individual buffer regions overlap a wider buffer region is created by mixing the two buffering agents.

## Buffer Capacity

Buffer capacity is a quantitative measure of the resistance

of a buffer solution to $pH$ change on addition of hydroxide ions. It can be defined as follows.

$$\text{buffer capacity} = \frac{dn}{d(pH)}$$

where dn is an infinitesimal amount of added base and $d(pH)$ is the resulting infinitesimal change in $pH$. With this definition the buffer capacity can be expressed as

$$\frac{dn}{d(pH)} = 2.303\left(\frac{K_w}{H^+} + [H^+) + \frac{C_A K_a [H^+]}{(K_a + [H^+])^2}\right),$$

where $K_w$ is the self-ionization constant of water and $C_A$ is the analytical concentration of the acid, equal to [HA] + [A⁻].

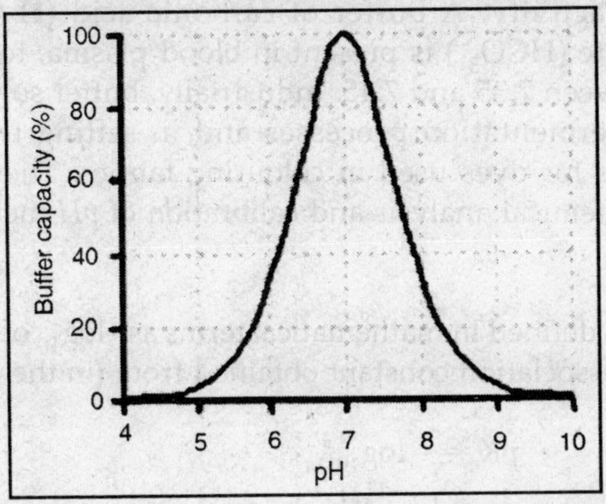


**Fig.** Buffer Capacity for $pK_a = 7$ as Percentage of Maximum

The term $K_w/[H^+]$ becomes significant at $pH$ greater than about 11.5 and the second term becomes significant at $pH$ less than about 2. Both these terms are properities of water and are independent of the weak acid. Considering the third term, it follows that

1. Buffer capacity of a weak acid reaches its maximum value when $pH = pK_a$
2. At $pH = pK_a \pm 1$ the buffer capacity falls to 33% of the maximum value. This is the approximate range within which buffering by a weak acid is effective. Note: at $pH = pK_a - 1$, The Henderson-Hasselbalch equation shows that the ratio [HA] : [A⁻] is 10 : 1.

3. Buffer capacity is directly proportional to the analytical concentration of the acid.

## Applications

Their resistance to changes in pH makes buffer solutions very useful for chemical manufacturing and essential for many biochemical processes. The ideal buffer for a particular *pH* has a $pK_a$ equal to that *pH*, since such a solution has maximum buffer capacity. Buffer solutions are necessary to keep the correct *pH* for enzymes in many organisms to work. Many enzymes work only under very precise conditions; if the *pH* strays too far out of the margin, the enzymes slow or stop working and can denature, thus permanently disabling its catalytic activity. A buffer of carbonic acid ($H_2CO_3$) and bicarbonate ($HCO_3^-$) is present in blood plasma, to maintain a pH between 7.35 and 7.45. Industrially, buffer solutions are used in fermentation processes and in setting the correct conditions for dyes used in colouring fabrics. They are also used in chemical analysis and calibration of *pH* meters.

## $pK_A$

$pK_a$ is defined in mathematical terms as $-\log_{10}$ of $K_a$, where $K_a$ is the association constant obtained from (in the case of an Acid):

$$pK_a = -\log_{10} K_a$$

$$K_a = \frac{[A^{-1}][H^+]}{[AH]}$$

Since $pH = -\log_{10}[H^+]$, the relationship between $pK_a$ and *pH* can be expressed as follows,

$$pH = pK_a - \log \frac{[AH]}{[A^-]}$$

From this equation, it is clear that in a practical sense a $pK_a$ value refers to the *pH* at which you have exactly the same proportion of Neutral and Ionized species present in solution (for a simple monoprotic compound).

The relationship between $pK_a$ and *pH* is most clearly expressed via a Distribution of Species Plot:

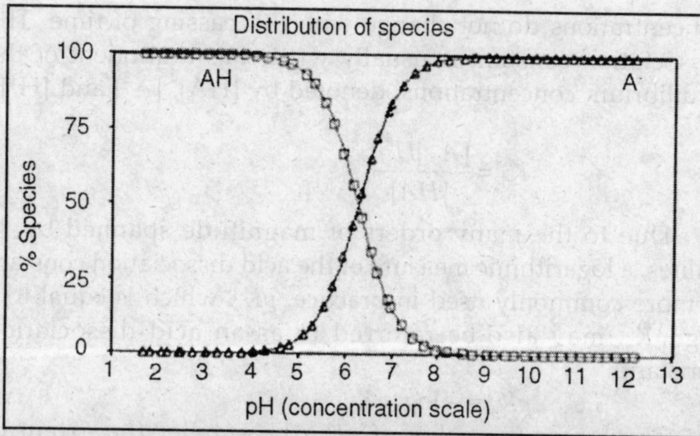


This example shows the Distribution of Species that one might expect for a monoprotic acid with a $pK_a$ of 6.27 (for example: Flumequine). At high *pH*, above the $pK_a$, it exists mostly in its ionized form [$A^-$]. At low pH, below the $pK_a$, it exists mostly in its neutral form [AH]. Where the *pH* is exactly 6.27, there is exactly 50% of both forms present in solution. Sirius Analytical are renowned experts in the measurement of $pK_a$, we offer measurements by pH-metric, *pH-UV* and high throughput methods and we can fully resolve even the most complex cases of ionic equilibria.

## ACID DISSOCIATION CONSTANT

An acid dissociation constant, $K_a$, (also known as acidity constant, or acid-ionization constant) is a quantitative measure of the strength of an acid in solution. It is the equilibrium constant for a chemical reaction known as dissociation in the context of acid-base reactions. The equilibrium can be written symbolically as

$$HA \rightleftharpoons A^- + H^+,$$

where HA is a generic acid which dissociates by splitting into $A^-$, known as the conjugate base of the acid, and the hydrogen ion or proton, $H^+$, which, in the case of aqueous solutions, exists as a solvated hydronium ion. HA represents acetic acid, and $A^-$ the acetate ion. The chemical species HA, $A^-$ and $H^+$ are said to be in equilibrium when their concentrations do not change with the passing of time. The dissociation constant is usually written as a quotient of the equilibrium concentrations, denoted by [HA], $[A^-]$ and $[H^+]$:

$$K_a = \frac{[A^-][H^+]}{[HA]}$$

Due to the many orders of magnitude spanned by $K_a$ values, a logarithmic measure of the acid dissociation constant is more commonly used in practice. $pK_a$, which is equal to $-\log_{10} K_a$, may also be referred to as an acid dissociation constant:

$$pK_a = -\log_{10} K_a$$

The larger the value of $pK_a$, the smaller the extent of dissociation. A weak acid has a $pK_a$ value in the approximate range –2 to 12 in water. Acids with a $pK_a$ value of less than about –2 are said to be strong acids; a strong acid is almost completely dissociated in aqueous solution, to the extent that the concentration of the undissociated acid becomes undetectable. $pK_a$ values for strong acids can, however, be estimated by theoretical means or by extrapolating from measurements in non-aqueous solvents in which the dissociation constant is smaller, such as acetonitrile and dimethyl sulphoxide.

The acid dissociation constant for an acid is a direct consequence of the underlying thermodynamics of the dissociation reaction; the $pK_a$ value is directly proportional to the standard Gibbs free energy change for the reaction. The value of the $pK_a$ changes with temperature and can be understood qualitatively based on Le Chatelier's principle: when the reaction is endothermic, the $pK_a$ decreases with increasing temperature; the opposite is true for exothermic reactions. The underlying structural factors that influence the magnitude of the acid dissociation constant include Pauling's rules for acidity constants, inductive effects, mesomeric effects, and hydrogen bonding.

The quantitative behaviour of acids and bases in solution can only be understood if their $pK_a$ values are known. In particular, the $pH$ of a solution can be predicted when the analytical concentration and $pK_a$ values of all acids and bases are known; conversely, it is possible to calculate the equilibrium concentration of the acids and bases in solution when the pH is known. These calculations find application in many different areas of chemistry, biology, medicine, and geology.

For example, many compounds used for medication are weak acids or bases, and a knowledge of the $pK_a$ values, together with the water–octanol partition coefficient, can be used for estimating the extent to which the compound enters the blood stream.

Acid dissociation constants are also essential in aquatic chemistry and chemical oceanography, where the acidity of water plays a fundamental role. In living organisms, acid-base homeostasis and enzyme kinetics are dependent on the $pK_a$ values of the many acids and bases present in the cell and in the body.

In chemistry, a knowledge of $pK_a$ values is necessary for the preparation of buffer solutions and is also a prerequisite for a quantitative understanding of the interaction between acids or bases and metal ions to form complexes. Experimentally, $pK_a$ values can be determined by potentiometric ($pH$) titration, but for values of $pK_a$ less than about 2 or more than about 11 spectrophotometric or NMR measurements may be required due to practical difficulties with $pH$ measurements.

## Equilibrium Constant

An acid dissociation constant is a particular example of an equilibrium constant. For the specific equilibrium between a monoprotic acid, HA and its conjugate base $A^-$, in water,

$$HA + H_2O \rightleftharpoons A^- + H_3O^+$$

the thermodynamic equilibrium constant, $K^{\varnothing}$ can be defined by

$$K^{\varnothing} = \frac{\{A^-\}\{H_3O^+\}}{\{HA\}\{H_2O\}}$$

where $\{A\}$ is the activity of the chemical species $A$ etc. $K^{\varnothing}$ is dimensionless since activity is dimensionless. Activities of the products of dissociation are placed in the numerator, activities of the reactants are placed in the denominator.

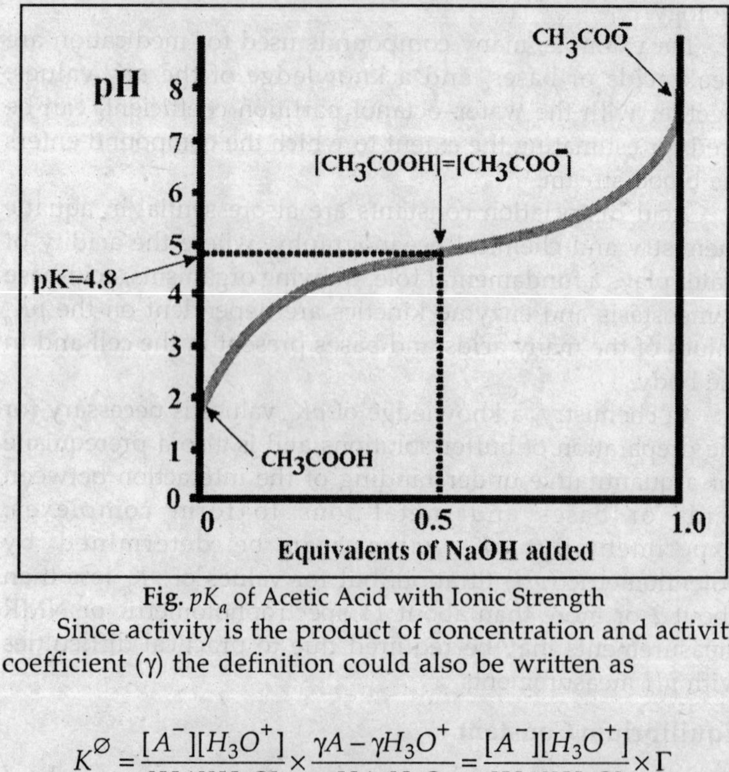


**Fig.** $pK_a$ of Acetic Acid with Ionic Strength

Since activity is the product of concentration and activity coefficient (γ) the definition could also be written as

$$K^{\varnothing} = \frac{[A^-][H_3O^+]}{[HA][H_2O]} \times \frac{\gamma A - \gamma H_3O^+}{\gamma HA \gamma H_2O} = \frac{[A^-][H_3O^+]}{[HA][H_2O]} \times \Gamma$$

where $[HA]$ represents the concentration of $HA$ and $\Gamma$ is a quotient of activity coefficients.

To avoid the complications involved in using activities, dissociation constants are determined, where possible, in a medium of high ionic strength, that is, under conditions in which $\tilde{A}$ can be assumed to be always constant. For example, the medium might be a solution of 0.1 M sodium nitrate or 3 M potassium perchlorate (1 M = 1 mol·dm$^{-3}$, a unit of molar concentration). Furthermore, in all but the most concentrated solutions it can be assumed that the concentration of water, $[H_2O]$, is constant, approximately 55 mol·dm$^{-3}$. On dividing $K^{\varnothing}$ by the constant terms and writing $[H^+]$ for the concentration of the hydronium ion the expression

$$K_a = \frac{[A^-][H^+]}{[HA]}$$

is obtained. This is the definition in common use. p$K_a$ is defined as $-\log_{10} K_a$. Dissociation constant values refer to the specific ionic medium used in their determination and that different values are obtained with different conditions for acetic acid in the illustration above. Although $K_a$ appears to have the dimension of concentration it must in fact be dimensionless or it would not be possible to take its logarithm. The illusion is the result of omitting the constant term $[H_2O]$ from the defining expression. Nevertheless it is not unusual, particularly in texts relating to biochemical equilibria.

## Water Self-ionization

Water has both acidic and basic properties. The equilibrium constant for the equilibrium

$$2\,H_2O \rightleftharpoons OH^- + H_3O^+$$

is given by

$$K_a = \frac{[H_3O^+][OH^-]}{[H_2O]^2}$$

When, as is usually the case, the concentration of water can be assumed to be constant, this expression may be replaced by

$$K_w = [H_3O^+]\,[OH^-]$$

The value of $K_w$ at SATP is $1.0 \times 10^{-14}$. The self-ionization constant of water, $K_w$, is thus just a special case of an acid dissociation constant.

## Bases

Historically the equilibrium constant $K_b$ for a base was defined as the *association* constant for protonation of the base, *B*, to form the conjugate acid, $HB^+$.

$$B + H_2O \rightleftharpoons HB^+ + OH^-$$

Using similar reasoning to that used before

$$K_b = \frac{[HB^+][OH^-]}{[B]}$$

In water, the concentration of the hydroxide ion, $[OH^-]$, is related to the concentration of the hydrogen ion by

$$K_w = [H^+][OH^-],$$

therefore

$$[OH^-] = \frac{K_w}{[H^+]}$$

Substitution of the expression for $[OH^-]$ into the expression for $K_b$ gives

$$K_b = \frac{[HB^+]K_w}{[B][H^+]} = \frac{K_w}{K_a}$$

When $K_a$, $K_b$ and $K_w$ are determined under the same conditions of temperature and ionic strength, it follows, taking cologarithms, that $pK_b = pK_w - pK_a$. In aqueous solutions at 25 °C, $pK_w$ is 13.9965, so $pK_b \sim 14 - pK_a$. In effect there is no need to define $pK_b$ separately from $pK_a$, but it is done here because $pK_b$ values can be found in the older literature.

## Temperature Dependence

All equilibrium constants vary with temperature according to the van't Hoff equation

$$\frac{d \ln K}{dT} = \frac{\Delta H^{\varnothing}}{RT^2}$$

$R$ is the gas constant and $T$ is the temperature in Kelvin. Thus, for exothermic reactions, (the standard enthalpy change, $\Delta H^{\varnothing}$, is negative) $K$ decreases with temperature, but for endothermic reactions ($\Delta H^{\varnothing}$ is positive) $K$ increases with temperature.

## SOLUBILITY

Solubility is the physical property referring to the ability of a given substance, the solute, to dissolve in a solvent. It is measured in terms of the maximum amount of solute dissolved in a solvent at equilibrium. The resulting solution is called a saturated solution. Certain liquids are soluble in all proportions with a given solvent, such as ethanol in water. This property is known as miscibility. Under certain conditions the equilibrium solubility can be exceeded to give a so-called supersaturated solution, which is metastable.

In a solution, the solvent is generally a liquid, which can be a pure substance or a mixture. The species that dissolves, the solute, can be a gas, another liquid, or a solid. Solubilities range widely, from infinitely soluble such as ethanol in water, to poorly soluble, such as silver chloride in water. The term *insoluble* is often applied to poorly soluble compounds, although in some cases insolubility means that a compound is very poorly soluble.

## Molecular View

Solubility occurs under dynamic equilibrium, which means that solubility results from the simultaneous and opposing processes of dissolution and precipitation. The solubility equilibrium occurs when the two processes proceed at a constant rate.

The solubility equilibrium is relatively straightforward for covalent substances such as benzene. When dissolved in water, the benzene molecules remain intact but interact with and are generally surrounded by molecules of water. When, however, an ionic compound such as sodium chloride (NaCl) dissolves in water, the sodium chloride lattice dissociates into individual ions that are solvated or surrounded by water molecules. Nonetheless, NaCl is said to dissolve in water, because evaporation of the solvent returns crystalline NaCl.

The term "dissolving" is sometimes applied to an irreversible chemical reaction, as with iron in nitric acid, but in such a case the thermodynamic concept of solubility does not apply.

When a solute dissolves, it may form several species in the solution. For example, an aqueous suspension of ferrous hydroxide, $Fe(OH)_2$, will contain the series $[Fe(H_2O)_{6-x}(OH)_x]^{(2-x)+}$ as well as other oligomeric species. Furthermore, the solubility of ferrous hydroxide and the composition of its soluble components depends on *pH*.

In general, solubility in the solvent phase can be given only for a specific solute which is thermodynamically stable, and the value of the solubility will include all the species in the solution (in the example above, all the iron-containing complexes).

## Criteria of Solubility

Solubility is defined for specific phases. For example, the solubility of aragonite and calcite in water are expected to differ, even though they are both polymorphs of calcium carbonate, and have the same chemical formula.

The solubility of one substance dissolving in another is determined by the balance of intermolecular forces between the solvent and solute, and the entropy change that accompanies the solvation. Factors such as temperature and pressure will alter this balance, thus changing the solubility.

Solubility may also strongly depend on the presence of other species dissolved in the solvent, for example, complex-forming anions (ligands) in liquids. Solubility will also depend on the excess or deficiency of a common ion in the solution, a phenomenon known as the common-ion effect. To a lesser extent, solubility will depend on the ionic strength of liquid solutions. The last two effects can be quantified using the equation for solubility equilibrium.

For a solid that dissolves in a redox reaction, solubility is expected to depend on the potential (within the range of potentials under which the solid remains the thermodynamically stable phase). For example, solubility of gold in high-temperature water is observed to be almost an order of magnitude higher when the redox potential is controlled using a highly-oxidizing $Fe_3O_4 - Fe_2O_3$ redox buffer than with a moderately-oxidizing Ni – NiO buffer.

Solubility (metastable) also depends on the physical size of the crystal or droplet of solute (or, strictly speaking, on the specific or molar surface area of the solute). For highly defective crystals, solubility may increase with the increasing degree of disorder. Both of these effects occur because of the dependence of solubility constant on the Gibbs energy of the crystal. The last two effects, although often difficult to measure, are of practical importance. For example, they provide the driving force for precipitate aging (the crystal size spontaneously increasing with time).

## Temperature

The solubility of a given solute in a given solvent typically depends on temperature. For many solids dissolved in liquid water, the solubility increases with temperature up to 100°C. In liquid water at high temperatures, (e.g., that approaching the critical temperature), the solubility of ionic solutes tends to decrease due to the change of properties and structure of liquid water; the lower dielectric constant results in a less polar solvent.

Gaseous solutes exhibit more complex behaviour with temperature. As the temperature is raised, gases usually become less soluble in water, but more soluble in organic solvents.

The chart shows solubility curves for some typical solid inorganic salts. Many salts behave like barium nitrate and disodium hydrogen arsenate, and show a large increase in solubility with temperature. Some solutes (e.g. NaCl in water) exhibit solubility which is fairly independent of temperature. A few, such as cerium(III) sulfate, become less soluble in water as temperature increases. This is sometimes referred to as "retrograde" or "inverse" solubility. Occasionally, a more complex pattern is observed, as with sodium sulfate, where the less soluble decahydrate crystal loses water of crystallization at 32°C to form a more soluble anhydrous phase.

The solubility of organic compounds nearly always increases with temperature. The technique of recrystallization, used for purification of solids, depends on a solute's different solubilities in hot and cold solvent. A few exceptions exist, such as certain cyclodextrins.

**Pressure**

For condensed phases (solids and liquids), the pressure dependence of solubility is typically weak and usually neglected in practice. Assuming an ideal solution, the dependence can be quantified as:

$$\left(\frac{\partial \ln N_i}{\partial P}\right)_T = -\frac{V_{i,aq} - V_{i,cr}}{RT}$$

where the index *i* iterates the components, $N_i$ is the mole fraction of the $i^{th}$ component in the solution, *P* is the pressure, the index *T* refers to constant temperature, $V_{i,aq}$ is the partial molar volume of the $i^{th}$ component in the solution, $V_{i,cr}$ is the partial molar volume of the $i^{th}$ component in the dissolving solid, and *R* is the universal gas constant.

The pressure dependence of solubility does occasionally have practical significance. For example, precipitation fouling of oil fields and wells by calcium sulfate (which decreases its solubility with decreasing pressure) can result in decreased productivity with time.

## SOLUBILITY OF GASES

Henry's law is used to quantify the solubility of gases in solvents. The solubility of a gas in a solvent is directly proportional to the partial pressure of that gas above the solvent. This relationship is written as:

$$p = kc$$

where k is a temperature-dependent constant (for example, 769.2 L • atm/mol for dioxygen ($O_2$) in water at 298 K), *p* is the partial pressure (atm), and *c* is the concentration of the dissolved gas in the liquid (mol/L).

## POLARITY

A popular aphorism used for predicting solubility is "*like dissolves like*". This indicates that a solute will dissolve best in a solvent that has a similar polarity to itself. This is a rather simplistic view, since it ignores many solvent-solute interactions, but it is a useful rule-of-thumb. For example, a very polar (hydrophilic) solute such as urea is very soluble in highly polar water, less soluble in fairly polar methanol, and practically insoluble in non-polar solvents such as benzene. In contrast, a non-polar or lipophilic solute such as naphthalene is insoluble in water, fairly soluble in methanol, and highly soluble in non-polar benzene.

Liquid solubilities also generally follow this rule. Lipophilic plant oils, such as olive oil and palm oil, dissolve in non-polar solvents such as alkanes, but are less soluble in polar liquids such as water.

Synthetic chemists often exploit differences in solubilities to separate and purify compounds from reaction mixtures, using the technique of liquid-liquid extraction.

## RATE OF DISSOLUTION

Dissolution is not always an instantaneous process. It is fast when salt and sugar dissolve in water but much slower for a tablet of aspirin or a large crystal of hydrated copper(II) sulfate. The speed at which a solid dissolves may depend on its crystalline properties (i.e. whether it is crystalline or amorphous), and the crystal size) and the presence of polymorphism. This is important in many practical systems, for example in designing methods for controlled drug delivery. Critically, the dissolution rate depends on the presence of mixing and other factors that determine the degree of undersaturation in the liquid solvent film immediately adjacent to the solid solute crystal. In some cases, solubility equilibria can take a long time to establish (hours, days, months, or many years; depending on the nature of the solute and other factors). In practice, it means that the amount of solute in a solution is not always determined by its thermodynamic solubility, but may depend on kinetics of dissolution (or precipitation).

The rate of dissolution and solubility should not be confused as they are different concepts, kinetic and thermodynamic, respectively.

### Quantification of Solubility

Solubility is commonly expressed as a concentration,

either by mass (g of solute per kg of solvent, g per dL (100 mL) of solvent), molarity, molality, mole fraction or other similar descriptions of concentration. The maximum equilibrium amount of solute that can dissolve per amount of solvent is the solubility of that solute in that solvent under the specified conditions. The advantage of expressing solubility in this manner is its simplicity, while the disadvantage is that it can strongly depend on the presence of other species in the solvent (for example, the common ion effect).

Solubility constants are used to describe saturated solutions of ionic compounds of relatively low solubility. The solubility constant is a special case of an equilibrium constant. It describes the balance between dissolved ions from the salt and undissolved salt. The solubility constant is also "applicable" (i.e. useful) to precipitation, the reverse of the dissolving reaction. As with other equilibrium constants, temperature can affect the numerical value of solubility constant. The solubility constant is not as simple as solubility, however the value of this constant is generally independent of the presence of other species in the solvent.

The Flory-Huggins solution theory is a theoretical model describing the solubility of polymers. The Hansen Solubility Parameters and the Hildebrand solubility parameters are empirical methods for the prediction of solubility. It is also possible to predict solubility from other physical constants such as the enthalpy of fusion.

The partition coefficient (Log P) is a measure of differential solubility of a compound in a hydrophobic solvent (octanol) and a hydrophilic solvent (water). The logarithm of these two values enables compounds to be ranked in terms of hydrophilicity (or hydrophobicity).

## Applications

Solubility is of fundamental importance in a large number of scientific disciplines and practical applications, ranging from ore processing, to the use of medicines, and the transport of pollutants.

Solubility is often said to be one of the "characteristic properties of a substance," which means that solubility is commonly used to describe the substance, to indicate a substance's polarity, to help to distinguish it from other substances, and as a guide to applications of the substance. For example, indigo is described as "insoluble in water, alcohol, or ether but soluble in chloroform, nitrobenzene, or concentrated sulfuric acid".

Solubility of a substance is useful when separating mixtures. For example, a mixture of salt (sodium chloride) and silica may be separated by dissolving the salt in water, and filtering off the undissolved silica. The synthesis of chemical compounds, by the milligram in a laboratory, or by the ton in industry, both make use of the relative solubilities of the desired product, as well as unreacted starting materials, byproducts, and side products to achieve separation.

Another example of this is the synthesis of benzoic acid from phenylmagnesium bromide and dry ice. Benzoic acid is more soluble in an organic solvent such as dichloromethane or diethyl ether, and when shaken with this organic solvent in a separatory funnel, will preferentially dissolve in the organic layer. The other reaction products, including the magnesium bromide, will remain in the aqueous layer, clearly showing that separation based on solubility is achieved. This process, known as liquid-liquid extraction, is an important technique in synthetic chemistry.

## Solubility of Ionic Compounds in Water

Some ionic compounds (salts) dissolve in water, which arises because of the attraction between positive and negative charges. For example, the salt's positive ions (i.e. $Ag^+$) attract the partially-negative oxygens in $H_2O$. Likewise, the salt's negative ions (i.e. $Cl^-$) attract the partially-positive hydrogens in $H_2O$. Note: oxygen is partially-negative because it is more electronegative than hydrogen, and vice-versa.

$$AgCl_{(s)} \rightleftharpoons Ag^+_{(aq)} + Cl^-_{(aq)}$$

However, there is a limit to how much salt can be dissolved in a given volume of water. This amount is given by the solubility product, $K_{sp}$. This value depends on the type of salt (AgCl vs. NaI, for example), temperature, and the common ion effect.

One can calculate the amount of AgCl that will dissolve in 1 liter of water, some algebra is required.

$K_{sp} = [Ag^+] \times [Cl^-]$ (definition of solubility product)

$K_{sp} = 1.8 \times 10^{-10}$ (from a table of solubility products)

$[Ag^+] = [Cl^-]$, in the absence of other silver or chloride salts,

$[Ag^+]^2 = 1.8 \times 10^{-10}$

$[Ag^+] = 1.34 \times 10^{-5}$

The result: 1 liter of water can dissolve $1.34 \times 10^{-5}$ moles of $AgCl_{(s)}$ at room temperature. Compared with other types of salts, AgCl is poorly soluble in water. In contrast, table salt (NaCl) has a higher $K_{sp}$ and is, therefore, more soluble.

*Main Article: Solubility chart*

| *Soluble* | *Insoluble* |
|---|---|
| Group I and $NH_4^+$ compounds | carbonates (except Group I, $NH_4^+$ and uranyl compounds) |
| nitrates | sulfites (except Group I and $NH_4^+$ compounds) |
| acetates (ethanoates) (except $Ag^+$ compounds) | phosphates (except Group I and $NH_4^+$ compounds) |
| chlorides, bromides and iodides (except $Ag^+$, $Pb^{2+}$, $Cu^+$ and $Hg_2^{2+}$) | hydroxides and oxides (except Group I, $NH_4^+$, $Ba^{2+}$, $Sr^{2+}$ and $Tl^+$) |
| sulfates (except $Ag^+$, $Pb^{2+}$, $Ba^{2+}$, $Sr^{2+}$ and $Ca^{2+}$) | sulfides (except Group I, Group II and $NH_4^+$ compounds) |

## SOLUBILITY OF ORGANIC COMPOUNDS

The principle outlined above under polarity, that *like dissolves like,* is the usual guide to solubility with organic systems. For example, petroleum jelly will dissolve in gasoline; both of which are lipophilic. This is because petroleum jelly consists of long carbon chains, as does the gasoline. It will not, on the other hand, dissolve in alcohol or water, since the polarity of these solvents is too high. Sugar will not dissolve in gasoline, since sugar is too polar in comparison with gasoline. A mixture of gasoline and sugar can therefore be separated by filtration, or extraction with water.

### Solid Solution

This term is often used in the field of metallurgy to refer to the extent that an alloying element will dissolve into the base metal without forming a separate phase. The solubility line (or curve) is the line (or lines) on a phase diagram which give the limits of solute addition. That is, the lines show the maximum amount of a component that can be added to another component and still be in solid solution. In microelectronic fabrication, solid solubility refers to the maximum concentration of impurities one can place into the substrate.

## INCONGRUENT DISSOLUTION

Many substances dissolve congruently, i.e., the composition of the solid and the dissolved solute stoichiometrically match. However, some substances may dissolve incongruently, whereby the composition of the solute in solution does not match that of the solid. This is accompanied by alteration of the "primary solid" and possibly formation of a secondary solid phase. However, generally, some primary solid also remains and a complex solubility equilibrium establishes. For example, dissolution of albite may result in formation of gibbsite.

$$NaAlSi_3O_8(s) + H^+ + 7H_2O = Na^+ + Al(OH)_3(s) + 3H_4SiO_4.$$

In this case, the solubility of albite is expected to depend on the solid-to-solvent ratio. This kind of solubility is of great importance in geology, where it results in formation of metamorphic rocks.

## HYDROPHOBICITY

In chemistry, hydrophobicity (from the combining form of water in Attic Greek *hydro-* and for fear *phobos*) refers to the physical property of a molecule (known as a hydrophobe) that is repelled from a mass of water.

Hydrophobic molecules tend to be non-polar and thus prefer other neutral molecules and nonpolar solvents. Hydrophobic molecules in water often cluster together forming micelles. Water on hydrophobic surfaces will exhibit a high contact angle. Examples of hydrophobic molecules include the alkanes, oils, fats, and greasy substances in general. Hydrophobic materials are used for oil removal from water, the management of oil spills, and chemical separation processes to remove non-polar from polar compounds.

Hydrophobic is often used interchangeably with lipophilic, "fat loving." However, the two terms are not synonymous. While hydrophobic substances are usually lipophilic, there are exceptions — such as the silicones and fluorocarbons.

According to thermodynamics, matter seeks to be in a low-energy state, and bonding reduces chemical energy. Water is electrically polarized, and is able to form hydrogen bonds internally, which gives it many of its unique physical properties. But, since hydrophobes are not electrically polarized, and because they are unable to form hydrogen bonds, water repels hydrophobes, in favour of bonding with itself. It is this effect that causes the hydrophobic interaction — which in itself is incorrectly named as the energetic force comes from the hydrophilic molecules. Thus the two immiscible phases (hydrophilic vs. hydrophobic) will change so that their corresponding interfacial area will be minimal. This effect can be visualized in the phenomenon called phase separation.

## SUPERHYDROPHOBICITY

Superhydrophobic materials such as the leaves of the lotus plant have surfaces that are extremely difficult to wet, with water contact angles in excess of 150° and slide angles less than 20°.

### Theory

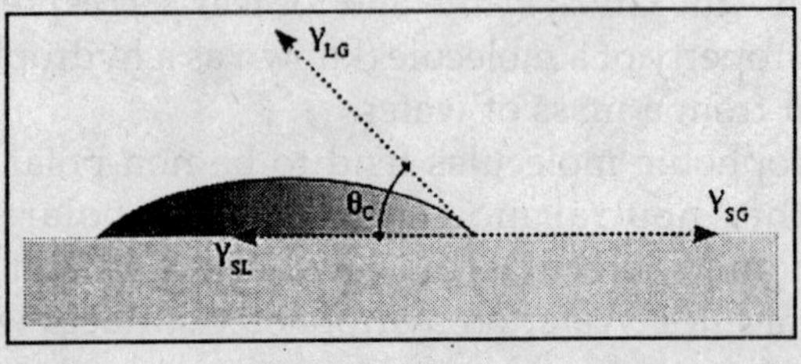

In 1805 Thomas Young defined the contact angle è by analyzing the forces acting on a fluid droplet resting on a solid surface surrounded by a gas.

A liquid droplet rests on a solid surface and is surrounded by gas. The contact angle, $\theta_C$, is the angle formed by a liquid at the three phase boundary where the liquid, gas, and solid intersect.

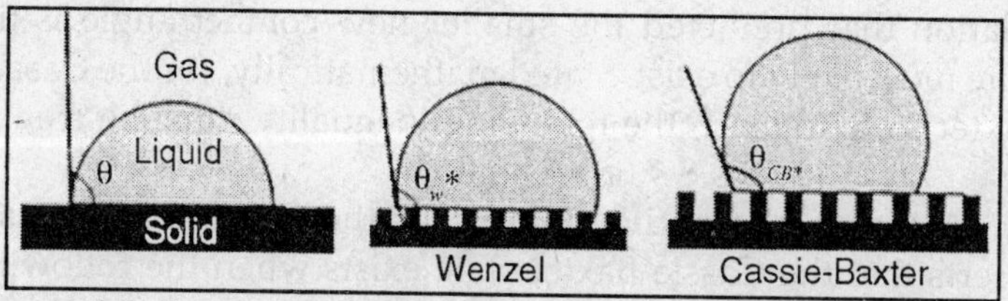


A droplet resting on a solid surface and surrounded by a gas forms a characteristic contact angle θ. If the solid surface is rough, and the liquid is in intimate contact with the solid asperities, the droplet is in the Wenzel state. If the liquid rests on the tops of the asperities, it is in the Cassie-Baxter state.

$$\gamma_{SG} = \gamma_{SL} + \gamma_{LG} \cos\theta$$

where

$\gamma_{SG}$ = Interfacial tension between the solid and gas

$\gamma_{SL}$ = Interfacial tension between the solid and liquid

$\gamma_{LG}$ = Interfacial tension between the liquid and gas

θ can be measured using a contact angle goniometer.

Wenzel determined that when the liquid is in intimate contact with a microstructured surface, θ will change to $\theta_W{}^*$

$$\cos \theta_W{}^* = r \cos \theta$$

where *r* is the ratio of the actual area to the projected area. Wenzel's equation shows that microstructuring a surface amplifies the natural tendency of the surface. A hydrophobic surface (one that has an original contact angle greater than 90°) becomes more hydrophobic when microstructured – its new contact angle becomes greater than the original. However, a hydrophilic surface (one that has an original contact angle less than 90°) becomes more hydrophilic when microstructured – its new contact angle becomes less than the original.

Cassie and Baxter found that if the liquid is suspended on the tops of microstructures, θ will change to $\theta_{CB}{}^*$

$$\cos \theta_{CB}^* = \phi(\cos \theta + 1) - 1$$

where $\phi$ is the area fraction of the solid that touches the liquid. Liquid in the Cassie-Baxter state is more mobile than in the Wenzel state.

We can predict whether the Wenzel or Cassie-Baxter state should exist by calculating the new contact angle with both equations. By a minimization of free energy argument, the relation that predicted the smaller new contact angle is the state most likely to exist. Stated mathematically, for the Cassie-Baxter state to exist, the following inequality must be true.

$$\cos \theta < (\varphi - 1)/(r - \varphi)$$

A recent alternative criteria for the Cassie-Baxter state asserts that the Cassie-Baxter state exists when the following 2 criteria are met: 1) Contact line forces overcome body forces of unsupported droplet weight and 2) The microstructures are tall enough to prevent the liquid that bridges microstructures from touching the base of the microstructures.

Contact angle is a measure of static hydrophobicity, and contact angle hysteresis and slide angle are dynamic measures. Contact angle hysteresis is a phenomenon that characterizes surface heterogeneity. When a pipette injects a liquid onto a solid, the liquid will form some contact angle. As the pipette injects more liquid, the droplet will increase in volume, the contact angle will increase, but its three phase boundary will remain stationary until it suddenly advances outward. The contact angle the droplet had immediately before advancing outward is termed the advancing contact angle.

The receding contact angle is now measured by pumping the liquid back out of the droplet. The droplet will decrease in volume, the contact angle will decrease, but its three phase boundary will remain stationary until it suddenly recedes inward. The contact angle the droplet had immediately before receding inward is termed the receding contact angle.

The difference between advancing and receding contact angles is termed contact angle hysteresis and can be used to characterize surface heterogeneity, roughness, and mobility. Surfaces that are not homogeneous will have domains which impede motion of the contact line.

## HYDROPHILICITY

Hydrophile, a physical property of a molecule that can transiently bond with water ($H_2O$) through hydrogen bonding. This is thermodynamically favorable, and makes these molecules soluble not only in water, but also in other polar solvents. There are hydrophilic and hydrophobic parts of the cell membrane.

A hydrophilic molecule or portion of a molecule is one that is typically charge-polarized and capable of hydrogen bonding, enabling it to dissolve more readily in water than in oil or other hydrophobic solvents. A hydrophilic is made up of alcohol and fatty acyl chains. Hydrophilic and hydrophobic molecules are also known as polar molecules and nonpolar molecules, respectively. Some hydrophilic substances don't dissolve. This type of mixture is called a colloid. Soap has a hydrophilic head and a hydrophobic tail, which allows it to dissolve in both waters and oils, therefore allowing the soap to clean a surface.

### Dielectric Constant of Water

A relatively small number of studies on the measurement of the dielectric constant of water as a function of temperature (T) and pressure (P) have been reported in the literature. Most of the experimental data available is for a relatively large range of temperatures (up to 670 K) at relatively low pressures (under 200 MPa).

There are few measurements available for pressures between 200MPa and 500MPa. The accuracy of most of the high temperature data is better than 1%, while in the lower temperature limit (below 70°C) the accuracy of most of the data is around 0.1%. As a consequence of this lack of data, attempts to estimate the properties of aqueous species at high temperature and/or high pressure rely on estimated or extrapolated values of the dielectric constant of water.

High temperature and pressure conditions are important for engineering processes and geothermal studies. For this reason, several equations have been proposed for different ranges of temperature and pressure. Almost all of them are entirely empirical. Although those equations may yield unphysical results if extrapolated too far outside the range of the experimental data, many of them were formulated to give some confidence about the extrapolated values. Most of those equations are based on the Kirkwood equation, in which the parameter g can be expressed as a function of experimental variables, like density and temperature, and adjusted to fit experimental data. An example of this approach has been recently reported in the derivation of an equation for the dielectric constant of water at temperatures from 238 K to 873 K and at pressures up to 1200 MPa.

Those authors used an empirical 12-parameter form for the g-factor as a function of the independent variables, temperature and density. Their formula correlates well a selected set of data from a collection of experimental data assembled by the authors. Another approach uses a hard sphere model for the water molecule and a modified Ornstein-Zernike equation, where the parameters are fit to the experimental data for water. One of the simplest but effective description of the dielectric constant dependence on the pressure and temperature, proposed by Bradley and Pitzer, uses an equation suggested by Tait in 1880 for volumetric data.

Bradley's equation was expressed in terms of the static dielectric constant of water (EPS) and its parameters were adjusted to reproduce experimental data. The resulting equation is valid in the range 0-70°C between 0.1-2000MPa, and 70-350°C between 0.1-5000MPa. A number of molecular dynamics simulations of the static dielectric constant of water. There are simulations at low pressure and temperatures below 373K, over a wide range of density and temperature values along the liquid-vapour coexistence curve, and under high pressure and temperature conditions. The SPC/E potential for water molecule was used in two of those simulations. The simulations carried by Wasserman covered temperatures ranging from 324K to 1278K and densities from 0.257 g/cm$^3$ to 1.1108 g/cm$^3$. Their simulations are in good agreement within the pressure-temperature range of the experimental data used. They showed that the SPC/E model tends to underestimate the values of the dielectric constant for densities greater than 1.0g/cm$^3$. The SPC/E potential used by those authors was parameterized to yield correct values of pressure and energy for a density of 0.998g/cm$^3$ at 300K and was not adjusted to experimental values of the dielectric constant. In the last years a considerable number of papers appeared in the literature addressing the important problem of proteins' folding/unfolding mechanism, using molecular dynamics simulations.

Different proteins have been studied under very distinct conditions of temperature and pressure. Since even the simplest proteins already have a large number of atoms, most simulations are carried out using a continuum dielectric medium (implicit solvent representation) to take the solvent molecules into account.

Among the several different conditions that have been employed in computer simulations, the use of hydrostatic pressure at relatively low temperatures seems to be very promising at revealing the details of folding/unfolding mechanism. We have recently analyzed the effects of pressure on the conformation of Myoglobin and showed that most of the experimental behaviour of this system is well reproduced by the simulations. However, very high pressures are usually required in order to conduct this type of study. Therefore, in order to simulate the effects of hydrostatic pressure on protein folding/ unfolding it is necessary to accurately describe the properties of the solvent (water) under the simulation conditions.

Two of these properties whose value depends on temperature and pressure are the static dielectric constant of water (EPS) and the density (DENS). The conditions for the study of pressure effects on unfolding are temperatures below 75°C and pressures up to 20kbar (2000MPa). Under these conditions, the water remains in the liquid state, and most proteins are thermally stable at that temperature. All the previous attempts at deriving equations describing the behaviour of EPS and DENS with $T$ and $P$, tried to cover relatively large ranges of temperature and pressure, and in order to do so, a large number of parameters are used. Bradley's is the only one among the equations available in the literature to cover the range of $T$ and $P$ needed in the pressure simulation studies. Although much simpler than the others, this equation is not accurate enough in the specific range of $T$ and $P$ (especially for low $T$) of interest for biological applications.

It would be ideal to have an equation analytically simple as Bradley's and accurate enough for the biologically relevant ranges of $T$ and $P$. However, Bradley's equation does not reproduce accurately enough the more recent experimental data for relatively low temperatures, most probably because of the fact that they attempted to cover a large range of $T$ and $P$.

Since the pressure simulation studies require the knowledge of EPS and DENS in a large range of P but at a limited set of temperatures, we adopted the following strategy: starting from an equation as simple as Bradley's, we tried to adjust its parameters as to reproduce as accurate as possible the available data for EPS and DENS at selected values of T, appropriated for the simulation studies. While it may be argued that such a procedure generates equations of limited applicability, this is not really the case because the fitting procedure is very simple and can be performed for any other desired temperature.

Besides, by selecting the temperature of interest we gain in accuracy. Therefore, although we exemplify its usage for the temperatures proper to our protein simulations, the equation is quite general and can be used in any simulation study where the knowledge of the water EPS and DENS as a function of $P$ at different values of $T$ is needed.

## Dielectric Constant

The equation suggested by Bradley and Pitzer for the region above saturation pressure and temperatures below 350°C was

$$EPS = EPS_{1000} + C \ln ((B + P)/(B + 1000),$$

where $P$ is the pressure in bars, $EPS$ is the dielectric

constant, and $EPS_{1000}$ was chosen arbitrarily as a reference value (*EPS* at 1000 bar). $EPS_{1000}$, *C* and *B* are temperature dependent parameters described by the equations (*T* in Kelvin):

$$EPS_{1000} = U_1 \exp [U_2T + U_3T^2]$$
$$C = U_4 + U_5/(U_6 + T)$$
$$B = U_7 + U_8/T + U_9T,$$

with

$U_1 = 3.4279\ E\ (02)$ $\quad U_6 = -1.8289\ E$
$U_2 = -\ 5.0866\ E\ (-03)$ $\quad U_7 = -8.0325\ E$
$U_3 = 9.4690\ E\ (-07)$ $\quad U_8 = 4.21452\ E$
$U_4 = -2.0525$ $\quad U_9 = 2.1417$
$U_5 = 3.1159\ E\ (03)$

The parameters showed in equation were optimized using a wide range of experimental data, including high temperature data. Although the experimental data below 70°C collected by the authors, had been fitted to better than 0.1%, the same agreement is not observed in the low temperature/ high pressure range.

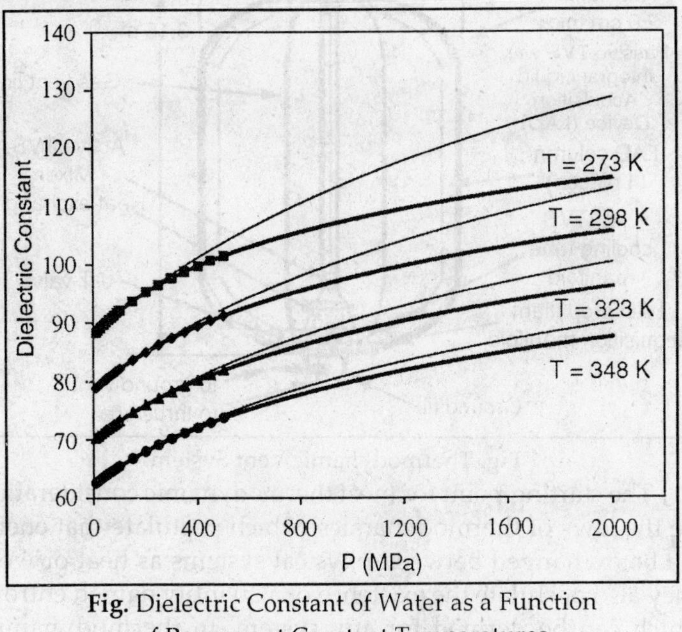


**Fig.** Dielectric Constant of Water as a Function of Pressure at Constant Temperatures

## THERMODYNAMICS

Thermodynamics is the study of the conversion of heat energy into different forms of energy (in particular, mechanical, chemical, and electrical energy); different energy conversions into heat energy; and its relation to macroscopic variables such as temperature, pressure, and volume. Its underpinnings, based upon statistical predictions of the collective motion of particles from their microscopic behaviour, is the field of statistical thermodynamics, a branch of statistical mechanics. Roughly, heat means "energy in transit" and dynamics relates to "movement"; thus, in essence thermodynamics studies the movement of energy and how energy instills movement. Historically, thermodynamics developed out of need to increase the efficiency of early steam engines.

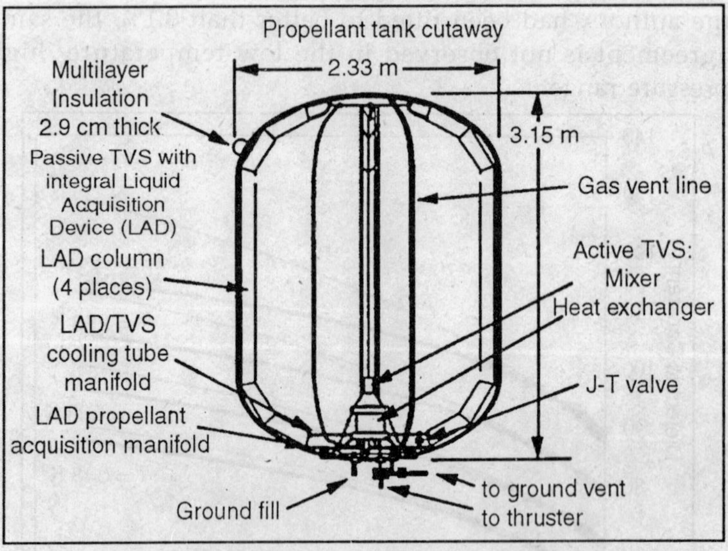


**Fig.** Thermodynamic Vent System

The starting point for most thermodynamic considerations are the laws of thermodynamics, which postulate that energy can be exchanged between physical systems as heat or work. They also postulate the existence of a quantity named entropy, which can be defined for any system. In thermodynamics, interactions between large ensembles of objects are studied and categorized.

Central to this are the concepts of *system* and *surroundings*. A system is composed of particles, whose average motions define its properties, which in turn are related to one another through equations of state. Properties can be combined to express internal energy and thermodynamic potentials, which are useful for determining conditions for equilibrium and spontaneous processes.

## THERMODYNAMIC FREE ENERGY

In thermodynamics, the term thermodynamic free energy refers to the amount of work that can be extracted from a system, and is helpful in engineering applications. It is a subtraction of the entropy of a system ("useless energy") from the total energy, yielding a thermodynamic state function which represents the "useful energy".

The free energy functions are Legendre transforms of the internal energy. For processes involving a system at constant pressure $p$ and temperature $T$, the Gibbs free energy is the most useful because, in addition to subsuming any entropy change due merely to heat flux, it does the same for the $pdV$ work needed to "make space for additional molecules" produced by various processes. (Hence its utility to solution-phase chemists, including biochemists.) The Helmholtz free energy has a special theoretical importance since it is proportional to the logarithm of the partition function for the canonical ensemble in statistical mechanics. (Hence its utility to physicists; and to gas-phase chemists and engineers, who do not want to ignore $pdV$ work.)

The (historically earlier) Helmholtz free energy is defined as $A = U - TS$, where $U$ is the internal energy, $T$ is the absolute temperature, and $S$ is the entropy. Its change is equal to the amount of reversible work done on, or obtainable from, a system at constant $T$. Thus its appellation "work content", and the designation $A$ from *Arbeit*, the German word for work. Since it makes no reference to any quantities involved in work (such as $p$ and $V$), the Helmholtz function is completely general: its decrease is the maximum amount of work which can be done *by* a system, and it can increase at most by the amount of work done *on* a system.

The Gibbs free energy $G = H - TS$, where $H$ is the enthalpy. ($H = U + pV$, where $p$ is the pressure and $V$ is the volume.)

There has been historical controversy:

- In physics, "free energy" most often refers to the *Helmholtz free energy*, denoted by $F$.
- In chemistry, "free energy" most often refers to the *Gibbs free energy*.

Since both fields use both functions, a compromise has been suggested, using $A$ to denote the Helmholtz function, with $G$ for the Gibbs function. While $A$ is preferred by IUPAC, $F$ is sometimes still in use, and the correct free energy function is often implicit in manuscripts and presentations.

## Application

The experimental usefulness of these functions is restricted to conditions where certain variables ($T$, and $V$ or *external* $p$) are held constant, although they also have theoretical importance in deriving Maxwell relations. Work other than $pdV$ may be added, e.g., for electrochemical cells, or $fdx$ work in elastic materials and in muscle contraction. Other forms of work which must sometimes be considered are stress-strain, magnetic, as in adiabatic demagnetization used in the approach to absolute zero, and work due to electric polarization. These are described by tensors. In most cases of interest there are internal degrees of freedom and processes, such as chemical reactions and phase transitions, which create entropy. Even for homogeneous "bulk" materials, the free energy functions depend on the (often suppressed) composition, as do all proper thermodynamic potentials (extensive functions), including the internal energy.

| *Name* | *Definition* | *Natural variables* |
|---|---|---|
| Helmholtz free energy | $A = U - TS$ | $T, V, \{N_i\}$ |
| Gibbs free energy | $G = U + pV - TS$ | $T, p, \{N_i\}$ |

$N_i$ is the number of molecules (alternatively, moles) of type $i$ in the system. If these quantities do not appear, it is impossible to describe compositional changes. The differentials for *reversible processes* are (assuming only $pV$ work)

$$dA = -pdV - SdT + \sum_i \mu_i dN_i$$

$$dG = VdP - SdT + \sum_i \mu_i dN_i$$

where $\mu_i$ is the chemical potential for the $i$-th component in the system. The second relation is especially useful at constant $T$ and $p$, conditions which are easy to achieve experimentally, and which approximately characterize living creatures.

$$(dG)_{T,p} = \sum_i \mu_i dN_i$$

Any decrease in the Gibbs function of a system is the upper limit for any isothermal, isobaric work that can be captured in the surroundings, or it may simply be dissipated, appearing as $T$ times a corresponding increase in the entropy of the system and/or its surrounding.

## ENTHALPY

In thermodynamics and molecular chemistry, the enthalpy (denoted as $H$, $h$, or rarely as $\chi$) is a quotient or description of thermodynamic potential of a system, which can be used to calculate the heat transfer during a quasi-static process taking place in a closed thermodynamic system under constant pressure. The term *enthalpy* was composed of the prefix *en-*, meaning "to put into" and the Greek word *-thalpein*, meaning "to heat", although the original definition is thought to have stemmed from the word "enthalpos".

This is the heat change which occurs when 1 mol of a substance reacts completely with oxygen to form products at 273.15 K and 1 atm. The function $H$ was introduced by the Dutch physicist Kamerlingh Onnes in early 20th century in the following form:

$$H = E + pV,$$

where $E$ represents the energy of the system. In the absence of an external field, the enthalpy may be defined, as it is generally known, by:

$$H = U + pV,$$

where (all units given in SI)

- $H$ is the enthalpy (joules)
- $U$ is the internal energy, (joules)
- $p$ is the pressure of the system, (pascals)
- $V$ is the volume, (cubic meters)

The form $pV$ is motivated by the following example of an isobaric process. Gas in a cylinder pushes a piston, maintaining constant pressure $p$. The force is calculated from the area $A$ of the piston and definition of pressure $p = F/A$: the force is $F = pA$. By definition, work $W$ done is $W = Fx$, where $x$ is the distance. Combining gives $W = pAx$, and the product $Ax$ is the volume traversed by the piston: $Ax = V$. Thus, the work done by the gas is $W = pV$, where $p$ is a constant pressure and $V$ the expansion of volume. Including this term allows the discussion of energy changes when not only temperature, but also volume or pressure are changed.

## Heats of Reaction

The total enthalpy of a system cannot be measured directly; the *enthalpy change* of a system is measured instead. Enthalpy change is defined by the following equation:

$$\Delta H = H_{\text{final}} - H_{\text{initial}}$$

where

$\Delta H$ is the *enthalpy change*

$H_{\text{final}}$ is the final enthalpy of the system, measured in joules. In a chemical reaction, $H_{\text{final}}$ is the enthalpy of the products.

$H_{\text{initial}}$ is the initial enthalpy of the system, measured in joules. In a chemical reaction, $H_{\text{initial}}$ is the enthalpy of the reactants.

For an exothermic reaction at constant pressure, the system's change in enthalpy is equal to the energy released in the reaction, including the energy retained in the system and lost through expansion against its surroundings. In a similar manner, for an endothermic reaction, the system's change in enthalpy is equal to the energy *absorbed* in the reaction, including the energy *lost by* the system and *gained* from compression from its surroundings. A relatively easy way to determine whether or not a reaction is exothermic or endothermic is to determine the sign of $\Delta H$. If $\Delta H$ is positive, the reaction is endothermic, that is heat is absorbed by the system due to the products of the reaction having a greater enthalpy than the reactants. On the other hand if $\Delta H$ is negative, the reaction is exothermic, that is the overall decrease in enthalpy is achieved by the generation of heat.

Although enthalpy is commonly used in engineering and science, it is impossible to measure directly, as enthalpy has no datum (reference point). Therefore enthalpy can only accurately be used in a closed system. However, few real world applications exist in closed isolation, and it is for this reason that two or more closed systems cannot be compared using enthalpy as a basis, although sometimes this is done erroneously.

### Standard Enthalpy Changes

Standard enthalpy changes describe the change in enthalpy observed in the constituents of a thermodynamic system when going between different states under standard conditions. The standard enthalpy change of vaporization, for example gives the enthalpy change when going from liquid to gas. These enthalpies are reversible; the enthalpy change of going from gas to liquid is the negative of the enthalpy change of vaporization. A common standard enthalpy change is the standard enthalpy change of formation, which has been determined for a large number of substances. The enthalpy change of any reaction under any conditions can be computed, given the standard enthalpy change of formation of all of the reactants and products.

## ENTROPY

The concept of entropy is particularly notable as it is applied across physics, information theory and mathematics.

In thermodynamics (a branch of physics), entropy, symbolized by *S*, is a measure of the unavailability of a system's energy to do work. It is a measure of the disorder of molecules in a system, and is central to the second law of thermodynamics and to the fundamental thermodynamic relation, both of which deal with physical processes and whether they occur unexpectedly. Spontaneous changes in isolated systems occur with an increase in entropy. Unexpected changes tend to average out differences in temperature, pressure, density, and chemical potential that may exist in a system, and entropy is thus a measure of how great the unexpected changes are.

Entropy is a function of a quantity of heat in a system which is capable of doing work. When heat is added to a system at high temperature, the increase in entropy is small. When heat is added to a system at low temperature, the increase in entropy is great. Thus, under maximum entropy, there is a minimum of energy available for doing work and, under minimum entropy, there is a maximum of energy available for doing work.

Entropy, *S*, is not defined directly, but rather by an equation relating the change in entropy of the system to the change in heat of the system. For a constant temperature, the change in entropy, $\Delta S$, is defined by the equation

$$\Delta S = \frac{\Delta Q}{T},$$

where $\Delta Q$ is the amount of heat absorbed in an isothermal and reversible process in which the system goes from one state to another, and $T$ is the absolute temperature at which the process is occurring. If the temperature of the system is not constant, then the relationship becomes a differential equation

$$dS = \frac{dQ}{T}.$$

To understand what this equation means, suppose the temperature $T$ can be expressed as a function $T(Q)$ of the heat $Q$. Then the total change in entropy as the heat-level varies is:

$$\Delta S = \int_A \frac{1}{T(Q)} dQ.$$

where $A$ is the set defining the range of heat values in the system.

Entropy is one of the factors that determines the free energy of the system. This thermodynamic definition of entropy is only valid for a system in equilibrium (because temperature is defined only for a system in equilibrium), while the statistical definition of entropy applies to any system. Thus the statistical definition is usually considered the fundamental definition of entropy.

Entropy increase has often been defined as a change to a more disordered state at a molecular level. In recent years, entropy has been interpreted in terms of the "dispersal" of energy. Entropy is an extensive state function that accounts for the effects of irreversibility in thermodynamic systems.

Chapter 2

# Carbohydrates

Carbohydrates are the most abundant class of organic compounds found in living organisms. They originate as products of photosynthesis, an endothermic reductive condensation of carbon dioxide requiring light energy and the pigment chlorophyll.

$$n\,CO_2 + n\,H_2O + \text{energy}\ C_nH_{2n}O_n + n\,O_2$$

| Complexity | Simple Carbohydrates | Complex Carbohydrates |
|---|---|---|
| | monosaccharides | disaccharides, oligosaccharides & polysaccharides |

| Size | Tetrose | Pentose | Hexose | Heptose | |
|---|---|---|---|---|---|
| | $C_4$ sugars | $C_5$ sugars | $C_6$ sugars | $C_7$ sugars | etc. |

| | |
|---|---|
| C=O Function | **Aldose**<br>sugars having an aldehyde function or an acetal equivalent.<br>**Ketose**<br>sugars having a ketone function or an acetal equivalent. |
| Reactivity | **Reducing**<br>sugars oxidized by Tollens' reagent (or Benedict's or Fehling's reagents).<br>**Non-reducing**<br>sugars not oxidized by Tollens' or other reagents. |

As noted here, the formulas of many carbohydrates can be written as carbon hydrates, $C_n(H_2O)_n$, hence their name. The carbohydrates are a major source of metabolic energy, both for plants and for animals that depend on plants for food. Aside from the sugars and starches that meet this vital nutritional role, carbohydrates also serve as a structural material (cellulose), a component of the energy transport compound ATP, recognition sites on cell surfaces and one of three essential components of DNA and RNA.

Carbohydrates are called saccharides or, if they are relatively small, sugars. Several classifications of carbohydrates have proven useful and are outlined in the above table.

## GLUCOSE

Carbohydrates have been given non-systematic names, although the suffix ose is generally used. The most common carbohydrate is glucose ($C_6H_{12}O_6$). Glucose is a monosaccharide, an aldohexose and a reducing sugar. The general structure of glucose and many other aldohexoses was established by simple chemical reactions. The following diagram illustrates the kind of evidence considered, although some of the reagents shown here are different from those used by the original scientists.

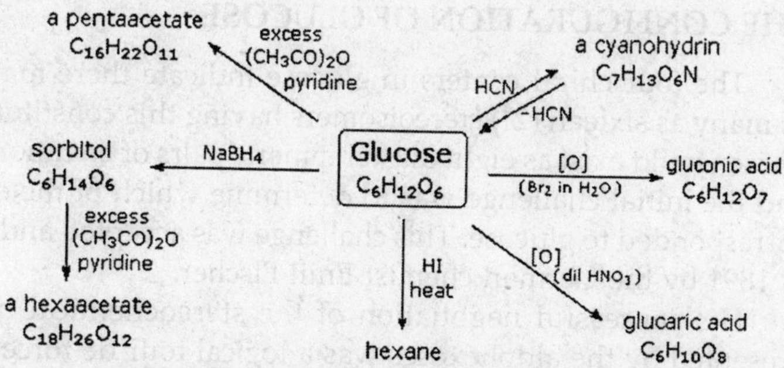


**Fig.** Glucose

Hot hydriodic acid (HI) was often used to reductively remove oxygen functional groups from a molecule and in the case of glucose this treatment gave hexane (in low yield). From this it was concluded that the six carbons are in an unbranched chain. The presence of an aldehyde carbonyl group was deduced from cyanohydrin formation, its reduction to the hexa-alcohol sorbitol, also called glucitol and mild oxidation to the mono-carboxylic acid, glucuronic acid. Somewhat stronger oxidation by dilute nitric acid gave the diacid, glucaric acid, supporting the proposal of a six-carbon chain. The five oxygens remaining in glucose after the aldehyde was accounted for were thought to be in hydroxyl groups, since a penta-acetate derivative could be made. These hydroxyl groups were assigned, one each, to the last five carbon atoms, because geminal hydroxyl groups are normally unstable relative to the carbonyl compound formed by loss of water. The four middle carbon atoms in the glucose chain are centers of chirality and are colored red.

Glucose and other saccharides are extensively cleaved by periodic acid, thanks to the abundance of vicinal diol moieties in their structure. This oxidative cleavage, known as the Malaprade reaction is particularly useful for the analysis of selective O-substituted derivatives of saccharides, since ether functions do not react. The stoichiometry of aldohexose cleavage is shown in the following equation.

$$HOCH_2(CHOH)_4CHO + 5\ HIO_4 \rightarrow H_2C$$
$$= O + 5\ HCO_2H + 5\ HIO_3$$

## THE CONFIGURATION OF GLUCOSE

The four chiral centers in glucose indicate there may be as many as sixteen ($2^4$) stereoisomers having this constitution. These would exist as eight diastereomeric pairs of enantiomers and the initial challenge was to determine which of the eight corresponded to glucose. This challenge was accepted and met in 1891 by the German chemist Emil Fischer.

His successful negotiation of the stereochemical maze presented by the aldohexoses was a logical tour de force and it is fitting that he received the 1902 Nobel Prize for chemistry for this accomplishment. One of the first tasks faced by Fischer was to devise a method of representing the configuration of each chiral cenre in an unambiguous manner. To this end, he invented a simple technique for drawing chains of chiral centers, that we now call the Fischer projection formula. At the time Fischer undertook the glucose project it was not possible to establish the absolute configuration of an enantiomer. Consequently, Fischer made an arbitrary choice for (+)-glucose and established a network of related aldose configurations that he called the D-family. The mirror images of these configurations were then designated the L-family of aldoses. To illustrate using present day knowledge, Fischer projection formulas and names for the D-aldose family (three to six-carbon atoms) are shown below, with the asymmetric carbon atoms (chiral centers) colored red.

The last chiral cenre in an aldose chain (farthest from the aldehyde group) was chosen by Fischer as the D / L designator site. If the hydroxyl group in the projection formula pointed to the right, it was defined as a member of the D-family. A left directed hydroxyl group then represented the L-family.

Fischer's initial assignment of the D-configuration had a 50:50 chance of being right, but all his subsequent conclusions concerning the relative configurations of various aldoses were soundly based. In 1951 x-ray fluorescence studies of (+)-tartaric acid, carried out in the Netherlands by Johannes Martin Bijvoet (pronounced "buy foot"), proved that Fischer's choice was correct. It is important to recognize that the sign of a compound's specific rotation (an experimental number) does not correlate with its configuration (D or L). It is a simple matter to measure an optical rotation with a polarimeter. Determining an absolute configuration usually requires chemical interconversion with known compounds by stereospecific reaction paths.

## OXIDATION

Sugars may be classified as reducing or non-reducing based on their reactivity with Tollens', Benedict's or Fehling's reagents. If a sugar is oxidized by these reagents it is called reducing, since the oxidant (Ag(+) or Cu(+2)) is reduced in the reaction, as evidenced by formation of a silver mirror or precipitation of cuprous oxide. The Tollens' test is commonly used to detect aldehyde functions; and because of the facile interconversion of ketoses and aldoses under the basic conditions of this test, ketoses such as fructose also react and are classified as reducing sugars.

When the aldehyde function of an aldose is oxidized to a carboxylic acid the product is called an aldonic acid. Because of the 2º hydroxyl functions that are also present in these compounds, a mild oxidizing agent such as hypobromite must be used for this conversion.

If both ends of an aldose chain are oxidized to carboxylic acids the product is called an aldaric acid. By converting an aldose to its corresponding aldaric acid derivative, the ends of the chain become identical (this could also be accomplished by reducing the aldehyde to CH2OH, as noted below). Such an operation will disclose any latent symmetry in the remaining molecule. Thus, ribose, xylose, allose and galactose yield achiral aldaric acids which are, of course, not optically active. Other aldose sugars may give identical chiral aldaric acid products, implying a unique configurational relationship. Remember, a Fischer projection formula may be rotated by 180º in the plane of projection without changing its configuration.

## REDUCTION

Sodium borohydride reduction of an aldose makes the ends of the resulting alditol chain identical, $HOCH_2(CHOH)_nCH_2OH$, thereby accomplishing the same configurational change produced by oxidation to an aldaric acid. Thus, allitol and galactitol from reduction of allose and galactose are achiral and altrose and talose are reduced to the same chiral alditol. A summary of these redox reactions and derivative nomenclature is given in the following table.

**Derivatives of $HOCH_2(CHOH)_nCHO$**

| | | |
|---|---|---|
| HOBr Oxidation | → | $HOCH_2(CHOH)_nCO_2H$ an Aldonic Acid |
| $HNO_3$ Oxidation | → | $H_2OC(CHOH)_nCO_2H$ an Aldaric Acid |
| $NaBH_4$ Reduction | → | $HOCH_2(CHOH)_nCH_2OH$ an Alditol |

## OSAZONE FORMATION

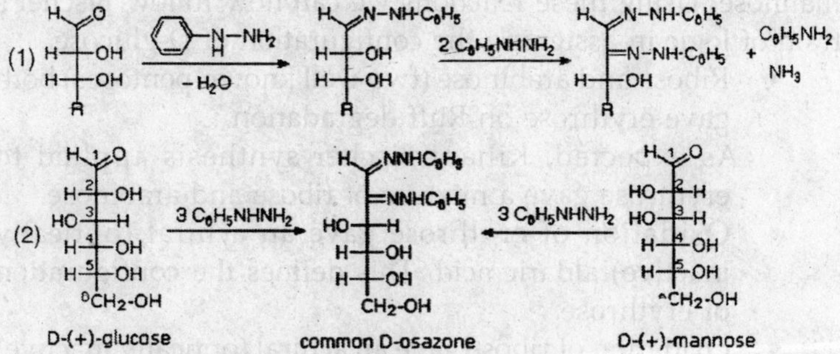


The osazone reaction was developed and used by Emil Fischer to identify aldose sugars differing in configuration only at the alpha-carbon. The upper equation shows the general form of the osazone reaction, which effects an alpha-carbon oxidation with formation of a bis-phenylhydrazone, known as an osazone. Application of the osazone reaction to D-glucose and D-mannose demonstrates that these compounds differ in configuration only at C-2.

## CHAIN SHORTENING AND LENGTHENING

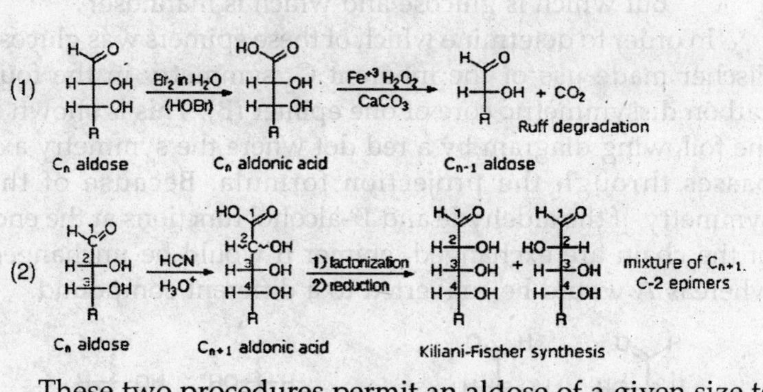


These two procedures permit an aldose of a given size to be related to homologous smaller and larger aldoses. The importance of these relationships may be seen in the array of aldose structures presented earlier, where the structural connections are given by the dashed blue lines. Thus Ruff degradation of the pentose arabinose gives the tetrose erythrose. Working in the opposite direction, a Kiliani-Fischer synthesis applied to arabinose gives a mixture of glucose and mannose. Using these reactions we can now follow Fischer's train of logic in assigning the configuration of D-glucose.

- Ribose and arabinose (two well known pentoses) both gave erythrose on Ruff degradation.

  As expected, Kiliani-Fischer synthesis applied to erythrose gave a mixture of ribose and arabinose.
- Oxidation of erythrose gave an achiral (optically inactive) aldaric acid. This defines the configuration of erythrose.
- Oxidation of ribose gave an achiral (optically inactive) aldaric acid. This defines the configuration of both ribose and arabinose.
- Ruff shortening of glucose gave arabinose and Kiliani-Fischer synthesis applied to arabinose gave a mixture of glucose and mannose.
- Glucose and mannose are therefore epimers at C-2, a fact confirmed by the common product from their osazone reactions.
- A pair of structures for these epimers can be written, but which is glucose and which is mannose?

In order to determine which of these epimers was glucose, Fischer made use of the inherent $C_2$ symmetry in the four-carbon dissymmetric core of one epimer (B). This is shown in the following diagram by a red dot where the symmetry axis passes through the projection formula. Because of this symmetry, if the aldehyde and 1º-alcohol functions at the ends of the chain are exchanged, epimer B would be unchanged; whereas A would be converted to a different compound.

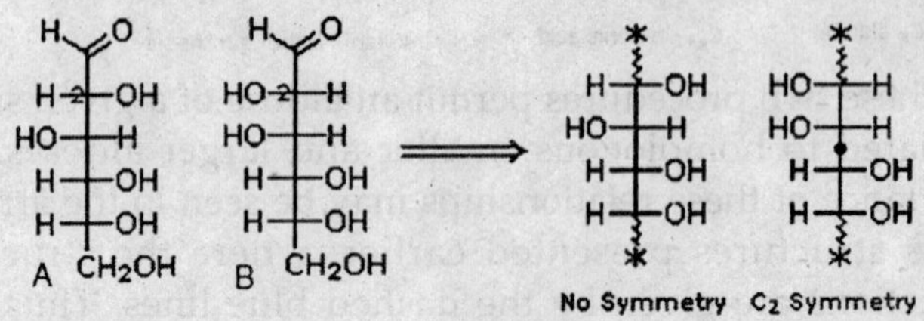


Fischer looked for and discovered a second aldohexose that represented the end group exchange for the epimer lacking the latent $C_2$ symmetry (A). This compound was L-(+)-gulose and its exchange relationship to D-(+)-glucose was demonstrated be oxidation to a common aldaric acid product.

## KETOSES

If a monosaccharide has a carbonyl function on one of the inner atoms of the carbon chain it is classified as a ketose. Dihydroxyacetone may not be a sugar, but it is included as the ketose analog of glyceraldehyde.

The carbonyl group is commonly found at C-2, as illustrated by the following examples (chiral centers are colored red). As expected, the carbonyl function of a ketose may be reduced by sodium borohydride, usually to a mixture of epimeric products. D-Fructose, the sweetest of the common natural sugars, is for example reduced to a mixture of D-glucitol (sorbitol) and D-mannitol, named after the aldohexoses from which they may also be obtained by analogous reduction. Mannitol is itself a common natural carbohydrate.

Although the ketoses are distinct isomers of the aldose monosaccharides, the chemistry of both classes is linked due to their facile interconversion in the presence of acid or base catalysts. This interconversion and the corresponding epimerization at sites alpha to the carbonyl functions, occurs by way of an enediol tautomeric intermediate.

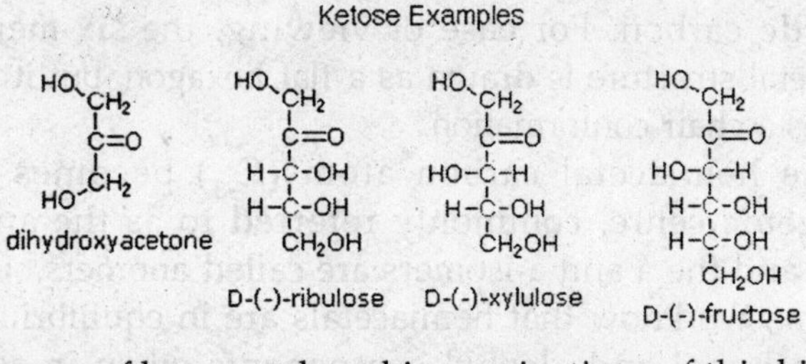


Because of base-catalyzed isomerizations of this kind, the Tollens' reagent is not useful for distinguishing aldoses from ketoses or for specific oxidation of aldoses to the corresponding aldonic acids. Oxidation by HOBr is preferred for the latter conversion.

## ANOMERIC FORMS OF GLUCOSE

Fischers brilliant elucidation of the configuration of glucose did not remove all uncertainty concerning its structure. Two different crystalline forms of glucose were reported in 1895. Each of these gave all the characteristic reactions of glucose and when dissolved in water equilibrated to the same mixture. This equilibration takes place over a period of many minutes and the change in optical activity that occurs is called mutarotation. These facts are summarized in the diagram below.

When glucose was converted to its pentamethyl ether (reaction with excess $CH_3I$ & AgOH), two different isomers were isolated and neither exhibited the expected aldehyde reactions. Acid-catalyzed hydrolysis of the pentamethyl ether derivatives, however, gave a tetramethyl derivative that was oxidized by Tollen's reagent and reduced by sodium borohydride, as expected for an aldehyde.

The search for scientific truth often proceeds in stages and the structural elucidation of glucose serves as a good example. Somehow a new stereogenic cenre must be created and the aldehyde must be deactivated in the pentamethyl derivative. A simple solution to this dilemma is achieved by converting the open aldehyde structure for glucose into a cyclic hemiacetal, called a glucopyranose.

The linear aldehyde is tipped on its side and rotation about the $C_4$-$C_5$ bond brings the C5-hydroxyl function close to the aldehyde carbon. For ease of viewing, the six-membered hemiacetal structure is drawn as a flat hexagon, but it actually assumes a chair conformation.

The hemiacetal carbon atom ($C_{-1}$) becomes a new stereogenic cenre, commonly referred to as the anomeric carbon and the á and â-isomers are called anomers.

First, we know that hemiacetals are in equilibrium with their carbonyl and alcohol components when in solution. Consequently, fresh solutions of either alpha or beta-glucose crystals in water should establish an equilibrium mixture of both anomers, plus the open chain chain form.

Second, a pentamethyl ether derivative of the pyranose

structure converts the hemiacetal function to an acetal. Acetals are stable to base, so this product should not react with Tollen's reagent or be reduced by sodium borohydride. Acid hydrolysis of acetals regenerates the carbonyl and alcohol components and in the case of the glucose derivative this will be a tetramethyl ether of the pyranose hemiacetal. This compound will, of course, undergo typical aldehyde reactions.

## CYCLIC FORMS OF MONOSACCHARIDES

The preferred structural form of many monosaccharides may be that of a cyclic hemiacetal. Five and six-membered rings are favored over other ring sizes because of their low angle and eclipsing strain.

Cyclic structures of this kind are termed furanose (five-membered) or pyranose (six-membered), reflecting the ring size relationship to the common heterocyclic compounds furan and pyran shown on the right. Ribose, an important aldopentose, commonly adopts a furanose structure by convention for the D-family, the five-membered furanose ring is drawn in an edgewise projection with the ring oxygen positioned away from the viewer.

The anomeric carbon atom (colored red here) is placed on the right. The upper bond to this carbon is defined as beta, the lower bond then is alpha. The cyclic pyranose forms of various monosaccharides are often drawn in a flat projection known as a Haworth formula, after the British chemist, Norman Haworth.

As with the furanose ring, the anomeric carbon is placed on the right with the ring oxygen to the back of the edgewise view. In the D-family, the alpha and beta bonds have the same orientation defined for the furanose ring. These Haworth formulas are convenient for displaying stereochemical relationships, but do not represent the true shape of the molecules. We know that these molecules are actually puckered in a fashion we call a chair conformation. Examples of four typical pyranose structures are shown below, both as Haworth projections and as the more representative chair conformers. The anomeric carbons are colored red.

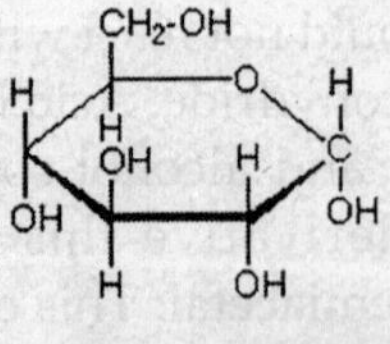

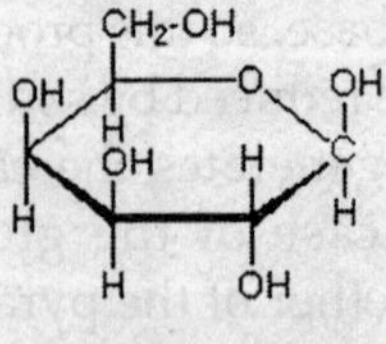

α-D-glucopyranose β-D-galactopyranose

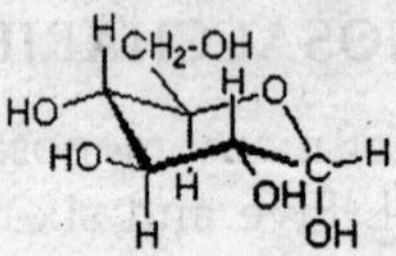

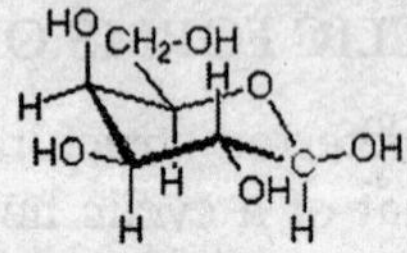

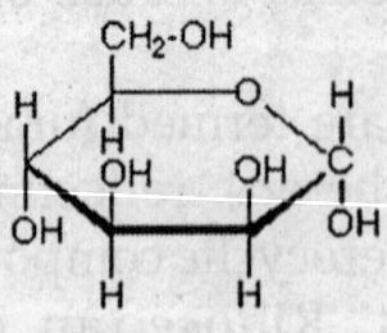

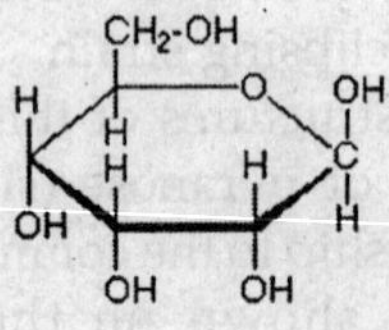

α-D-mannopyranose β-D-allopyranose

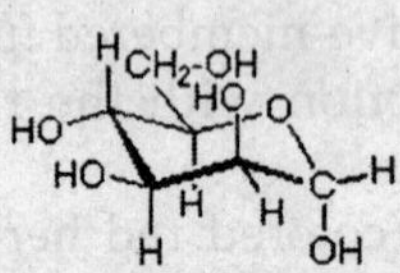

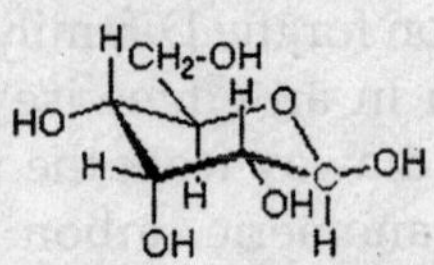

The size of the cyclic hemiacetal ring adopted by a given sugar is not constant, but may vary with substituents and other structural features. Aldolhexoses usually form pyranose rings and their pentose homologs tend to prefer the furanose form, but there are many counter examples. The formation of acetal derivatives illustrates how subtle changes may alter this selectivity. A pyranose structure for D-glucose is drawn in the rose-shaded box on the left. Acetal derivatives have been prepared by acid-catalyzed reactions with benzaldehyde and acetone. As a rule, benzaldehyde forms six-membered cyclic acetals, whereas acetone prefers to form five-membered acetals. The top equation shows the formation and some reactions of the 4,6-O-benzylidene acetal, a commonly employed protective group.

A methyl glycoside derivative of this compound leaves the C-2 and C-3 hydroxyl groups exposed to reactions such as the periodic acid cleavage, shown as the last step. The formation of an isopropylidene acetal at C-1 and C-2, cenre structure, leaves the C-3 hydroxyl as the only unprotected function. Selective oxidation to a ketone is then possible. Finally, direct di-O-isopropylidiene derivatization of glucose by reaction with excess acetone results in a change to a furanose structure in which the C-3 hydroxyl is again unprotected. However, the same reaction with D-galactose, shown in the blue-shaded box, produces a pyranose product in which the C-6 hydroxyl is unprotected. Both derivatives do not react with Tollens' reagent. This difference in behaviour is attributed to the cis-orientation of the C-3 and C-4 hydroxyl groups in galactose, which permits formation of a less strained five-membered cyclic acetal, compared with the trans-C-3 and C-4 hydroxyl groups in glucose. Derivatizations of this kind permit selective reactions to be conducted at different locations in these highly functionalized molecules.

## GLYCOSIDES

Acetal derivatives formed when a monosaccharide reacts with an alcohol in the presence of an acid catalyst are called glycosides. This reaction is illustrated for glucose and methanol in the diagram below. In naming of glycosides, the "oseΔ suffix of the sugar name is replaced by "osideΔ and the alcohol group name is placed first. As is generally true for most aldols, glycoside formation involves the loss of an equivalent of water. The diether product is stable to base and alkaline oxidants such as Tollen's reagent. Since acid-catalyzed aldolization is reversible, glycosides may be hydrolyzed back to their alcohol and sugar components by aqueous acid.

The anomeric methyl glucosides are formed in an equilibrium ratio of 66% alpha to 34% beta. In the case of glucose, the substituents on the beta-anomer are all equatorial, whereas the C-1 substituent in the alpha-anomer changes to axial. Since substituents on cyclohexane rings prefer an equatorial location over axial (methoxycyclohexane is 75% equatorial), the preference for alpha-glycopyranoside formation is unexpected and is referred to as the anomeric effect. Glycosides abound in biological systems. By attaching a sugar moiety to a lipid or benzenoid structure, the solubility and other properties of the compound may be changed substantially. Because of the important modifying influence of such derivatization, numerous enzyme systems, known as glycosidases, have evolved for the attachment and removal of sugars from alcohols, phenols and amines. Chemists refer to the sugar component of natural glycosides as the glycon and the alcohol component as the aglycon. Salicin, one of the oldest herbal remedies known, was the model for the synthetic analgesic aspirin.

A large class of hydroxylated, aromatic oxonium cations called anthocyanins provide the red, purple and blue colors of many flowers, fruits and some vegetables. Peonin is one example of this class of natural pigments, which exhibit a pronounced pH colour dependence. The oxonium moiety is only stable in acidic environments and the colour changes or disappears when base is added. The complex changes that occur when wine is fermented and stored are in part associated with glycosides of anthocyanins. Finally, amino derivatives of ribose, such as cytidine play important roles in biological phosphorylating agents, coenzymes and information transport and storage materials.

## DISACCHARIDES

When the alcohol component of a glycoside is provided by a hydroxyl function on another monosaccharide, the compound is called a disaccharide. Four examples of disaccharides composed of two glucose units are shown in the following diagram. The individual glucopyranose rings are labled A and B and the glycoside bonding is circled in light blue. Notice that the glycoside bond may be alpha, as in maltose and trehalose, or beta as in cellobiose and gentiobiose. Acid-catalyzed hydrolysis of these disaccharides yields glucose as the only product.

Enzyme-catalyzed hydrolysis is selective for a specific glycoside bond, so an alpha-glycosidase cleaves maltose and trehalose to glucose, but does not cleave cellobiose or gentiobiose. A beta-glycosidase has the opposite activity. In order to draw a representative structure for cellobiose, one of the glucopyranose rings must be rotated by 180º, but this feature is often omitted in favour of retaining the usual perspective for the individual rings.

The bonding between the glucopyranose rings in cellobiose and maltose is from the anomeric carbon in ring A to the C-4 hydroxyl group on ring B. This leaves the anomeric carbon in ring B free, so cellobiose and maltose both may assume alpha and beta anomers at that site . Gentiobiose has a beta-glycoside link, originating at C-1 in ring A and terminating at C-6 in ring B. Its alpha-anomer is drawn in the diagram. Because cellobiose, maltose and gentiobiose are hemiacetals they are all reducing sugars (oxidized by Tollen's reagent). Trehalose, a disaccharide found in certain mushrooms, is a bis-acetal and is therefore a non-reducing sugar. A systematic nomenclature for disaccharides exists, but as the following examples illustrate, these are often lengthy.

*Cellobiose:* 4-O-β-D-Glucopyranosyl-D-glucose
*Maltose:* 4-O-α-D-Glucopyranosyl-D-glucose
*Gentiobiose:* 6-O-β-D-Glucopyranosyl-D-glucose
*Trehalose:* α-D-Glucopyranosyl-α-D-glucopyranoside

Although all the disaccharides shown here are made up of two glucopyranose rings, their properties differ in interesting ways. Maltose, sometimes called malt sugar, comes from the hydrolysis of starch. It is about one third as sweet as cane sugar (sucrose), is easily digested by humans and is fermented by yeast. Cellobiose is obtained by the hydrolysis of cellulose. It has virtually no taste, is indigestible by humans and is not fermented by yeast. Some bacteria have beta-glucosidase enzymes that hydrolyze the glycosidic bonds in cellobiose and cellulose. The presence of such bacteria in the digestive tracts of cows and termites permits these animals to use cellulose as a food. Finally, it may be noted that trehalose has a distinctly sweet taste, but gentiobiose is bitter.

Disaccharides made up of other sugars are known, but glucose is often one of the components.. Lactose, also known as milk sugar, is a galactose-glucose compound joined as a beta-glycoside. It is a reducing sugar because of the hemiacetal function remaining in the glucose moiety. Many adults, particulaly those from regions where milk is not a dietary staple, have a metabolic intolerance for lactose. Infants have a digestive enzyme which cleaves the beta-glycoside bond in lactose, but production of this enzyme stops with weaning. Cheese is less subject to the lactose intolerance problem, since most of the lactose is removed with the whey. Sucrose, or cane sugar, is our most commonly used sweetening agent. It is a non-reducing disaccharide composed of glucose and fructose joined at the anomeric carbon of each by glycoside bonds (one alpha and one beta). In the formula shown here the fructose ring has been rotated 180º from its conventional perspective.

## POLYSACCHARIDES

As the name implies, polysaccharides are large high-molecular weight molecules constructed by joining monosaccharide units together by glycosidic bonds. They are sometimes called glycans. The most important compounds in this class, cellulose, starch and glycogen are all polymers of glucose. This is easily demonstrated by acid-catalyzed hydrolysis to the monosaccharide. Since partial hydrolysis of cellulose gives varying amounts of cellobiose, we conclude the glucose units in this macromolecule are joined by beta-glycoside bonds between C-1 and C-4 sites of adjacent sugars. Partial hydrolysis of starch and glycogen produces the disaccharide maltose together with low molecular weight dextrans, polysaccharides in which glucose molecules are joined by alpha-glycoside links between C-1 and C-6, as well as the alpha C-1 to C-4 links found in maltose.

Over half of the total organic carbon in the earth's biosphere is in cellulose. Cotton fibres are essentially pure cellulose and the wood of bushes and trees is about 50% cellulose. As a polymer of glucose, cellulose has the formula (C6H10O5)n where n ranges from 500 to 5,000, depending on the source of the polymer. The glucose units in cellulose are linked in a linear fashion, as shown in the drawing below. The beta-glycoside bonds permit these chains to stretch out and this conformation is stabilized by intramolecular hydrogen bonds. A parallel orientation of adjacent chains is also favored by intermolecular hydrogen bonds. Although an individual hydrogen bond is relatively weak, many such bonds acting together can impart great stability to certain conformations of large molecules.

Most animals cannot digest cellulose as a food and in the diets of humans this part of our vegetable intake functions as roughage and is eliminated largely unchanged. Some animals (the cow and termites, for example) harbour intestinal microorganisms that breakdown cellulose into monosacharide nutrients by the use of beta-glycosidase enzymes.

Cellulose is commonly accompanied by a lower molecular weight, branched, amorphous polymer called hemicellulose. In contrast to cellulose, hemicellulose is structurally weak and is easily hydrolyzed by dilute acid or base. Also, many enzymes catalyze its hydrolysis. Hemicelluloses are composed of many D-pentose sugars, with xylose being the major component. Mannose and mannuronic acid are often present, as well as galactose and galacturonic acid.

Starch is a polymer of glucose, found in roots, rhizomes, seeds, stems, tubers and corms of plants, as microscopic granules having characteristic shapes and sizes. Most animals, including humans, depend on these plant starches for nourishment. The structure of starch is more complex than that of cellulose. The intact granules are insoluble in cold water, but grinding or swelling them in warm water causes them to burst. The released starch consists of two fractions. About 20% is a water soluble material called amylose. Molecules of amylose are linear chains of several thousand glucose units joined by alpha C-1 to C-4 glycoside bonds.

Amylose solutions are actually dispersions of hydrated helical micelles. The majority of the starch is a much higher molecular weight substance, consisting of nearly a million glucose units and called amylopectin. Molecules of amylopectin are branched networks built from C-1 to C-4 and C-1 to C-6 glycoside links and are essentially water insoluble. The branching in this diagram is exaggerated, since on average, branches only occur every twenty five glucose units. Hydrolysis of starch, usually by enzymatic reactions, produces a syrupy liquid consisting largely of glucose. When cornstarch is the feedstock, this product is known as corn syrup. It is widely used to soften texture, add volume, prohibit crystallization and enhance the flavour of foods.

Glycogen is the glucose storage polymer used by animals. It has a structure similar to amylopectin, but is even more highly branched (about every tenth glucose unit). The degree of branching in these polysaccharides may be measured by enzymatic or chemical analysis.

## SYNTHETIC MODIFICATION OF CELLULOSE

Cotton, probably the most useful natural fibre, is nearly pure cellulose. The manufacturie of textiles from cotton involves physical manipulation of the raw material by carding, combing and spinning selected fibers. For fabrics the best cotton has long fibers and short fibers or cotton dust are removed. Crude cellulose is also available from wood pulp by dissolving the lignan matrix surrounding it. These less desirable cellulose sources are widely used for making paper.

In order to expand the ways in which cellulose can be put to practical use, chemists have devised techniques for preparing solutions of cellulose derivatives that can be spun into fibers, spread into a film or cast in various solid forms. A key factor in these transformations are the three free hydroxyl groups on each glucose unit in the cellulose chain, —$[C_6H_7O(OH)_3]_n$. Esterification of these functions leads to polymeric products having very different properties compared with cellulose itself.

Cellulose Nitrate, first prepared over 150 years ago by treating cellulose with nitric acid, is the earliest synthetic polymer to see general use. The fully nitrated compound, —$[C_6H_7O(ONO_2)_3]_n$, called guncotton, is explosively flammable and is a component of smokeless powder. Partially nitrated cellulose is called pyroxylin. Pyroxylin is soluble in ether and at one time was used for photographic film and lacquers. The high flammability of pyroxylin caused many tragic cinema fires during its period of use. Furthermore, slow hydrolysis of pyroxylin yields nitric acid, a process that contributes to the deterioration of early motion picture films in storage.

Cellulose Acetate, $-[C_6H_7O(OAc)_3]_n$, is less flammable than pyroxylin and has replaced it in most applications. It is prepared by reaction of cellulose with acetic anhydride and an acid catalyst.

The properties of the product vary with the degree of acetylation. Some chain shortening occurs unavoidably in the preparations. An acetone solution of cellulose acetate may be forced through a spinnerette to generate filaments, called acetate rayon, that can be woven into fabrics.

Viscose Rayon, is prepared by formation of an alkali soluble xanthate derivative that can be spun into a fibre that reforms the cellulose polymer by acid quenching. The following general equation illustrates these transformations. The product fibre is called viscose rayon.

$$\underset{\text{cellulose}}{ROH} \overset{NaOH}{\rightleftarrows} RO^{(-)}Na^{(+)} + S{=}C{=}S \rightarrow \underset{\text{viscose solution}}{RO\text{-}CS_2^{(-)}Na^{(+)}} \overset{H_3O^{(+)}}{\longrightarrow} \underset{\text{rayon}}{ROH}$$

## STRUCTURE OF CARBOHYDRATES

## MONOSACCHARIDES STRUCTURES

Sugars can be defined as polyhydroxy aldehydes or ketones. Hence the simplest sugars contain at least three carbons. The most common are the aldo- and keto-trioses, tetroses, pentoses and hexoses. The simplest 3C sugars are glyceraldehye and dihydroxyacetone.

Glucose, an aldo-hexose, is a central sugar in metabolism. It and other 5 and 6C sugars can cyclize through intramolecular nucleophilic attack of one of the OH's on the carbonyl C of the aldehyde or ketone. Such intramolecular reactions occur if stable 5 or 6 member rings can form. The resulting rings are labeled furanose (5 member) or pyranose (6 member) based on their similarity to furan and pyran. On nucleophilic attack to form the ring, the carbonyl O becomes an OH which points either below the ring (a anomer) or above the ring (β anomer).

Monosaccharides in solution exist as equilbrium mixtures of the straight and cyclic forms. In solution, glucose (Glc) is mostly in the pyranose form, fructose is 67% pyranose and 33% furanose and ribose is 75% furanose and 25% pyranose. However, in polysaccharides, Glc is exclusively pyranose and fructose and ribose are furanoses. Sugars can be drawn in the straight chain form as either Fisher projections or perspective structural formulas.

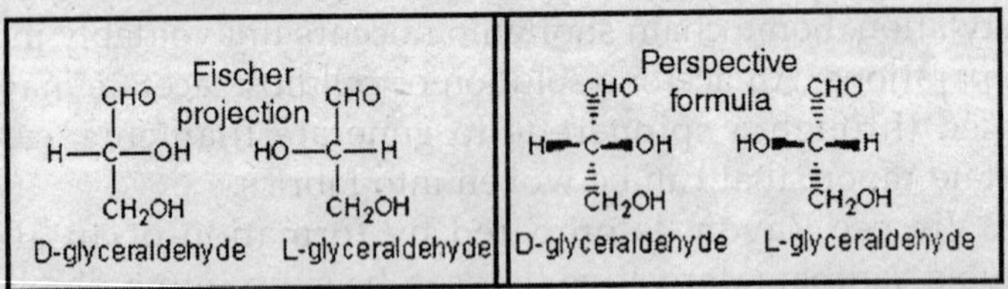


Cyclic forms can be drawn either as the Haworth projections, which shows the molecule as cyclic and planar with substituents above or below the ring) or the more plausible bent forms (showing Glc in the chair or boat conformations, for example). b-D-glucopyranose is the only aldohexose which can be drawn with all its bulky substituents (OH and $CH_2OH$) in equatorial positions, which probably accounts for its widespread prevalence in nature.

Haworth projections are more realistic than the Fisher projections, but you should be able to draw both structures. In general, if a substituents points to the right in the Fisher structure, it points down in the Haworth. if it points left, it points up. In general, the OH on the a-anomer points down (ants down) while on the b-anomer it points up (butterflys up).

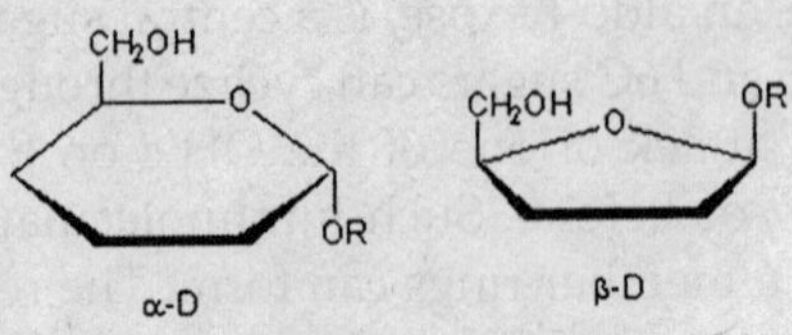


**Fig.** A more Rigorous view of the Relationship between the Anomeric OH and the OH on the Last Chiral C of a Sugar

The most common monosaccharides (other than glyceraldehyde and dihydroxyacetone) which you need to know are shown below.

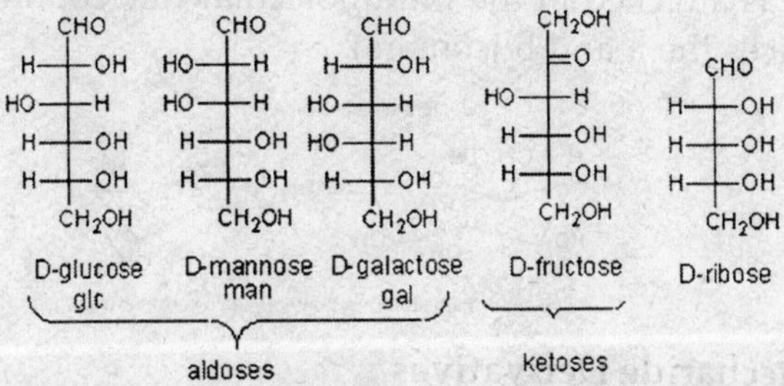


The mirror image of D-Glc is L-Glc. For common sugars, the prefix D and L refer to the cenre of asymmetry most remote from the aldehyde or ketone. By convention, all chiral centers are related to D- glyceraldehyde, so sugar isomers related to D-glyceraldehyde at their last asymmetric cenre are D sugars.

## ISOMERS

Sugars can exists as either configurational isomers (interconverted only by breaking covalent bonds) and conformational isomers.

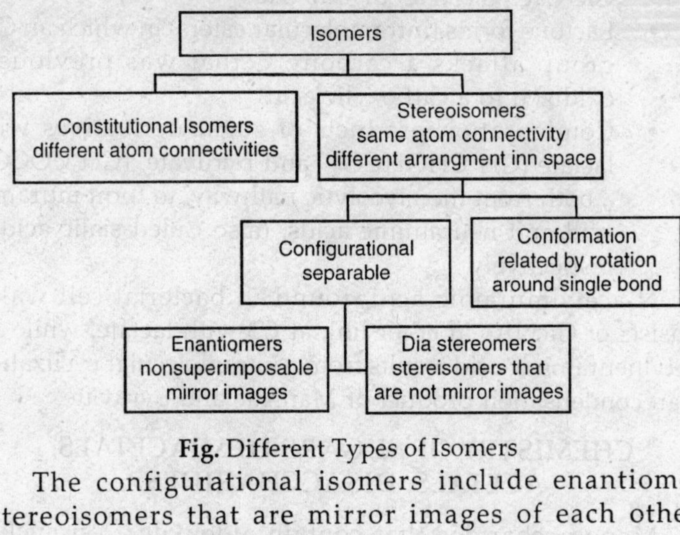


**Fig.** Different Types of Isomers

The configurational isomers include enantiomers (stereoisomers that are mirror images of each other), diastereomers (stereoisomers that are not mirror images), epimers (diastereomers that differ at one stereocenter) and anomers (a special form of stereoisomer, diastereomer and epimer that differ only in the configuration around the carbon which was attacked in the intramolecular nucleophilic attack to produces the a and b isomers).

Conformational Isomers

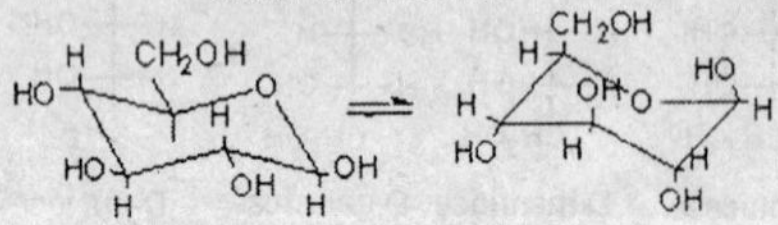

β-D-glucopyranose: chairnforms, boat form also is conformer

## Monosaccharide Derivatives

Many derivatives of monosaccharides are found in nature. These include

- Oxidized forms in which the aldehyde and/or alcohol functional groups are oxidized to carboxylic acids
- Phosporylated forms in which phosphate is added by ATP to form phosphoester derivatives
- Amine derivatives such as glucosamine or galactosamine
- Acetylated amine derivatives such as N-Acetyl-GlcNAc (GlcNAc) or GalNAc
- Lactone forms (intramolecular esters) in which an OH group attacks a carbony C that was previously oxidized to a carboxylic acid
- Condensation products of sugar derivatives with lactate ($CH_3CHOHCO_2^-$) and pyruvate, ($CH_3COCO_2^-$), both from the glycolytic pathway, to form muramic acid and neuraminic acids, (also called sialic acids), respectively.

N-acetylmuramic acid, found in bacterial cell walls, consists of GlcNAc in ether link at C3 with lactate, while N-acetylneuraminic acid results from a intramolecular cyclization of an condensation product of ManNac and pyruvate.

## CHEMISTRY OF SUGARS-HEMIACETALS, ACETALS, DISACCHARIDES,

Monosaccharides that contain aldehydes can cyclize

through intramolecular nucleophilic attack of an OH at the carbonyl carbon in an addition reaction to form a hemiacetal (hemiketal if attack on a ketone). On the addition of acid (which protonates the anomeric OH, forming water as a potential leaving group), another alcohol can add forming an acetal (or ketal from a ketone) with water leaving.

Aldehydes, Hemiacetals, and Acetals

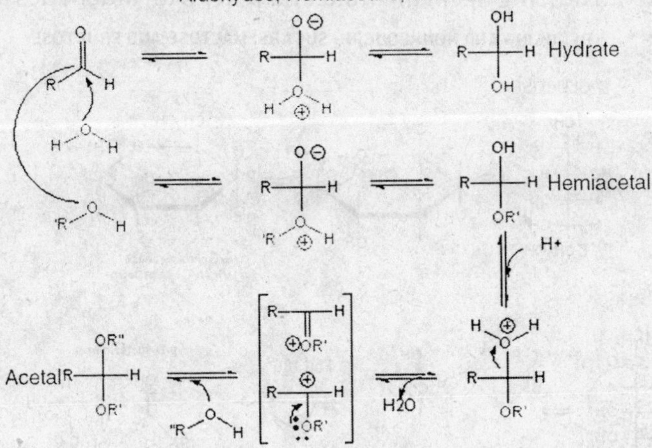


If the other alcohol is a second monosaccharide, a dissacharide results. The acetal (or ketal) link bonding to the two monosaccharides is called a glycosidic link. Links between the two sugars can be either a (if the OH on C1 involved in the glycosidic link is pointing down) or b (if the O on C1 involved in the glycosidic link is pointing up). Since sugars contain so many OH groups which can act as the "secondΔ alcohol in acetal (or ketal) formation, links between sugars can be quite diverse. These include a and b forms of 1-2, 1-3, 1-4, 1-5, 1-6, 2-2, etc. links. For example:

- *Lactose:* Gal(β 1->4)Glc Since Glc is attached to Gal through the OH on C4, its anomeric carbon, C1, could revert to the noncyclic aldehyde form. This aldehyde is susceptible to oxidation by reagents (Benedicts Solution - with citrate, Fehlings reagent - with tartrate) which are subsequently reduced. In both reagents, reducing sugars reduce a basic blue solution of $CuSO_4$

($Cu^{2+}$) to form a brick red precipate of $Cu_2O$ ($Cu^+$). Sugars (monosaccharides, dissacharides and polysaccharides which can form an aldehyde at C1 or have an a-hydroxymethyl ketone group which can isomerize to an aldehyde under basic conditions (such as fructose) are called reducing sugars. These oxidizing agent are mild and react with aldehydes and

**REDUCING AND NONREDUCING SUGARS: MALTOSE AND FRUCTOSE**

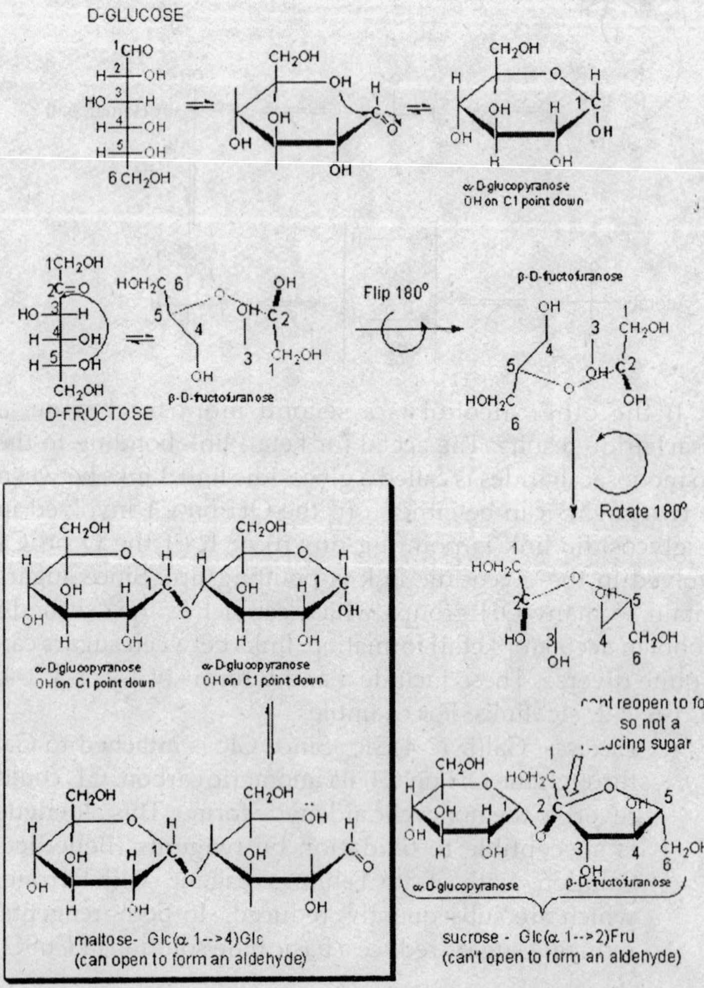

- *Sucrose:* Glc(α 1->2)Fru. Since Fru is attached through the anomeric OH of this ketose, the Fru is not in equilibrium with its straight chain keto form and hence sucrose is a nonreducing sugar.

The glycosidic (acetal or ketal) link can be cleaved by hydrolysis, just as the peptide bond in proteins.

## CLASSIFICATION OF CARBOHYDRATE

Carbohydrates are classified generally according to their degree of complexity. Hence, the free sugars such as glucose and fructose are termed monosaccharides; sucrose and maltose, disaccharides; and the starches and celluloses, polysaccharides. Carbohydrates of short chain lengths such as raffinose, stachyose and verbascose, which are three, four and five sugar polymers respectively, are classified as oligosaccharides.

## PENTOSES

Pentoses are five-carbon sugars seldom found in the free state in nature. In plants they occur in polymeric forms and are collectively known as pentosans. Thus, xylose and arabinose are the constituents of pentosans present in plant fibres and vegetable gums, respectively. As the sugar moieties in nucleic acids and riboflavin, ribose and deoxyribose are indispensable constituents of the life process. D-ribose has the following chemical structure.

## HEXOSES

The hexoses comprise a large group of sugars. Principal among these are: glucose, fructose, galactose and mannose. While glucose and fructose are found free in nature, galactose and mannose occur only in combined form. The hexoses are divided into aldoses and ketoses according to whether they possess aldehydic or ketonic groups. Thus, glucose is an aldo sugar and fructose is a keto sugar.

The presence of aymmetric centres in all sugars with three or more carbon atoms gives rise to stereoisomers. Galactose and mannose are stereoisomers of glucose which, theoretically, is only one of 16 stereoisomers. Because the ketohexoses have only three asymmetric centres, fructose is one of eight stereoisomers.

A general phenomenon, known as mutarotation, is observed in a variety of pentoses and hexoses as well as in certain disaccharides. For example, it has been established that two isomers of D-glucose exist, hence requiring an additional asymmetric centre in this sugar. It became apparent that D-glucose and most other sugars have cyclic structures.

The position of the hydroxyl group in relation to the ring oxygen characterizes this additional configurations modification. By convention, the positioning of the hydroxyl group on carbon atom 1 on the same side of the structure as the oxygen ring -modification; and, the positioning of the same hydroxyl group on the opposite side of the ring oxygen b-modification.

Carbohydrases, which catalyse the hydrolysis of glycosidic linkages of simple glycosides, oligosaccharides and polysaccharides often exhibit specificity with regard to substrate configuration. The specificity for enzyme hydrolysis of certain oligosaccharides helps to explain the poor utilization of this class of carbohydrates in fish nutrition.

Sugars containing the aldo or the keto group are capable of reducing copper in alkaline solutions to produce the brick-red colouration of cuprous ions. These sugars are called reducing sugars and the reaction, although not specific for reducing sugars, has use for both qualitative and quantitative determinations.

Glucose is widely distributed in small amounts in fruits, plant juices and honey. It is commercially produced by the acid or enzyme hydrolysis of grain and root starches. Glucose is of special interest in nutrition because it is the end-product of carbohydrate digestion in all non-ruminant animals including fish. Fructose is the only important ketohexose and is found in the free state alongside glucose in ripening fruits and honey. Combined with glucose it forms sucrose. Fructose is somewhat sweeter than sucrose and is produced in increasing quantities commercially as a sweetener.

Galactose occurs in milk in combination with glucose. It is also present in oligo-saccharides of plant origin, in combination with both glucose and fructose.

Mannose is present in some plant polysaccharides collectively termed mannans.

## DISACCHARIDES

Disaccharides are condensation products of two molecules of monosaccharides. Sucrose is the predominant disaccharide occurring in the free form and is the principal substance of sugar cane and sugar beet. It is also formed during germination of legume seeds. Other common disaccharides are maltose and lactose. Maltose is a dimer of glucose and lactose is a copolymer of galactose and glucose. The two molecules of glucose in maltose are held together in an a -1,4 glycosidic linkage whereas the two hexose entities of galactose are linked at the b -1,4 position. Glucose and fructose are combined in an a -1,2 linkage in sucrose. The abbreviated name of sucrose is D-Glu-(a, 1 $\rightarrow$ 2)-D-Fru.

## OLIGOSACCHARIDES

The Oligosaccharides raffinose, stachyose and verbascose are present in significant quantities in legume seeds. Raffinose, which is the most widespread among the three, consists of one molecule of glucose linked to a molecule of sucrose at the $\alpha$-1, 6 position. Its abbreviated chemical name is $\alpha$-D-Gal (1$\rightarrow$6) -$\alpha$ - D -Glu - (1 $\rightarrow$ 2) - $\beta$-D-Fru. Further chain elongation at the galactose end with another galactose molecule will yield stachyose. These galactose-galactose linkages are all at the $\alpha$-l,6 position and digestion of these Oligosaccharides by animals requires a highly specific enzyme not elaborated by the animals themselves but by certain bacteria present in the animals guts. The gradual disappearance of oligosaccharides from the cotelydons of legume seeds during germination is part of an intricate process beginning with uptake of water by the seed. This uptake of moisture releases gibberellic acid which in turn activates the DNA in the seed, thereby triggering the life cycle of the plant. The DNA directs the production of α-galactosidase which is required for the hydrolysis of these Oligosaccharides. Any interference of the DNA transcription process blocks enzyme production and will be evidenced by continued senescence of the seed and persistence of oligosaccharides in the seed cotelydons.

## POLYSACCHARIDES

The polysaccharides represent a large group of complex carbohydrates which are condensation products of undetermined numbers of sugar molecules. The various subgroups are rather ill-defined and there is a lack of agreement on their classification. Most polysaccharides are insoluble in water. Upon hydrolysis with acids or enzymes they eventually yield their constituent monosaccharides.

Starch is a high molecular weight polymer of D-glucose and is the principal reserve carbohydrate in plants. Most starches consist of a mixture of two types of polymers, namely; amylose and amylopectin. The proportion of amylose and amylopectin is generally one part of amylose and three parts of amylopectin. Enzymes capable of catalyzing the hydrolysis of starch are present in the digestive secretions of animals and fish within their cells.

The a-amylases which are found virtually in all living cells cleave the α-D-(1→4) linkages at random and bring about an eventual total conversion of the starch molecule into the reducing sugars. The principal α-amylases of animal origin are those produced in the salivary gland and the pancreas. Starch is insoluble in water and is stained blue by iodine.

Glycogen is the only complex carbohydrate of animal origin. It exists in limited quantities in liver and muscle tissues and acts as a readily available energy source. Dextrins are intermediate compounds resulting from incomplete hydrolysis or digestion of starch. The presence of α-D-(1→6) linkages in amylopectin and the inability of α-amylase to cleave these bonds give rise to low molecular weight carbohydrate segments called limit dextrins. These residues are acted upon primarily by acidophilic bacteria in the digestive tract.

Cellulose is made up of long chains of glucose units held

together by β-D-(1→4) linkages. The enzymes which cleave these linkages are not ordinarily present in the digestive secretions of animals and fish although some species of shellfish are believed to elaborate cellulase, the enzyme which catalyzes the hydrolysis of cellulose. Cellulase producing micro-organisms present in the gut of herbivorous animals and fish impart to their host animals the ability to utilize as food the otherwise indigestible cellulose.

Other complex polysaccharides in common occurrence are the hemicelluloses and pentosans. Hemicellulose represents a group of carbohydrates including araban, xylan, certain hexosans and polyuronides. These substances are generally less resistant to chemical treatment and undergo some degree of enzymatic hydrolysis during normal digestive processes. Pentosans are polymers of either xylose or arabinose as constituents of plant structural material and vegetable gums, respectively. Arbohydrates are classified generally according to their degree of complexity. Hence, the free sugars such as glucose and fructose are termed monosaccharides; sucrose and maltose, disaccharides; and the starches and celluloses, polysaccharides. Carbohydrates of short chain lengths such as raffinose, stachyose and verbascose, which are three, four and five sugar polymers respectively, are classified as oligosaccharides.

## GLYCOPROTEINS

For many years glycoproteins have been a subject of interest. However, it is in the second half of this century that they have aroused the interest of biochemists and biologists from a wide range of fields. This increased interest is partly due to the fact that glycoproteins were discovered to be abundant in living organisms. It is also due to the diverse functions of glycoproteins; glycoproteins appear in nearly every biological process studied. Many glycoproteins have structural functions. One of many instances is their role as a constituent of the cell wall. Glycoproteins also form connective tissues such as collagen. They are also found in gastrointestinal mucus secretions. Glycoproteins are used as protective agents and lubricants. They are also found abundantly in the blood plasma where they serve many functions.

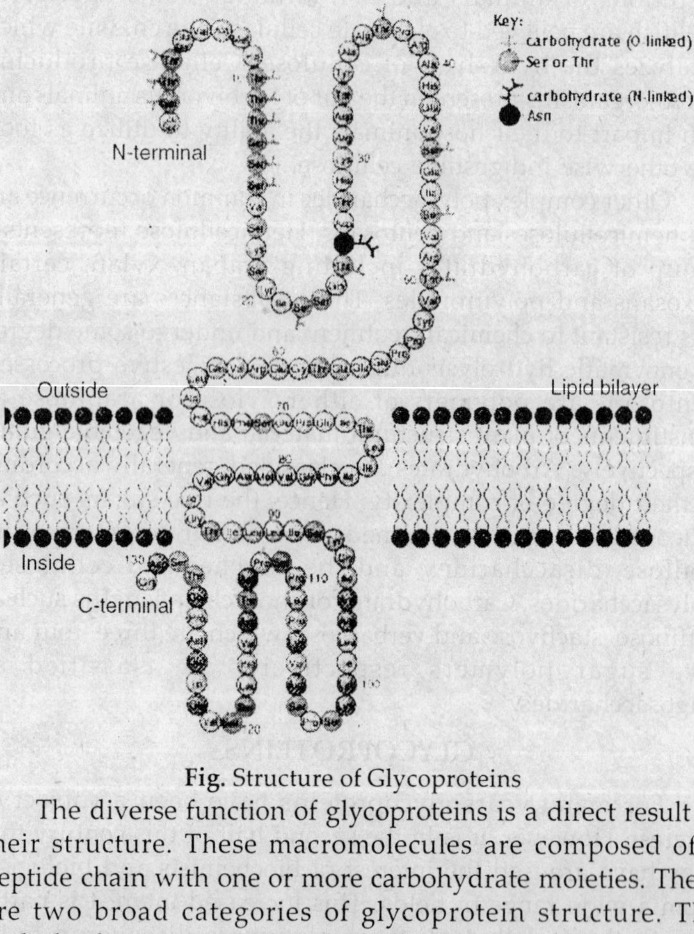


**Fig.** Structure of Glycoproteins

The diverse function of glycoproteins is a direct result of their structure. These macromolecules are composed of a peptide chain with one or more carbohydrate moieties. There are two broad categories of glycoprotein structure. The carbohydrates are either linked N-glycosidically or O-glycosidically to their constituent protein.

Within these broader categories, there can be fine structural differences which account for the large diversity of functions among glycoproteins. Controlling of glycoproteins is achieved through synthesis and degradation. Those processes are controlled by very specific enzymes. Not so much is yet researched on glycoprotein regulation in general.

## STRUCTURE

Structurally, glycoproteins consist of a polypeptide covalently bonded to a carbohydrate moiety. The carbohydrate can make up anywhere from less than one percent to more than 80 percent of the total protein mass. The saccharide chains, referred to as glycans, can be linked to the polypeptide in two major ways. The first class of glycoproteins are the O-linked glycans. These usually contain an N-acetylgalactosamine which is attached through a glycosidic bond to the O-terminus of either threonine or serine. The other class of glycoproteins are the N-linked glycans. These involve a glycosidic bond between N-acetylglucosamine and the N-terminus of an asparagine residue.

O-linked glycans consist of N-acetylgalactosamine attached to the O-terminus of a threonine or serine residue. N-acetylgalactosamine is simply a galactose molecule with an amine group covalently bonded to the second carbon. This amine group is bonded to a carboxyl group. N-acetylgalactosamine attaches to the carboxyl group of the amino acid through the hydroxyl group of its anomeric carbon. Another type of O-linked glycan consists of a galactose or a glucosyl-galactose disaccharide linked to the hydroxyl of hydroxylysine. Yet another type of O-linkage involves the binding of arabinose to the hydroxyl of hydroxyproline. In all of the 0-linked glycans, there can be a variety of different monosaccharide or polysaccharide chains attached to the sugar that is bonded to the amino acid.

The other class of glycoproteins are N-linked glycans. These molecules consist of an N-acetylglucosamine bonded to the amide nitrogen of an asparagine molecule. N-acetylglucosamine is simply a glucose molecule which is bonded to an amine group. This amine group, in turn, is bonded to a hydroxyl group. The N-acetylglucosamine is bonded to the asparagine through its anomeric carbon. The asparagine must be surrounded by a specific amino acid sequence, or sequon. This sequence is -X-Asn-X-Thr; the X can be any amino acid. A large variety of polysaccharide side chains can be linked to the N-acetylglucosamine. A typical polysaccharide chain is Man2 a(1-6)-Man B(1-4)-GlcNAc B(1-4)-GlcNAc B(1-N) Asn. Adding on to this structure can create many different N-linked glycans.

The carbohydrate chains of glycoproteins can play a role in the structure of the polypeptide. For example, in human immunoglobulins, the carbohydrate chain wraps around one of the protein domains. By doing so it prevents contact of this domain with the neighboring domain. An experiment that was done by Koide et al illustrates how the carbohydrate can change the overall structure of the polypeptide. The carbohydrate side chains of a rabbit antibody were removed through glycosidase digestion. The result was that the domain where the carbohydrate had previously been attached could no longer perform its ordinary function. Because an immunoglobulin's function is determined to a large extent by its structure, it can be concluded that removing the carbohydrate affected the structure of the molecule.

## FUNCTION

Because carbohydrates and proteins by themselves serve in a vast number of biological functions, it should not be surprising that linking the two together results in a macromolecule with an extremely large number of functions. Because of this and their biologically ubiquitous nature, the best way to go about exploring glycoprotein function is to break it down into categories that are fairly general. The following is an attempt to do this.

*Structural:* Glycoproteins are found throughout matrices. They act as receptors on cell surfaces that bring other cells and proteins together giving strength and support to a matrix

*Proteoglycan:* Linking glycoproteins cross links proteoglycan molecules and is involved in the formation of the ordered structure within cartilage tissue. In nerve tissue glycoproteins are abundant in gray matter and appear to be associated with synaptosomes, axons and microsomes. Prothrombin, thrombin and fibrinogen are all glycoproteins that play an intricate role in the blood clotting mechanism . In certain bacteria the slime layer that surrounds the outermost components of cell walls are made up of glycoproteins of high molecular weight. In addition to forming these s-layers, glycoproteins also function as bacterial flagella. These are made up of bundles of glycoproteins protruding from the cell's surface. Their rotation provides propulsion. In plants, glycoproteins have roles in cell wall formation, tissue differentiation, embryogenesis and sexual adhesion

*Protection:* High molecular weight polymers called mucins are found on internal epithelial surfaces. They form a highly viscous gel that protects epithelium form chemical, physical and microbial disturbances. Examples of mucin sites are the human digestive tract, urinary tract and respiratory tracts. "Cervical mucinΔ is a glycoprotein found in the cervix of animals that regulates access of spermatozoa to the upper reproductive tract. Recently it was discovered that mucins may be responsible for aiding in metastasis of transformed cancer cells Mucins are also found on the outer body surfaces of fish to protect the skin Not only does mucin serve the function of protection, but it also acts as a lubricant. Human lacrimal glands produce a glycoprotein which protects the corneal epithelium from desiccation and foreign particles. Human sweat glands secrete glycoproteins which protect the skin from the other excretory products that could harm the skin

*Reproduction:* Glycoproteins found on the surface of spermatozoa appear to increase a sperm cell's attraction for the egg by altering the electrophoretic mobility of the plasma membrane. Actual binding of the sperm cell to the egg is mediated by) -linked glycoproteins serving as receptors on the surface of each the two membranes. The zona pellucida is an envelope made of glycoprotein that surrounds the egg and prevents polyspermy from occurring after the first sperm cell has penetrated the egg's plasma membrane Hen ovalbumin is a glycoprotein found in egg white that serves as a food storage unit for the embryo

*Adhesion:* Glycoproteins serve to adhere cells to cells and cells to substratum. Cell-cell adhesion is the basis for the development of functional tissues in the body. The interactions between cells is mediated by the glycoproteins on those cell's surfaces. In different domains of the body, different glycoproteins act to unite cells. For example, nerve cells recognize and bind to one another via the glycoprotein N-CAM (nerve cell adhesion molecule). N-CAM is also found on muscle cells indicating a role in the formation of myoneural junctions. With cell-substratum adhesion, glycoproteins serve as cell surface receptors for certain adhesion ligands that mediate and coordinate the interaction of cells. Substrates with the appropriate receptor will bind to the cell related to that receptor. For example, a substrate containing the glycoprotein fibronectin will be recognized and adhered to by fibroblasts. The fibroblasts will then secrete adhesion molecules and continue to spread, producing a pericellular matrix .

*Hormones:* There are many glycoproteins that function as hormones such as human chorionic gonadotropin (HCG) which is present in human pregnancy urine. Another example is erythropoietin which regulates erythrocyte production

*Enzymes:* Glycoprotein enzymes are of three types. These are oxidoreductases, transferases and hydrolases

*Carriers:* Glycoproteins can bind to certain molecules and serve as vehicles of transport. They can bind to vitamins, hormones, cations and other substances.

*Inhibitors:* Many glycoproteins in blood plasma have shown antiproteolytic activity. For example, the glycoprotein a1-antichymotrypsin inhibits chymotrypsin.

*Defence:* In beetles pygidial glands secrete a glycoprotein disinfecting paste that covers the body and hardens. This shell provides protection against attack by bacteria and fungi

*Freezing-point depression:* Glycoproteins were found in the sera of antarctic fishes to decrease the freezing point due to their apparent interaction with water

Vision: In bovine visual pigment a glycoprotein forms the outer membranes of retinal rods

*Immunological:* The interaction of blood group substances with antibodies is determined by the glycoproteins on erythrocytes. Adding or removing just one monosaccharide from a blood group structure, the antigenicity and therefore a person's blood type can be altered Many immunoglobulins are actually glycoproteins Soluble immune mediators such as helper, suppressor and activator cell have been shown to bind to glycoproteins found on the surface of their target cells. B and T cells contain surface glycoproteins that attract bacteria to these sites and bind them. In much the same manner, glycoproteins can direct phagocytosis. Because the HIV virus recognizes the receptor protein CD4, it binds to helper T cells which contain it.

## REGULATION/CONTROL

Regulation and control of glycoproteins is not as straight forward as some might think. "If someone was to understand regulation in glycoproteins he would have to look at the enzymes that are involved in the biosynthesis pathway of these molecules; what is affecting the enzyme activity (could be hormonal). The big interest in this area is in the cloning of the genes, the cloning of the enzymes that are involved in the synthesis of the oligosaccharide on the glycoproteins.

## SYNTHESIS

The protein part of the glycoprotein is formed at the ribosomes, where all proteins are synthesized, on template represented by RNA and DNA. As a result, its structure can change only through the mutation of the genetic material of the cell. On the other hand, the carbohydrate component of a glycoprotein is not a product of the ribosome; it is synthesized somewhere in the cytoplasm, but the exact site of synthesis has not yet been established. Since it is not directly genetically controlled, the oligosaccharide part shows a much greater variation

In contrast with O-linked glycoproteins, where oligosaccharide assemble occurs on the polypeptide chain, N-linked glycoproteins assemble their oligosaccharide portions on a lipid linked intermediate, dolichol phosphate. The first step in oligosaccharide synthesis is the formation of that intermediate. In subsequent steps, sugars, the first being always N-acetylglucosamine are chain-like connected to dolichol phosphate. One more GlcNAc and three more mannose sugars are linked to the first GlcNAc to form the typical core of N-linked glycoproteins. Addition of any other sugar can be in any possible combination, according to the desired resulting functions. The specificity of the enzymes is very important in the synthesis process. Every sugar added, is catalyzed by a different enzyme.

These group of enzymes are called glycosyltransferases The first addition of GlcNAc to dolichol phosphate is catalyzed by the specific enzyme example) that cleave peptide bonds and glycosidases enzymes that remove sugars one at a time from the end of an oligosaccharide chain. Both of these groups are contained in lysozomes. The lysosome attaches to a phagocyte, which has engulfed a substance that needs to be broken down and releases its enzymes in it Next, these enzymes begin their catalytic action. In degradation, ad in synthesis, the enzymes involved are very specific. After the glycoprotein is broken down, its amino acid and sugar components are either metabolized or can be used in the formation of another glycoprotein. Enzymatic degradation can provide much information about the structure of the oligosaccharide chains, as well as about the carbohydrate peptide linkage. For example, if a glycoprotein is treated with mannosidase (removes mannose) and mannose is released, one can conclude that mannose residues were located at the periphery of the molecule since glycosidases remove sugars from the end of the oligosaccharide chain

Regulation and control is the organism's ultimate tool to monitor and adjust the production or degradation of different molecules. Like a factory, it would be useless and inefficient to produce excess products when there is no need for them. On the other hand, shortage of products could be a problem too. Hopefully, further research will put some light in the area of glycoprotein regulation.

## GLYCOPROTEIN STRUCTURE AND STEROID HORMONE FUNCTION

Glycoproteins play an important part in hormone function. The action of hormones depends on the initial binding of the hormone to a protein receptor molecule. In many cases this molecule is a glycoprotein. Many hormones bind to receptors in the cell membrane; these hormones never actually enter the cell.

Steroids, on the other hand, bind to an intracellular protein receptor. There is still controversy over whether the steroid hormone receptor is found in the nucleus or the cytosol. However, it is clear that after the steroid binds, the hormone-receptor complex moves to the nucleus.

The hormone binding domain of the receptor is found at the C-terminus. The amino acid sequence of this region is highly diverse; it is not conserved from one protein to the next. Binding of the hormone stimulates a conformational change in the hormone receptor. This change allows the hormone-receptor complex to bind to the DNA.

The DNA binding domain is highly conserved and is found within the central core of the protein. The central core contains a very basic amino acid sequence. In the cellular environment, these bases will tend to pick up a hydrogen and become positively charged. The positive charge will attract the hormone-receptor complex to the negatively charged DNA. Binding of the complex to the DNA stimulates transcription

## SIMILARITIES BETWEEN GLYCOPROTEINS, CYTOCHROMES AND METALLOENZYMES

Cytochromes have features that are quite similar to transferrins, a particular class of glycoproteins. First of all, cytochromes make up some of the integral proteins found in membranes, as do glycoproteins. Second of all, the main function of cytochromes is the binding of ion cations. Via oxidoreduction with these cations of 2+state, cytochromes bind iron to a porphyrin prosthetic group, forming a heme group The function is to transport cations in and out of the cell. Transferrin is an iron binding protein as well.

It interacts with iron cations of the 2+state which also bind to a porphyrin group, Protoporphyrin IX, to form a heme group. Transferrins carry ions to many different locations in the body including in and out of cells Another group of molecules that serve a function related to glycoproteins are the metalloenzymes. One class of glycoprotein enzymes, the oxidoreductases, contain an enzyme called chloroperoxidase. Like transferrin, chloroperoxidase contains protophyrin IX and forms the heme group with iron 3+ Eicosanoids, are also related to glycoproteins because they recognize the receptors on cells which are glycoproteins.

## INHIBITION OF ASPARAGINE LINKED GLYCOPROTEINS BY ANTIBIOTICS

Synthesis of N-linked glycoproteins can be inhibited with the use of antibiotics (function of antibiotics). Examples are monensin, mevinolin and tunicamycin. Monensin, a monocovalent ionophore, inhibits the transport of proteins through the endoplasmic reticulum- Golgi complex . Mevinolin on the other hand, is an inhibitor of mevalonate synthesis. Mevalonate is the precursor of dolichol and other isoprenoid lipids Asparagine linked glycosylation involves the first step of addition of GlcNAc to dolichol diphosphate.

This reaction is catalyzed by the enzyme UDP-GlcNAc:dolichol phosphate N-acetylglucosamine-1-phosphate transferase or GPT. Tunicamycin is a specific inhibitor of GPT, therefore it can inhibit the synthesis of all N-linked glycoproteins.

To give just an example of the effectiveness of tunicamycin, when it was injected to rats' hearts, it killed 12, 24, 50 and 60 h after the injection . Another connection could be seen with the function of the ribosomes, since the polypeptide chain of all glycoproteins.

## PEPTIDOGLYCAN

Peptidoglycan, also known as murein, is a polymer consisting of sugars and amino acids that forms a mesh-like layer outside the plasma membrane of eubacteria. The sugar component consists of alternating residues of β-(1,4) linked N-acetylglucosamine and N-acetylmuramic acid residues. Attached to the N-acetylmuramic acid is a peptide chain of three to five amino acids.

The peptide chain can be cross-linked to the peptide chain of another strand forming the 3D mesh-like layer. Some Archaea have a similar layer of pseudopeptidog-lycan. Peptidoglycan serves a structural role in the bacterial cell wall, giving structural strength, as well as counteracting the osmotic pressure of the cytoplasm.

A common misconcep-tion is that peptidoglycan gives the cell its shape; however, whereas peptidoglycan helps maintain the structure of the cell, it is actually the MreB protein that facilitates cell shape. Peptidoglycan is also involved in binary fission during bacterial cell reproduction. The peptidoglycan layer is substantially thicker in Gram-positive bacteria (20 to 80 nm) than in Gram-negative bacteria (7 to 8 nm), with the attachment of the S-layer. Peptidoglycan forms around 90% of the dry weight of Gram-positive bacteria but only 10% of Gram-negative strains. In Gram-positive strains, it is important in attachment roles and sterotyping purposes. For both Gram-positive and Gram-negative bacteria, particles of approximately 2 nm can pass through the peptidoglycan.

## ANTIBIOTIC INHIBITION

Some antibacterial drugs such as penicillin interfere with the production of peptidoglycan by binding to bacterial enzymes known as penicillin-binding proteins or transpe-ptidases. Penicillin-binding proteins form the bonds between oligopeptide crosslinks in peptidoglycan.

For a bacterial cell to reproduce through binary fission, more than a million peptidoglycan subunits (NAM-NAG+oligopeptide) must be attached to existing subunits. Mutations in transpeptidases that lead to reduced interactions with an antibiotic are a significant source of emerging antibiotic resistance.

Considered the human body's *own antibiotic*, lysozymes found in tears work by breaking the â-(1,4)-glycosidic bonds in peptidoglycan and thereby destroying many bacterial cells. Antibiotics such as penicillin commonly target bacterial cell wall formation (of which peptidoglycan is an important component) because animal cells do not have cell walls.

## Structure

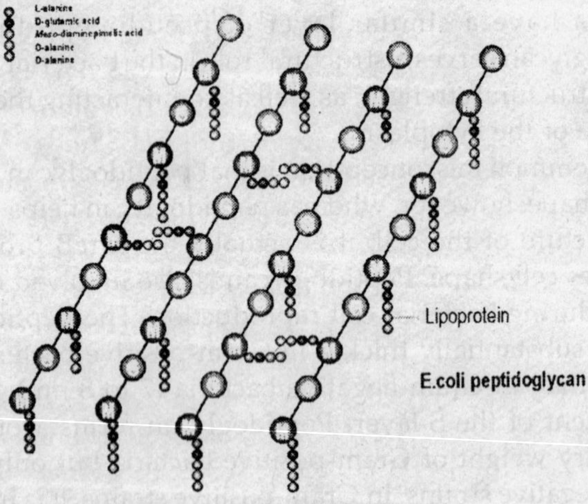


**Fig.** Structure of Peptidoglycan

The peptidoglycan layer in the bacterial cell wall is a crystal lattice structure formed from linear chains of two alternating amino sugars, namely *N*-acetylglucosamine (GlcNAc or NAG) and *N*-acetylmuramic acid (MurNAc or NAM). The alternating sugars are connected by a â-(1,4)-glycosidic bond. Each MurNAc is attached to a short (4- to 5-residue) amino acid chain, normally containing D-alanine, D-glutamic acid and mesodiaminopimelic acid. These three amino acids do not occur in proteins and are thought to help protect against attacks by most peptidases. Cross-linking between amino acids in different linear amino sugar chains by an enzyme known as transpeptidase result in a 3-dimensional structure that is strong and rigid. The specific amino acid sequence and molecular structure vary with the bacterial species.

## LIPOPOLYSACCHARIDES

## LIPOPOLYSACCHARIDES

These complex compounds (LPS) are the endotoxic O-antigens found in the cell walls of Gram-negative bacteria

(S-lipopolysaccharides) but are also present in one fungus, they may cause several pathophysiological symptoms, such as fever, diarrhea, blood pressure decrease, septic shock and death. A lipid part (Lipid A) forms a complex with a core polysaccharide part through a glycosidic linkage. The core part is linked to a third external region of a highly immunogenic and variable O-chain polysaccharide or O-antigen made up of repeating oligosaccharide units. The latter region of the LPS molecule is responsible for bacterial serological strain specificity.

Lipid A consists of a backbone of b-1,6-glucosaminyl-glucosamine with two phosphoester groups in the 1-position of glucosamine I and in the 4-position of glucosamine II. The 3-position of glucosamine II forms the acid-labile glycosidic linkage to the long-chain polysaccharide. The other groups are substituted with hydroxylated fatty acids as hydroxymyristate (two ester-linked and two amide-linked) and normal fatty acids. The fatty acid profiles of lipopolysaccharides differ markedly from those of the phospholipids from the same species.

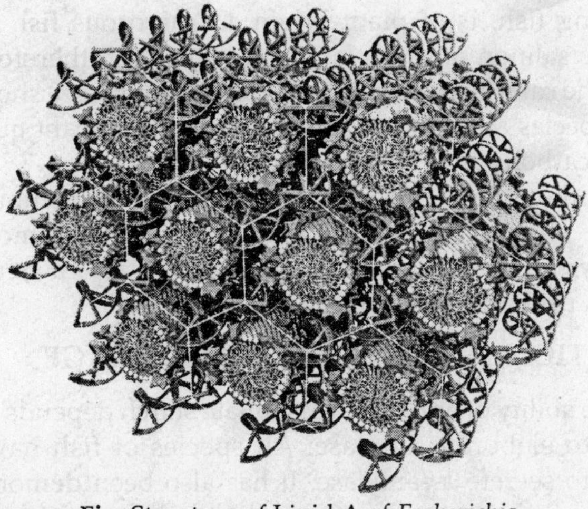

**Fig.** Structure of Lipid A of *Escherichia*

Although the hydroxylated fatty acids are generally 3-hydroxyalkanoic acids, 2-hydroxyalkanoic acids are also common. Exceptionally, some bacteria seem to lack hydroxy acids such as some species of *Gardnerella, Thermus, Thiobacillus, Bacteroides, Sorangium, Achromobacter* and *Brucella*. In most organisms, the major hydroxy acids are straight-chain, even-carbon compounds (from 12 to 18 carbon atoms). Odd-carbon acids are minor components (13 to 17 carbon atoms) in some enterobacteria, some *Vibrio*, some Pseudomonads and *Bordetella*. They are also found in *Pectinatus, Veillonella* and *Selenomonas*. Branched-chain hydroxy fatty acids are also present in *Pseudomonas, Moraxella, Desulfovibrio, Capnocytophaga, Flavobacterium, Legionella* species (3-OH i-11:0 up to3-OH i-17:0).

Lipopolysaccharides are the major toxins of Gram-negative bacteria (endotoxin). The only Gram-positive bacteria that possesses LPS is *Listeria monocytogenes*, an infective agent in milk. While the polysaccharide is non-toxic, the lipid A part is responsible for the toxic activities of these bacteria which result in septic shock.

## CARBOHYDRATE METABOLISM IN FISH

Much of the carbohydrates that enter the diets of animals, including fish, is of plant origin. Carnivorous fish like the Atlantic salmon and the Japanese yellowtail, therefore, deal with little carbohydrate,. Indeed, experiments have shown that these species are ill-equipped to handle significant quantities of raw carbohydrate, in their diets.

On the other hand, omnivores such as the common carp and the channel catfish are able to digest fair amounts of carbohydrates in their diets. The grass carp, a herbivore, subsists primarily on a vegetarian diet.

## DIGESTION, ABSORPTION AND STORAGE

The ability of animals to assimilate starch depends on their ability to elaborate amylase. All species of fish have been shown to secrete a -amylase. It has also been demonstrated that activity of this enzyme was greatest in herbivores. In carnivores such as the rainbow trout and sea perch, amylase is primarily of pancreatic origin whereas in herbivores the enzyme is widespread throughout the entire digestive tract.

In Tilapia mossambica the pancreas has been shown to be the site of greatest amylase activity followed by the upper intestine. Although the digestion of starch and dextrin by the carnivorous rainbow trout was shown to decrease progressively as levels of the carbohydrates were increased beyond the 20 percent level, the fish could effectively utilize up to 60 percent glucose, sucrose or lactose in the diet. This demonstrates that, contrary to earlier belief, carnivorous fish are capable of efficiently utilizing simple carbohydrate as a primary energy source.

The crystalline structure of starch appears also to influence its attack by amylase as evidenced by the two-fold increase in metabolizable energy content of fully cooked (gelatinized) maize in feeding trials with channel catfish. Rainbow trout have also been shown to have a higher tolerance for carbohydrate (present as wheat starch) in the diet when it was cooked. The process of gelatinization involves both heat and water. If an aqueous suspension of starch is heated, the granules do not change in appearance until a certain critical temperature is reached. At this point some of the starch granules swell and simultaneously lose their crystallinity. The critical temperature is that at which hydrogen bonds of the starch molecule loosen to permit complete hydration, leading to a phenomenon known as "swellingΔ.

Alpha-amylase, promotes a more or less random fragmentation of the starch molecule by hydrolyzing at the α -D-(l→4) glucosidic bonds in the inner and outer chains of the compound. The result of complete hydrolysis of the amylose component are maltose and D-glucose, while the amylopectin component is reduced to maltose, D-glucose and branched limit dextrins. As a consequence of these action patterns by a -amylase on starch, other enzymes are needed for complete hydrolysis of starch to D-glucose in fish. In this regard, it has been demonstrated that even the carnivorous sea bream possess the ability to digest maltose.

On the other hand, cellulase and a -galactosidase have not been shown to be secreted by fish although cellulase of bacterial origin is present in the gut of most species of carps.

The lack of a -galactosidase may partly explain the poor response by fish to dietary soybean meal which contains significant levels of the galactosidic oligosaccharides raffinose and stachyose. As has been pointed out earlier, these oligosaccharides do undergo enzymatic hydrolysis during the germination process to yield galactose and sucrose.

It would, therefore, appear that the nutritive value of soybean meal will be enhanced if the bulk of this indigestible starch is first transformed.

This can be achieved by soaking the beans for 48 hours prior to processing for meal production. It should also be pointed out that the nutritive value of pulses and other legume seeds can likewise be improved for fish since oligosaccharides constitute a large portion of the carbohydrates in legume seeds. Data on glucose absorption by fish are scanty. Work with goldfish has shown that active transport of glucose is coupled with $Na^+$ transport as in most mammals. It is generally believed that absorption takes place on the mucosal surface of intestinal cells.

The mono-saccharides which result from carbohydrate digestion consist primarily of glucose, fructose, galactose, mannose, xylose and arabinose. Although the rates of absorption of these sugars have been determined for many land mammals, similar information for fish is not available.

Glucose does not appear to be a superior energy source for fish over protein or fat although digestible carbohydrates do spare protein for tissue building. Also, unlike in mammals, glycogen is not a significant storage depot of energy despite evidence of an active and reversible Emden-Meyerhoff pathway in fish. The more efficient metabolism of amino acids over glucose for energy could be due to the ability of fish to excrete nitrogenous waste as ammonia from their gills without the high cost of energy in converting the waste to urea.

## OTHER FACTORS AFFECTING METABOLISM

Apart from genetic adaptation, climatic factors also play an important role in carbohydrate metabolism in fish. Acclimation in fish, in essence, reflects enzyme acclimation, since the animal's ability to survive depends largely upon its ability to carry out normal metabolic functions. Some enzymes for metabolic acclimation show good compensation while others do not.

The enzymes associated with energy liberation (enzymes of glycolysis, pentose shunt, tricarboxylic acid cycle, electron transport and fatty acid oxidation) exhibit temperature compensation whereas, those enzymes dealing largely with the degradation of metabolic products show poor or reverse compensation

**Table:** Enzymes Subject to Metabolic Acclimation

| *Enzymes exhibiting compensation* | *Enzymes exhibiting reverse or no compensation* |
|---|---|
| Phosphofructokinase | Catalase |
| Aldolase | Peroxidase |
| Lactic dehydrogenase | Acid phosphatase |
| 6-phosphogluconate dehydrogenase | D-amino acid oxidase |
| Succinic dehydrogenase | Mg-ATP ase |
| Malic dehydrogenase | Choline acetyl transferase |
| Cytochrome oxidase | Acetylcholine esterase |
| Succinate-cytochrome C reductase | Alkaline phosphatase |
| NAD-cytochrome C reductase | Allantoinase |
| Aminoacyl transferase | Uricase |
| Na-K-ATPase | Amylase |
| | Lipase |
| protease | Malic enzyme |
| | Glucose-6-phosphate dehydrogenase |

1. *Adapted from:* Comparative Animal Physiology, edited by C.L. Prosser, 1973

It is interesting to note that two key enzymes involved in carbohydrate metabolism, amylase and glucose-6-phosphate dehydrogenase, together with an enzyme involved in fat digestion, lipase, show no temperature compensation. It is not certain if this is in any way connected with the cessation of feeding by fish at low temperatures. The molecular mechanism of thermal acclimation are not well understood and may consist of changes in synthesis or amounts of a given enzyme. Differences in kinetics, changes in the proportion of isoenzymes suitable for particular temperatures and changes in co-factors such as lipids, co-enzymes, or other factors such as pH and ions may be important in the animal's adjustment to temperature changes.

## ENERGY TRANSFORMATION

Despite species differences in the tolerance of dietary carbohydrates it is generally believed that the principal end-product of carbohydrate digestion, glucose, is metabolized in a manner prevailing in all cells, i.e., via the reversible Emden-Meyerhoff pathway.

In this pathway, glucose has only one principal fate: phosphorylation to glucose-6-phosphate. The major metabolic transformations are depicted as follows: All transformations proceed with a loss of free energy. Thus, the formation of two moles of lactate from glucose-6-phosphate occurs with free energy change of D $G^{o}$ = -22000 cal/mole. The net result is the formation of four molecules of ATP. A functional reversal of this transformation can only occur via a different sequence requiring the input of six ATP molecules per mole of glucose-6-phosphate recovered.

Cells do not store glucose or glucose-6-phosphate. The readily available storage form is glycogen which is made from glucose-1-phosphate by one pathway and returned by another. Although in mammalian cells glucose-6-phosphate is transformed into fatty acids, such transformation does not appear to take place in fish.

Studies with the common carp indicate that the precursor for lipogenesis is citrate formed when amino acids are actively metabolized through the tricarboxylic acid cycle. The major form of utilizable energy in all cells is ATP. In most cells this energy currency is generated by the oxidation of NADH by the mitochondrial electron-transport systems. The reductants of $NAD^+$ for this process are intermediates derived from the TCA cycle and fatty acids.

## CARBOHYDRATE METABOLISM

Carbohydrate metabolism denotes the various biochemical processes responsible for the formation, breakdown and interconversion of carbohydrates in living organisms. The most important carbohydrate is glucose, a simple sugar (monosaccharide) that is metabolized by nearly all known organisms.

Glucose and other carbohydrates are part of a wide variety of metabolic pathways across species: plants synthesize carbohydrates from atmospheric gases by photosynthesis storing the absorbed energy internally, often in the form of starch or lipids. Plant components are eaten by animals and fungi, and used as fuel for cellular respiration. Oxidation of one gram of carbohydrate yields approximately 4 kcal of energy and from lipids about 9 kcal. Energy obtained from metabolism (eg, oxidation of glucose) is usually stored temporarily within cells in the form of ATP. Organisms capable of aerobic respiration metabolize glucose and oxygen to release energy with carbon dioxide and water as byproducts.

Carbohydrates are a superior short-term fuel for organisms because they are simpler to metabolize than fats or those amino acid portions of proteins that are used for fuel. In animals, the most important carbohydrate is glucose; so much so, that the level of glucose is used as the main control for the central metabolic hormone, insulin. Starch, and cellulose in a few animals (eg, termites, ruminants, and some bacteria), being both glucose polymers are disassembled during digestion and absorbed as glucose. Some simple carbohydrates have their own enzymatic oxidation pathways, as do only a few of the more complex carbohydrates. The disaccharide lactose, for instance, requires the enzyme lactase to be broken into into its monosaccharides components; many animals lack this enzyme in adulthood.

Carbohydrates are typically stored as long polymers of glucose molecules with Glycosidic bonds for structural support (e.g. chitin, cellulose) or for energy storage (e.g. glycogen, starch). However, the strong affinity of most carbohydrates for water makes storage of large quantities of carbohydrates inefficient due to the large molecular weight of the solvated water-carbohydrate complex.

In most organisms, excess carbohydrates are regularly catabolised to form Acetyl-CoA, which is a feed stock for the fatty acid synthesis pathway; fatty acids, triglycerides, and other lipids are commonly used for long-term energy storage. The hydrophobic character of lipids makes them a much more compact form of energy storage than hydrophilic carbohydrates.

## Glycolysis

Glycolysis (from *glycose*, an older term for glucose + *-lysis* degradation) is the metabolic pathway that converts glucose, $C_6H_{12}O_6$, into pyruvate, $C_3H_5O_3^-$. The free energy released in this process is used to form the high energy compounds, ATP (adenosine triphosphate) and NADH (reduced nicotinamide adenine dinucleotide). Glycolysis is a sequence of ten reactions involving ten intermediate compounds (one of the steps involves two intermediates). The intermediates provide entry points to glycolysis. For example, most monosaccharides, such as fructose, glucose, and galactose, can be converted to one of these intermediates. The intermediates may also be directly useful. For example, the intermediate, dihydroxyacetone phosphate, is a source of the glycerol that combines with fatty acids to form fat (triacylglycerides). Glycolysis results in the net production of 2 ATP and 2 NADH ions. If oxygen is present, the product of glycolysis, pyruvate, can enter the citric acid cycle, where it is converted into carbon dioxide. The free energy released in that process is used to produce considerably more ATP and NADH. Under anaerobic conditions, glycolysis may be the main energy source of an organism. This is the case in many prokaryotes, in eukaryotic cells devoid of mitochondria (e.g., mature erythrocytes), and in aerobic eukaryotic cells under low-oxygen conditions (e.g., heavily-exercising muscle or fermenting yeast).

Glycolysis is the archetype of a universal metabolic pathway. It occurs, with variations, in nearly all organisms, both aerobic and anaerobic. The wide occurrence of glycolysis indicates that it is one of the most ancient metabolic pathways known.Glycolysis takes place in the cytoplasm. However, some of the glycolytic reactions also occur in the Calvin cycle, which takes place in the chloroplasts of plant cells.

## TCA Cycle

The TCA cycle showing enzymes, substrates and products. The abbreviated enzymes are: IDH = *isocitrate dehydrogenase* and α-KGDH = *α-ketoglutarate dehydrogenase*. The GTP generated during the *succinate thiokinase* (succinyl-CoA synthetase) reaction is equivalent to a mole of ATP by virtue of the presence of *nucleoside diphosphokinase*. The 3 moles of NADH and 1 mole of $FADH_2$ generated during each round of the cycle feed into the oxidative phosphorylation pathway. Each mole of NADH leads to 3 moles of ATP and each mole of $FADH_2$ leads to 2 moles of ATP. Therefore, for each mole of pyruvate which enters the TCA cycle, 12 moles of ATP can be generated.

## CITRATE SYNTHASE (CONDENSING ENZYME)

The first reaction of the cycle is condensation of the methyl carbon of acetyl-CoA with the keto carbon (C-2) of oxaloacetate (OAA) to form citrate The standard free energy of the reaction, -8.0 kcal/mol, drives it strongly in the forward direction. Since the formation of OAA from its precursor is thermodynamically unfavorable, the highly exergonic nature of the *citrate synthase* reaction is of central importance in keeping the entire cycle going in the forward direction, since it *drives* oxaloacetate formation by mass action principals.

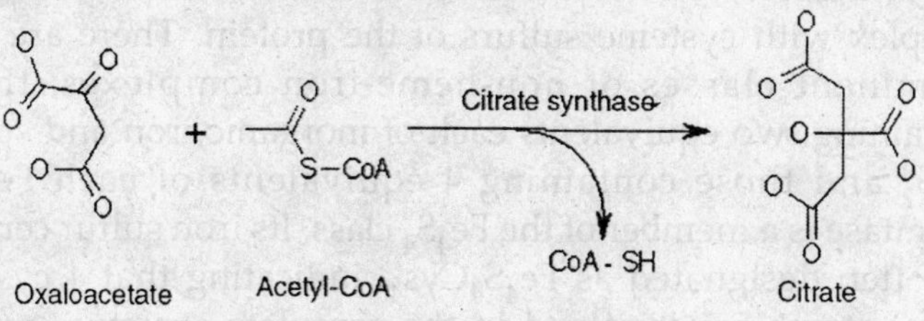


When the cellular energy charge increases the rate of flux through the TCA cycle will decline leading to a build-up of citrate. Excess citrate is used to transport acetyl-CoA carbons from the mitochondrion to the cytoplasm where they can be used for fatty acid and cholesterol biosynthesis. Additionally, the increased levels of citrate in the cytoplasm activate the key regulatory enzyme of fatty acid biosynthesis, acetyl-CoA carboxylase (ACC) and inhibit PFK-1. In non-hepatic tissues citrate is also required for ketone body synthesis.

## ACONITASE

The isomerization of citrate to isocitrate by ***aconitase*** is stereospecific, with the migration of the -OH from the central carbon of citrate (formerly the keto carbon of OAA) being always to the adjacent carbon which is derived from the methylene (-$CH_2$-) of OAA.During the reaction the intermediate linked to the enzyme *cis*-aconitate is produced.

The stereospecific nature of the isomerization determines that the $CO_2$ lost, as isocitrate is oxidized to succinyl-CoA, is derived from the oxaloacetate used in citrate synthesis.

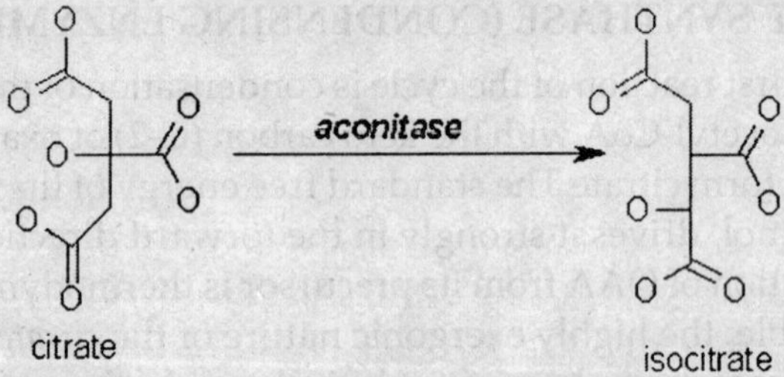


Aconitase is one of several mitochondrial enzymes known as non-heme-iron proteins. These proteins contain inorganic iron and sulfur, known as iron sulfur centers, in a coordination complex with cysteine sulfurs of the protein. There are two prominent classes of non-heme-iron complexes, those containing two equivalents each of inorganic iron and sulfur $Fe_2S_2$ and those containing 4 equivalents of each $Fe_4S_4$. Aconitase is a member of the $Fe_4S_4$ class. Its iron sulfur centers are often designated as $Fe_4S_4Cys_4$, indicating that 4 cystine sulfur atoms are involved in the complete structure of the complex. In iron sulfur compounds the iron is generally involved in oxidation-reduction events.

## ISOCITRATE DEHYDROGENASE

Isocitrate is oxidatively decarboxylated to a-ketoglutarate by isocitrate dehydrogenase, (IDH).

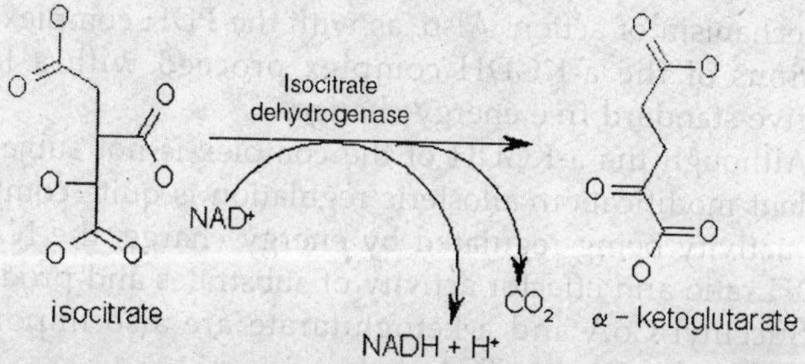


There are two different IDH enzymes. The IDH of the TCA cycle uses $NAD^+$ as a cofactor, whereas the other IDH uses $NADP^+$ as a cofactor. Unlike the NAD+-requiring enzyme, which is located only in the mitochondrial matrix, the $NADP^+$-requiring enzyme is found in both the mitochondrial matrix and the cytosol.

IDH catalyzes the rate-limiting step, as well as the first NADH-yielding reaction of the TCA cycle. The $CO_2$ produced by the IDH reaction is the original C-1 of the oxaloacetate used in the citrate synthase reaction. It is generally considered that control of carbon flow through the cycle is regulated at IDH by the powerful negative allosteric effectors NADH and ATP and by the potent positive effectors; isocitrate, ADP and AMP. From the latter it is clear that cell energy charge is a key factor in regulating carbon flow through the TCA cycle.

## α-KETOGLUTARATE DEHYDROGENASE COMPLEX

α-ketoglutarate is oxidatively decarboxylated to succinyl-CoA by the *a-ketoglutarate dehydrogenase* (α-KGDH) complex.

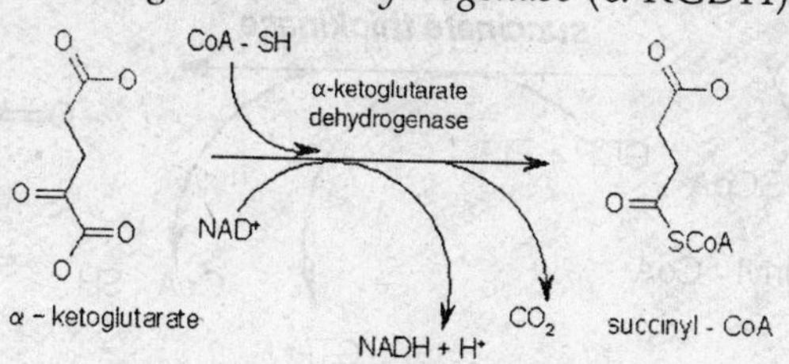

This reaction generates the second TCA cycle equivalent of $CO_2$ and NADH.

This multienzyme complex is very similar to the PDH complex in the intricacy of its protein makeup, cofactors and its mechanism of action. Also, as with the PDH complex, the reactions of the a-KGDH complex proceed with a large negative standard free energy change.

Although the a-KGDH of the complex is not subject to covalent modification, allosteric regulation is quite complex, with activity being regulated by energy charge, the $NAD^+$/ NADH ratio and effector activity of substrates and products.

Succinyl-CoA and a-ketoglutarate are also important metabolites outside the TCA cycle.

In particular,a-ketoglutarate represents a key anapleurotic metabolite linking the entry and exit of carbon atoms from the TCA cycle to pathways involved in amino acid metabolism. α-ketoglutarate is also important for driving the malate-aspartate shuttle.

Succinyl-CoA, along with glycine, contributes all the carbon and nitrogen atoms required for the synthesis of protoporphyrin heme biosynthesis and for non-hepatic tissue utilization of ketone bodies.

## SUCCINYL COA SYNTHETASE

The conversion of succinyl-CoA to succinate by succinyl CoA synthetase involves use of the high-energy thioester of succinyl-CoA to drive synthesis of a high-energy nucleotide phosphate, by a process known as substrate-level phosphorylation.

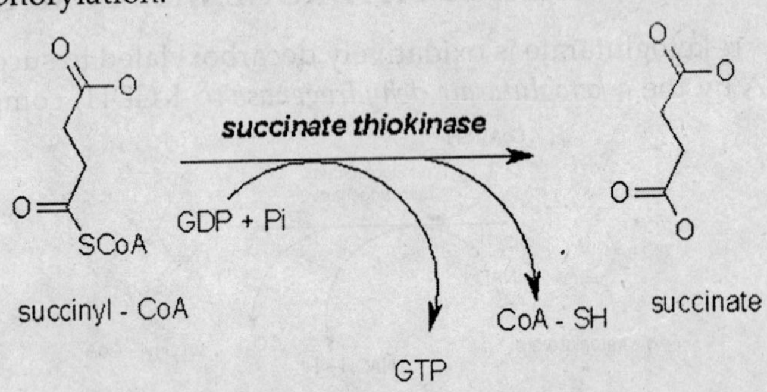

In this process a high energy succinyl-phosphate intermediate is formed, with the phosphate subsequently being transferred to GDP.

Mitochondrial GTP is used in a trans-phosphorylation reaction catalyzed by the mitochondrial enzyme nucleoside diphospho kinase to phosphorylate ADP, producing ATP and regenerating GDP for the continued operation of succinyl CoA synthetase.

## SUCCINATE DEHYDROGENASE (SDH)

*Succinate dehydrogenase* catalyzes the oxidation of succinate to fumarate with the sequential reduction of enzyme-bound FAD and non-heme-iron. In mammalian cells the final electron acceptor is coenzyme $Q_{10}$ ($CoQ_{10}$), a mobile carrier of reducing equivalents that is restricted by its lipophilic nature to the lipid phase of the mitochondrial membrane.

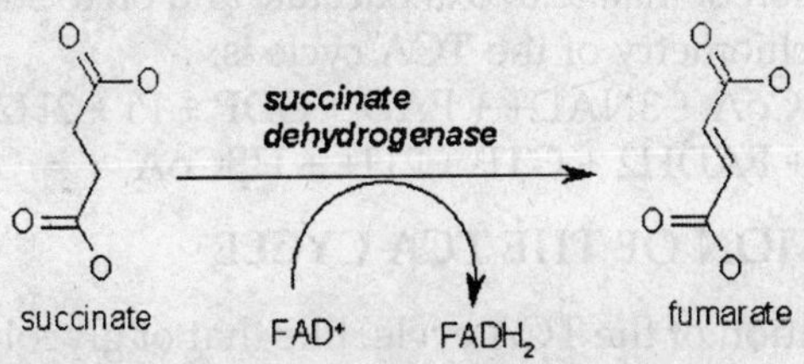


## FUMARASE (FUMARATE HYDRATASE)

The fumarase-catalyzed reactions specific for the *trans* form of fumarate.

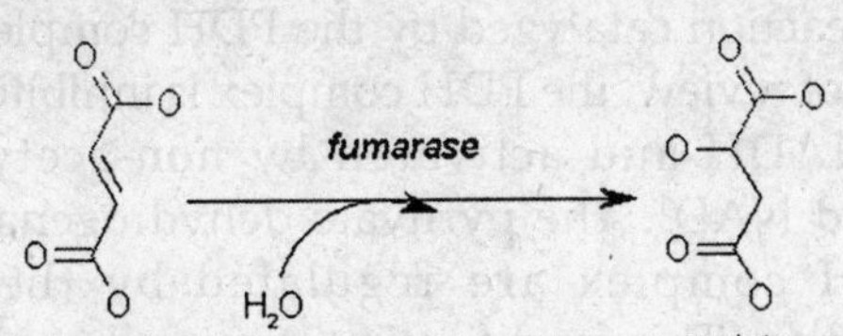


The result is that the hydration of fumarate proceeds stereospecifically with the production of L-malate.

## MALATE DEHYDROGENASE (MDH)

L-malate is the specific substrate for MDH, the final

enzyme of the TCA cycle. The forward reaction of the cycle, the oxidation of malate yields oxaloacetate(OAA).

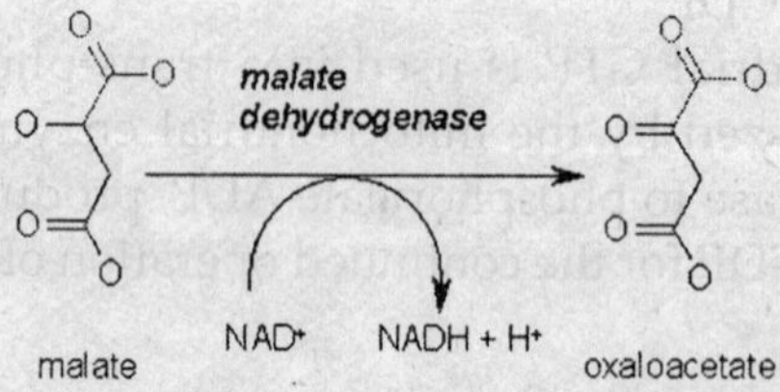


In the forward direction the reaction has a standard free energy of about +7 kcal/mol, indicating the very unfavorable nature of the forward direction. As noted earlier, the citrate synthase reaction that condenses oxaloacetate with acetyl-CoA has a standard free energy of about -8 kcal/mol and is responsible for pulling the MDH reaction in the forward direction. The overall change in standard free energy change is about -1 kcal/mol for the conversion of malate to oxaloacetate and on to succinate. The overall stoichiometry of the TCA cycle is:

acetyl-CoA + 3NAD+ + FAD + GDP + Pi + 2H2O → 2CO2 + 3NADH + FADH2 + GTP + 2H+ + HSCoA

## REGULATION OF THE TCA CYCLE

Regulation of the TCA cycle. like that of glycolysis, occurs at both the level of entry of substrates into the cycle as well as at the key reactions of the cycle. Fuel enters the TCA cycle primarily as acetyl-CoA. The generation of acetyl-CoA from carbohydrates is, therefore, a major control point of the cycle. This is the reaction catalyzed by the PDH complex.

By way of review, the PDH complex is inhibited by acetyl-CoA and NADH and activated by non-acetylated CoA (CoASH) and $NAD^+$. The pyruvate dehydrogenase activities of the PDH complex are regulated by their state of phosphorylation. This modification is carried out by a specific kinase (PDH kinase) and the phosphates are removed by a specific phosphatase (PDH phosphatase). The phosphorylation of PDH inhibits its activity and, therefore, leads to decreased oxidation of pyruvate. PDH kinase is activated by NADH and acetyl-CoA and inhibited by pyruvate, ADP, CoASH, $Ca^{2+}$ and

$Mg^{2+}$. The PDH phosphatase, in contrast, is activated by $Mg^{2+}$ and $Ca^{2+}$.

Since three reactions of the TCA cycle as well as PDH utilize $NAD^+$ as co-factor it is not difficult to understand why the cellular ratio of $NAD^+$/NADH has a major impact on the flux of carbon through the TCA cycle.

Substrate availability can also regulate TCA flux. This occurs at the *citrate synthase* reaction as a result of reduced availability of oxaloacetate. Product inhibition also controls the TCA flux, e.g. citrate inhibits *citrate synthase,* a-KGDH is inhibited by NADH and succinyl-CoA. The key enzymes of the TCA cycle are also regulated allosterically by $Ca^{2+}$, ATP and ADP.

## THE ELECTRON TRANSPORT CHAIN

2NADH + 2H+ + O2 ==> 2NAD+ + 2H2O

The oxidation of NADH occurs in a stepwise manner involving an electron transport chain. Reducing equivalents go via a series of carriers to oxygen. Electrons flow downhill, energetically. Energy released in this process is used to generate the Proton Motive Force which is used by the cell to make ATP.

## COMPONENTS OF THE ELECTRON TRANSPORT CHAIN (ETC)

The components of the ETC are membrane bound. Their redox potentials vary between NAD (Eo = -320mV) and O2 (+820mV). Note that the redox potentials of different cytochromes and flavoproteins etc. may differ even though they have the same prosthetic groups. The surrounding protein modifies the Eo of the heme and flavin groups. The mitochondrial ETC is fixed in composition and is longer and generates more energy than that of *E. coli*. In contrast, many bacteria, including *E. coli* have multiple alternative ETCs produced in response to different environmental conditions.

### NADH Dehydrogenase

Found in the cytoplasmic membrane with other

components of the ETC. *E. coli* contains two alternative NADH dehydrogenases: NADH-DH I contains one FAD, several FeS groups and one bound ubiquinone(UQ) per enzyme. This isoenzyme is coupled and generates PMF. Encoded by the gene *nuo* (NADH UQ oxidoreductase). Reducing equivalents go from NADH Ø FAD Ø FeS Ø UQ. NADH-DH II has FAD & UQ but contains no FeS clusters and is not coupled and cannot generate PMF. Encoded by *ndh.* It is used when there is surplus NADH to be oxidized but the cells have enough energy. The mitochondrial enzyme resembles NADH I except that it has FMN instead of FAD.

## Ubiquinone

All ETCs contain quinones. Quinones are hydrogen carriers and carry 2H at a time. The structure of the quinone nucleus varies: ubiquinone (respiration) & plastoquinone (photosynthesis) each have a single aromatic ring whereas menaquinone (anaerobic respiration) has a double ring. Ubiquinone is sometimes called coenzyme Q and its side chain (R-group) is made of 6-10 isoprenoid (C5) units depending on the organism. In*E. coli* n = 8 and in mitochondria n = 10.

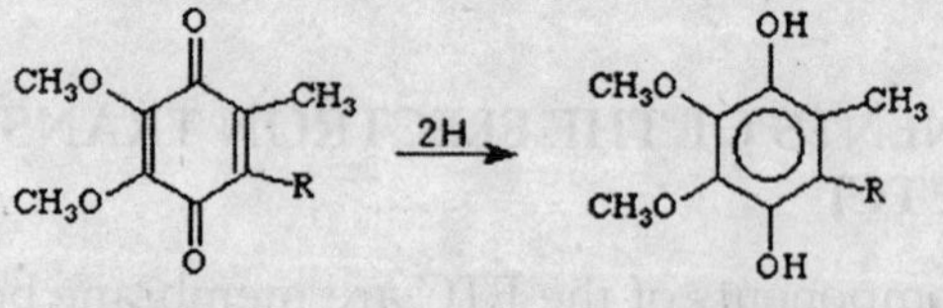


UBIQUINONE UBIQUINOL

There is more ubiquinone than all other ETCcomponents together. A small portion of the ubiquinone is tightly bound to ETC proteins and the rest is free to diffuse around in the membrane - the so-called ubiquinone pool.

The R-group = $-[CH_2 - CH = C(CH_3) - CH_2]^n - H$

## Iron Sulfur Proteins (FeS Proteins)

These have iron but no heme group, instead the iron is linked to acid-labile, i.e. inorganic, sulfur. Sometimes called non-heme-iron (NHI) proteins. The FeS groups are electron carriers. Each FeS group carries only one electron even though it may have 2 or 4 iron atoms. The electron is shared among the irons:

$$e + Fe^{2+} = Fe^{3+}$$

In *E. coli*, FeS groups are found in NADH-dehydrogenase-I and in the cytochrome b region of the chain. In mitochondria there are at least 7 FeS centers. Four are associated with mitochondrial NADH-dehydrogenase two are associated with cytochrome b and one with cytochrome $c_1$.

Several variant structures, the most common are the flat $Fe_2S_2$ and the cube shaped $Fe_4S_4$ groups. Both are bound to the protein by four cysteine residues.

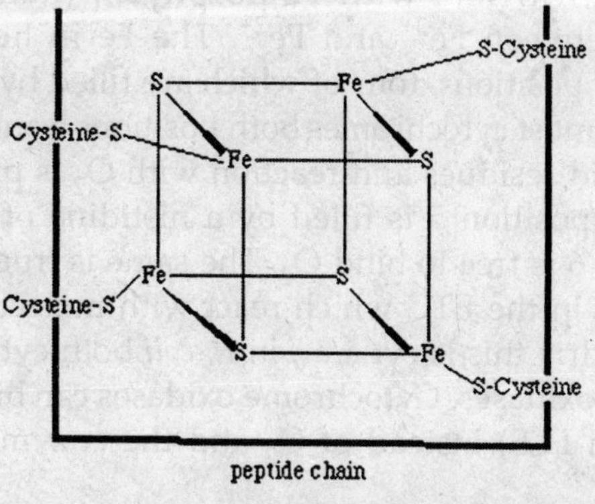


**Fig.** Peptide Chain

## FLAVOPROTEINS

The NADH-dehydrogenases are flavoproteins. Several other flavoproteins join the ETC as side branches at ubiquinone, thus missing coupling site #1. Examples are succinate dehydrogenase(SDH), lactate dehydrogenase(LDH) and glycerol-P dehydrogenase. Note that *E. coli* contains two types of LDH:

- Soluble, NAD-linked, converts pyruvate to lactate in fermentation
- Membrane-bound, FAD-linked, oxidizes lactate to pyruvate in air and feeds into ETC. There are actually two FAD-linked isoenzymes, one each for the D- & L-isomers of lactate.

Flavoproteins may also contain FeS groups (e.g., SDH,

NADH-DH I). Usually FAD or FMN is not bound covalently, but occasionally FAD may be bonded to a histidine (e.g. in SDH). Flavoprotein dehydrogenases are linked to the ETC and are not reoxidised by molecular oxygen. In contrast, flavoprotein oxidases are not linked to the ETC and may be reoxidized directly by $O_2$ e.g., aldehyde oxidase, glucose oxidase. Facultative anaerobes such as *E. coli* have few oxidase type flavoproteins.

## CYTOCHROMES

Electron carriers with Fe-porphyrin (heme) groups. Alternate between $Fe^{3+}$ and $Fe^{2+}$. The Fe in heme has six coordination positions, four of which are filled by N-atoms of the heme. In most cytochromes both positions 5 and 6 are filled by amino acid residues and reaction with $O_2$ is prevented. In hemoglobin position 5 is filled by a histidine of the protein and position 6 is free to bind $O_2$. The same is true of the final cytochromes in the ETC which react with molecular oxygen. In mitochondria this is cyt $a/a_3$, in *E. coli* both cyt d and cyt o are terminal oxidases. Cytochrome oxidases can bind CO, $CN^-$ or $HS^-$ (from $H_2S$) instead of $O_2$ and the enzyme activity is then killed.

In mitochondria and many obligately aerobic bacteria there is a full set of cytochromes: Electrons go from UQ to cyt b to cyt $c_1$ to cyt c to cyt $a/a_3$. Cyt $a/a_3$ is known as cytochrome c oxidase since it oxidizes cyt c at the expense of molecular oxygen. Cyt $a/a_3$ contains copper ions as well as heme.

In facultative anaerobes there is a shorter chain and cyt c & cyt $c_1$ are both missing. Electrons go from UQ to cyt b to cyt o or d.

In *E. coli* high aeration gives mostly the cytochrome o complex:

2 heme b (b555 & b562) and 2 copper ions.

Low aeration results in appearance of the cytochrome d complex:

2 heme b (b558 & b595) and 2 heme d, but no copper.

The Km values for $O_2$ are 0.2μM for cyt o and 0.02μM for cyt d which is consistent with the idea that cyt d is needed when the concentration of $O_2$ is low. Thes are sometimes called ubiquinol oxidases since they oxidize UQ at the expense of molecular oxygen. There is no cytochrome c, c1 or a in the *E.coli* aerobic respiratory chain. Cyt d used to be called cyt a1 but is not really an a-type. A c-type cytochrome is found in the anaerobic nitrite reductase of *E.coli.*

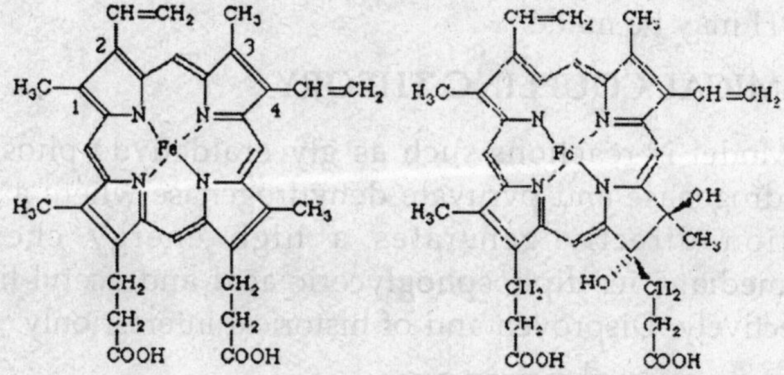

Protoheme IX = Heme b Heme d [a chlorin]

Porphyrin = Heme without the Fe atom and with 2H atoms instead. The four 5-membered rings each containing N are pyrrole rings. Cytochromes are classified according to the substituents at positions 2 and 4 of the porphyrin/heme ring:

*Cytochrome b:* vinyl groups at both 2 and 4 positions. This gives protoheme IX and is also found in hemoglobin, myoglobin, catalase and peroxidase. Cytochrome o of *E. coli* is a b-type.

*Cytochrome c:* positions 2 and 4 both are -CH($CH_3$)-S-Protein.

*Cytochrome a:* position 2 has a 17-carbon side chain, position 4 is vinyl and position 8 has -CHO (instead of the usual $CH_3$).

Chlorins are hemes in which one or more of the pyrrole rings are reduced. Heme d of *E.coli* cytochrome d (= cyt a1) and siroheme of sulfite reductase are chlorin derivatives.

## THE PROTON MOTIVE FORCE

Originally, it was thought that ATP was made by direct chemical phosphorylation at 3 places along the ETC where sufficient energy was available from the drop in redox potential. These are the three coupling sites. It is now known that 2 protons are expelled at each coupling site, so generating the Proton Motive Force (PMF). ATP is made indirectly using the PMF as a source of energy. Each pair of protons yields one ATP. In *E. coli* and many bacteria there is no cytochrome c or c1 and thus coupling site #2 is missing so that only 2 ATP per NADH may be made.

## CHEMICAL COUPLING THEORY

Model is reactions such as glyceraldehyde phosphate dehydrogenase and pyruvate dehydrogenase where a redox reaction directly generates a high energy chemical intermediate, ie: diphosphoglyceric acid and acetyl-lipoate respectively. Disproven and of historical interest only.

## CHEMIOSMOTIC THEORY

ATP synthesis is coupled indirectly to electron transport via the Proton Motive Force (PMF). In the chemiosmotic model, each of the coupling sites is responsible for extruding protons, so creating a proton concentration gradient across the membrane. The ATP synthase uses this gradient to energize the synthesis of ATP.

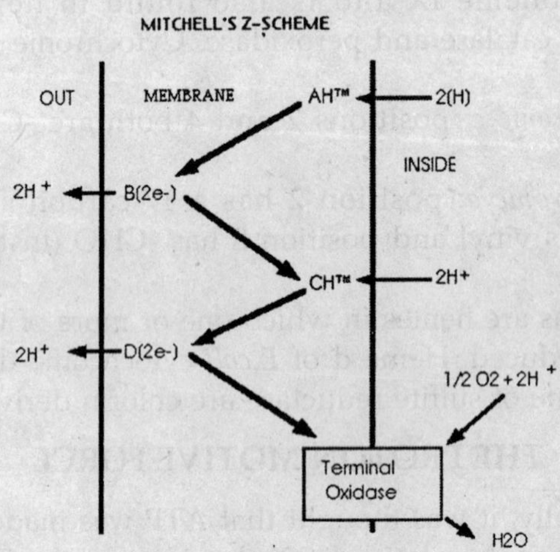


Fig. Mitchell's Z-scheme

The membrane must be impermeable to H+ (and OH-). Since H+ is pumped out, the exterior will become acidified and the cell interior should become alkaline relative to the outside. In practice the internal pH is held constant at 7.5 approx. The proton gradient may be all or partially converted to a charge difference by exchanging potassium ions for protons. The PMF thus consists of two components: a pH difference (Delta pH) and a charge difference (or membrane potential, Delta Psi) across the membrane.

## STOICHIOMETRY

Original theory of Mitchell has 2H+ extruded per site with 3 sites per mitochondrion or 2 sites per *E. coli* and hence P:O ratios of 3:1 or 2:1 assuming 2H+ can drive formation of one ATP. (P:O ratio = ATP per oxygen atom consumed = ATP per 2 electrons sent down ETC). Some data suggests 3 or 4 protons may be extruded at certain sites, but this only applies to mitochondria

## MITCHELL'S Z-SCHEME

In Mitchell's original Z-scheme the ETC is arranged in a series of loops with hydrogen carriers facing the inside of the bacterial cell or mitochondrion and electron carriers facing outwards. Each hydrogen carrier receives two hydrogen atoms –2 and it transfers $2e^-$ to the following electron carrier. Since 2 = $2H^+ + 2e^-$ this leaves two protons ($2H^+$) to dispose of and these are expelled so generating the PMF. The electron carrier transfers the electrons to the next hydrogen carrier. In order to get 2, two $2H^+$ must also be picked up from the inside of the bacterial cell. The ETC contains one proton extrusion loop for each coupling site.

## MITCHELLS EQUATION

Proton chemical potential: μ = 2.3RT(pHi – pHo) = 2.3RT Delta pH

Electrical potential: m = F(Psi i- Psi o) = F DeltaPsi

*i* refers to inside, o to outside, F = Faradays constant, Psi = charge

Electrochemical potential = m(chemical) + m(electrical):
$\mu(H+)$ = F DeltaPsi - 2.3RT Delta pH.
Proton Motive Force:
PMF = $\mu(H+)/F$ = DeltaPsi - 60 Delta pH

PMF is the proton motive force in millivolts. Conversion factor is approx. 60mV/pH unit at 37°C. Note that the relative contributions of DeltaPsi and Delta pH to PMF vary with the pH of the growth medium. Note also that PMF can only be maintained as long as membrane forms a sealed vesicle.

To generate 100mV of PMF requires the extrusion of about 60,000 protons/cell. The PMF generated by *E. coli* is 150-200 mV. How the intracellular pH is maintained at pH7.5 is unknown. It is known that at an external pH of 7.5 the Delta pH is replaced by a DeltaPsi by swapping K+ for H+.

In effect, *E. coli* is a tiny living battery. To convert ADP to ATP requires 7.3 kcal/mole energy input. This is equivalent to the energy stored in a pair of protons at a pH gradient of 3.5 pH units or at a charge difference of 240mV. Uncouplers are proton ionophores(= ion carriers).

They carry H+ across membranes and consequently allow the PMF to collapse. Both the protonated and unprotonated forms can diffuse.

Protons are picked up on the outside and released on the inside. eg: 2, 4-Dinitrophenol (2,4-DNP), carbonyl cyanide-p-trifluoromethoxyphenylhydrazone (FCCP) and the similar chloro compound CCCP.

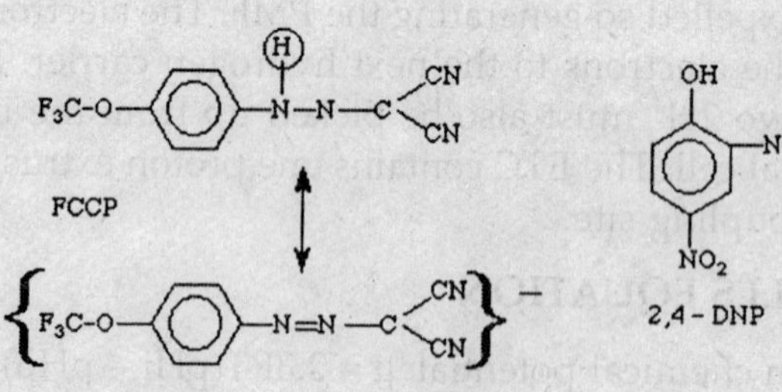

## Other Ionophores

Carriers exist for ions other than the proton eg:

- Valinomycin carries potassium($K^+$) or rubidium($Rb^+$)

ions and will therefore collapse the DeltaPsi but not the Delta pH

- Nigericin exchanges $Na^+$, $K^+$ or $Rb^+$ for $H^+$ so collapsing Delta pH but not DeltaPsi ie opposite effect to valinomycin
- Gramicidin is not a carrier but forms a channel in the membrane allowing through $H^+$, $Na^+$, $K^+$, & $Rb^+$ thus collapsing both Delta pH & DeltaPsi

## Measurements of DeltaPsi and Delta

Delta pH may be measured by the distribution of weak acids across the membrane. This depends on the Delta pH. Acids used eg: acetic acid, salicylic acid, DMO (5,5-dimethyloxazolidine-2,4-dione):

- Membrane permeable to undissociated acid(HA) but not to $H^+$ or $A^-$
- On each side of membrane HA $\infty$ $H^+ + A^-$
- The extent of dissociation depends on pH
- Thus extent of dissociation and hence the amount of HA differs on either side of membrane and this can be measured.

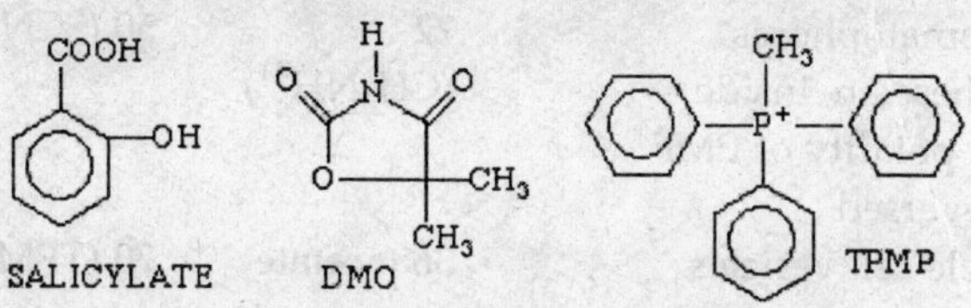


DeltaPsi may be measured indirectly by a similar method. A membrane permeable cation will be distributed according to DeltaPsi.

For example, the ionophore valinomycin plus K+ or Rb+, or a lipophilic cation such as TPMP+ (triphenylmethyl-phosphonium) can be used. DeltaPsi may be measured directly with a voltmeter and electrodes:

- Grow *E. coli* with mecillinam, a penicillin derivative which, in sublethal amounts, produces giant spherical cells (upto 6 microns diam).
- Use small electrodes! (0.08 to 0.2 microns at tip).

- Coat electrodes with phospholipid to get a good seal with the cell membrane - required to avoid PMF collapsing.
- Read DeltaPsi from voltmeter.

*Example Proton Motive Force Values*

| *Organism* | *pHo* | *-60DeltapH* | *Delta Psi* | *PMF* |
|---|---|---|---|---|
| Fermentative | | | | |
| *Staphylococcus lactis* | 5.0 | 58 (benzoate) | 95 (dye) | 158 |
| *Staphylococcus faecalis* | 5.0 | 54 (DMO) | 170 (DDA) | 224 |
| Fermentative/Respiratory | | | | |
| *Bacillus subtilis* | 5.0 | 120 (salicylate) | 30 (TPMP) | 150 |
| *Bacillus subtilis* | 7.5 | 0 (salicylate) | 140 (TPMP) | 140 |
| *Escherichia coli* | 5.5 | 72 (DMO) | 57 ($Rb^+$/Val) | 129 |
| *Escherichia coli* | 7.0 | 38 (DMO) | 122 ($Rb^+$/Val) | 160 |
| *Escherichia coli* | 5.5 | 120 (DMO) | 100 (electrodes) | 220 |
| *Escherichia coli* | 7.5 | 0 (DMO) | 142 (electrodes) | 142 |
| Photosynthetic | | | | |
| *Rhodospirillum rubrum* chromatophores* as these are inside out, polarity of PMF is reversed | | 72 ($CH_3NH^{4+}$) | 50 (SCN-) | 122 |
| whole cell vesicles | | 36 (acetate) | 70 (TPMP) | 106 |
| *Halobacterium halobium* | 6.0 | 73 (DMO) | 113 (TPMP) | 186 |
| *Halobacterium halobium* | 8.0 | 8 (DMO) | 151 (TPMP) | 159 |

## OXIDATIVE PHOSPHORYLATION

ATP synthesis is driven by the PMF. It can be demonstrated that only the PMF and the ATP synthase enzyme complex are required in the presence of sealed membrane vesicles. Note: Experiment is usually done "inside outΔ with the ATPase stuck to the outside.

- Make liposomes (i.e., lipid vesicles)

- Add purified ATPase and ADP and Pi
- Add excess $K^+$ plus Valinomycin to outside
- Positive charge builds up on inside - an artificial PMF
- The artificial DeltaPsi repels protons, derived from: $H_2O = H^+ + OH^-$
- Protons exit via ATP synthase and ATP is made

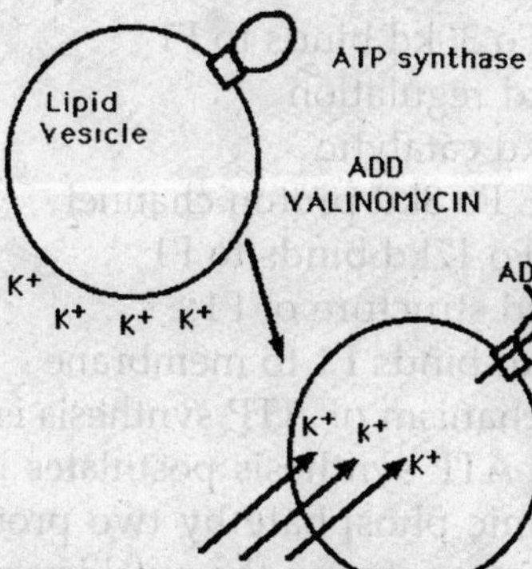


**Fig.** ATP synthase

ATPase and ATP synthase refer to the same enzyme - which direction the reaction goes depends on the conditions.

The ATP synthase consists of the headpiece, or $F_1$-complex, which sticks into the interior of the cell and the Fo-complex which is membrane-bound. The catalytic a and b subunits of the $F_1$-complex are responsible for synthesizing ATP.

The Fo complex consists of 3 proteins a, b and c (sometimes referred to as c, y, w). Fo forms a proton channel through the membrane.

Protein c is very hydrophobic (a "proteolipidΔ) and probably acts as the channel while a and b control assembly and binding to the $F_1$-headpiece.

Fo alone makes membranes permeable to protons. Addition of $F_1$ stops proton transport except to drive ATP synthesis. DCCD (dicyclohexyl carbodiimide) binds covalently to protein c and blocks the proton channel.

Oligomycin inhibits the ATPase reaction of the catalytic $F_1$ complex. Oligomycin halts the whole electron transport chain & ATP synthetic pathway when added alone. However, if both oligomycin and an uncoupler are added then the ETC runs but no ATP synthesis occurs.

## PROTEINS OF THE ATP SYNTHETASE:

Protein Ratio Gene Subunit MW Function
α 3 uncA F1 55kd catalytic
a (or c) 1 uncB Fo 30kd binds to F1
ε 1 uncC F1 15kd regulation
β 3 uncD F1 50kd catalytic
c (or w) 10 uncE Fo 8kd proton channel
b (or y) 2 uncF Fo 17kd binds to F1
γ 1 uncG F1 31kd structure of F1
δ 1 uncH F1 20kd binds F1 to membrane

The precise mechanism of ATP synthesis is obscure. The simplest scheme for ATP synthesis postulates removal of an oxygen from inorganic phosphate by two protons from the outside thus producing water on the outside and ATP on the inside.

The inorganic phosphate is closest to the proton channel with the ADP to the inside An alternative is the use of energy from the protons to alter the conformation of the catalytic a and β subunits. The high energy state then releases its energy by converting ADP to ATP.

## ATP VERSUS THE FORCE

The ATPase can interconvert PMF and ATP. In the absence of a PMF the ATPase will hydrolyse ATP and generate PMF. Hence ATP and PMF are interconvertible. If an uncoupler is added to *E. coli* or a mitochondrion the PMF is dissipated. The ATPase then tries to re-create the PMF by hydrolyzing all the ATP. Oligomycin inhibits the ATPase and will prevent hydrolysis of ATP under these conditions just as it prevents synthesis of ATP from PMF.

ATP is used for biosynthesis and also drives certain transport systems. Arsenate inhibits ATP driven transport systems, but not those energized directly by the proton motive force. In eukaryotes the PMF is generated in the mitochomdria or chloroplasts and is all converted to ATP. In bacteria the PMF is used directly to energize several membrane bound operations:-

## PMF DRIVEN TRANSPORT SYSTEMS :

Symport: e.g., Lactose Substrate and H+ enter together.
Antiport: e.g., $Na^+$ $H^+$ taken in and Na+ excreted.
Complex: e.g., Melibiose Overall Melibiose plus $H^+$ enter: Melibiose plus $Na^+$ enter together ii) Na+ out/$H^+$ in.

### Energy Linked Transhydrogenase

Normally NADH is converted to ATP whereas NADPH is used as reducing power in biosynthesis. NADPH may be generated directly by certain metabolic pathways which generate NADPH instead of NADH.

A proportion of the glucose (around 20%) goes via the pentose phosphate pathway instead of the Embden-Meyerhof pathway. In many bacteria the isocitrate dehydrogenase in the Krebs cycle also makes NADPH. Another source is the NADP linked malic enzyme.

However if their is a shortage of NADPH then it is possible to convert NADH to NADPH. This requires energy in order to shift the equilibrium in favour of NADPH. Energy linked transhydrogenase is energized directly by the PMF and converts NADH + NADP to NADPH + NAD. Mechanism unknown. It is particularly important in some photosynthetic organisms which do not have any glucose to send down the pentose pathway. It is present in *E. coli* but not important.

### Flagellar Motion

*E. coli* is powered by a proton drive - just like the ships of the galactic empire The M-ring of the basal body is located in the cytoplasmic membrane and rotates relative to the S- or stator ring of the basal body of the flagellum which is embedded in the cell wall.

Proton acceptor groups (e.g., $-NH_2$) are present on the rotor-M-ring and anionic groups (e.g., $COO^-$) are located over the exit channels for protons . Each time one $H^+$ enters the input channel the ring turns a fraction of the revolution as the newly formed -NH3+ group is attracted towards the -COO- group.

## ANAEROBIC RESPIRATION

Oxygen is the ultimate electron acceptor for the ETC when bacteria grow aerobically. However, bacteria can respire in the absence of oxygen if some other suitable oxidant is available to act as electron acceptor. Use of such alternative electron acceptors is known as anaerobic respiration. It requires a modified ETC in order to transport electrons to the new acceptor. *E. coli* can use nitrate (NO3-), amine oxides (R3NO), sulfoxides (R2SO), fumarate and nitrite (NO2). Many sulfur compounds can be used by a variety of bacteria. "Sulfide fermentationΔ is not actually fermentation but is an example of anaerobic respiration with sulfate (SO42-) as the electron acceptor. "Methane fermentationΔ is also anaerobic respiration, in this case with carbon dioxide ($CO_2$) as the electron acceptor.

Do not confuse anaerobic respiration with the oxidative metabolism of lithotrophs such as *Nitrobacter*. Oxygen has a more positive Eo than nitrite, therefore oxidation of nitrite by oxygen produces energy. Nitrate has a more positive Eo than 2 from NADH or formate. Therefore oxidation of NADH or formate by nitrate produces energy. Lithotrophs oxidize inorganic materials using $O_2$. A substance with an Eo between $O_2$ and NADH may be used either to oxidize NADH or to reduce O2. e.g. the NO3-/NO2- couple which has Eo of +420mV compared to +820mV for O2/H2O and -320mV for NAD+/ NADH.

## ANAEROBIC NITRATE RESPIRATION: $NO_3 + 2 = NO_{2-} + H_2O$

Nitrite oxidation by *Nitrobacter* $NO_{2-} + 1/2O_2 = NO_{3-}$

## NITRATE REDUCTASE

Nitrate reductase (NAR) of *E. coli* is a very large membrane bound protein (Total MW approx. 800,000). It is induced by nitrate and repressed by oxygen. When maximally induced it comprises upto 25% of the protein of the cytoplasmic membrane. Pyruvate formate lyase is present in anaerobic cells growing on nitrate and produces a lot of formate plus acetyl-CoA. Most of the acetyl-CoA is converted to acetate which is excreted. This is because the Krebs cycle is not complete during anaerobic growth because oxoglutarate dehydrogenase is not made. The formate is oxidized via formate dehydrogenase to give 2 plus $CO_2$. The reducing equivalents travel via the cytochrome b of the FDH complex to a quinone. NADH produced in glycolysis can also be reoxidized by nitrate. The quinone used anaerobically is mostly menaquinone (Vitamin K2). However, when grown on nitrate, which is a good oxidant, a mixture of ubiquinone and menaquinone is found.

Nitrate reductase contains 4 subunits:

- NarG protein x4 (= α) MW = 150,000, carries Mo & FeS groups
- NarH protein x4 (= β) MW = 60,000, carries FeS groups
- NarI protein (= γ) cytochrome b, often lost on purification
- NarJ protein membrane binding subunit

The Mo is present as part of the low molecular weight (approx 1000) molybdopterin cofactor. All Mo proteins share the same cofactor, (except for nitrogenase which has a different Mo/Fe cofactor).

## THE RESPIRATORY CHAIN TO NITRATE

Formate dehydrogenase is another large membrane protein. It contains molybdopterin, FeS groups and selenium. There are two types of FDH. The anaerobic respiratory FDH is found in anaerobic cells using nitrate or other alternative electron acceptors. The fermentative FDH is linked to hydrogenase and is produced anaerobically in the absence of alternative electron acceptors. The Se is present as selenocysteine ie cysteine in which sulfur has been replaced by selenium.

Most anaerobic reductases such as those for TMAO (trimethylamine oxide), DMSO (dimethyl sulfoxide) etc are molybdoproteins. Tungsten (W) is an analog of Mo. In the presence of tungstate the processing of molybdate into molybdopterin is inhibited and nitrate reductase and other Mo enzymes are inactivated. Excess molybdate will remedy this i.e., tungstate is a competitive inhibitor. The two anaerobic reductases which do not contain molybdenum are nitrite reductase - a heme protein and fumarate reductase - a flavoprotein.

Fumarate Reductase (FRD) FRD is repressed by both oxygen and nitrate but induced by fumarate. FRD contains FAD and FeS groups.

There are 4 subunits. A large flavoprotein, a medium sized FeS protein and two small membrane anchor proteins. FRD is connected via a cytochrome b to menaquinone.

FRD cannot use ubiquinone, it has to use menaquinone. No molydenum or selenium are found in FRD. Note that fumarate is an organic molecule and the 2 go to saturate a double bond i.e., no water is formed as with nitrate, TMAO or DMSO etc.

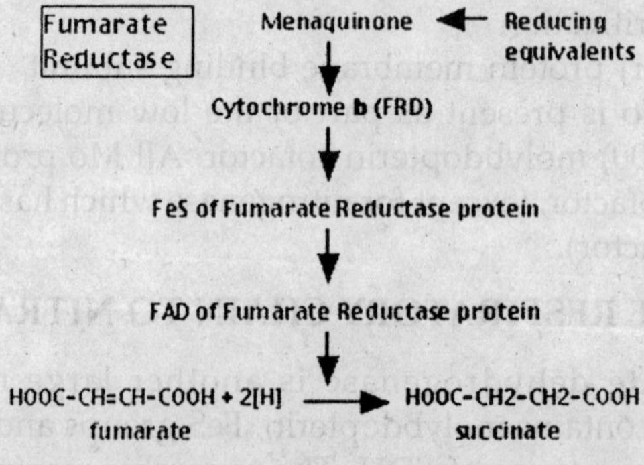


**Fig.** Respiratory Chain to Nitrate

Unlike $O_2$ and nitrate which both generate two pairs of protons per 2e- which travel down the electron transport chain and hence yield 2ATP, fumarate gives only one pair of protons and only one ATP.

When a cell changes from oxygen to fumarate as electron acceptor the PMF drops from around –200 mV to approx. –100 mV.

## OXIDATIVE PHOSPHORYLATION

The NADH and $FADH_2$ formed in glycolysis, fatty acid oxidation and the citric acid cycle are energy-rich molecules because each contains a pair of electrons having a high transfer potential. When these electrons are used to reduce molecular oxygen to water, a large amount of free energy is liberated, which can be used to generate ATP. *Oxidative phosphorylation is the process in which ATP is formed as a result of the transfer of electrons from NADH or $FADH_2$ to $O_2$ by a series of electron carriers.*

This process, which takes place in mitochondria, is the major source of ATP in aerobic organisms. For example, oxidative phosphorylation generates 26 of the 30 molecules of ATP that are formed when glucose is completely oxidized to $CO_2$ and $H_2O$.

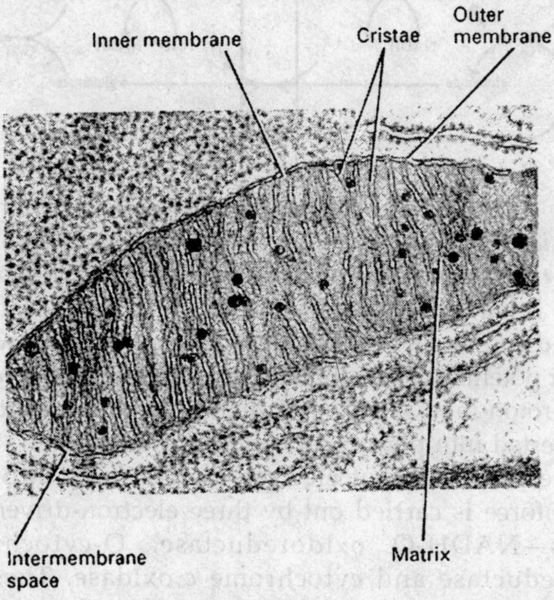


**Fig.** Electron Micrograph of a Mitochondrion

Oxidative phosphorylation is conceptually simple and mechanistically complex. Indeed, the unraveling of the mechanism of oxidative phosphorylation has been one of the most challenging problems of biochemistry.

The flow of electrons from NADH or $FADH_2$ to $O_2$ through protein complexes located in the mitochondrial inner membrane leads to the pumping of protons out of the mitochondrial matrix.

The resulting uneven distribution of protons generates a pH gradient and a transmembrane electrical potential that creates a *proton-motive force.* ATP is synthesized when protons flow back to the mitochondrial matrix through an enzyme complex. Thus, *the oxidation of fuels and the phosphorylation of ADP are coupled by a proton gradient across the inner mitochondrial membrane.*

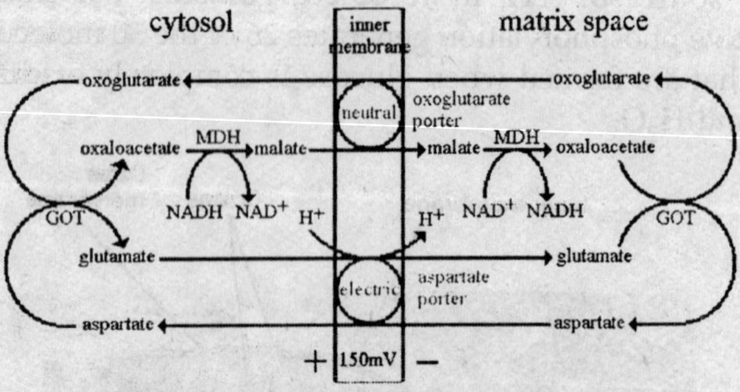


**Fig.** Essence of Oxidative Phosphorylation.

*Oxidative phosphorylation is the culmination of a series of energy transformations* that are called *cellular respiration* or simply *respiration* in their entirety. First, carbon fuels are oxidized in the citric acid cycle to yield electrons with high transfer potential. Then, this electron-motive force is converted into a proton-motive force and, finally, the proton-motive force is converted into phosphoryl transfer potential.

The conversion of electron-motive force into proton-motive force is carried out by three electron-driven proton pumps—NADH-Q oxidoreductase, Q-cytochrome *c* oxidoreductase and cytochrome *c* oxidase. These large transmembrane complexes contain multiple oxidation-reduction centers, including quinones, flavins, iron-sulfur clusters, hemes and copper ions. The final phase of oxidative phosphorylation is carried out by *ATP synthase,* an ATP-synthesizing assembly that is driven by the flow of protons back into the mitochondrial matrix. Components of this remarkable enzyme rotate as part of its catalytic mechanism. Oxidative phosphorylation vividly shows that *proton gradients are an interconvertible currency of free energy in biological systems.*

## PENTOSE PHOSPHATE PATHWAY

The pentose phosphate pathway is a process that serves to generate NADPH and the synthesis of pentose (5-carbon) sugars.

There are two distinct phases in the pathway. The first is the oxidative phase, in which NADPH is generated, and the second is the non-oxidative synthesis of 5-carbon sugars. This pathway is an alternative to glycolysis. While it does involve oxidation of glucose, its primary role is anabolic rather than catabolic. For most organisms it takes place in the cytosol; in plants most steps take place in plastids.

### Uses

The primary functions of the pathway are:

- To generate reducing equivalents, in the form of NADPH, for reductive biosynthesis reactions within cells.
- To provide the cell with ribose-5-phosphate (R5P) for the synthesis of the nucleotides and nucleic acids.
- Although not a significant function of the PPP, it can operate to metabolize dietary pentose sugars derived from the digestion of nucleic acids as well as to rearrange the carbon skeletons of dietary carbohydrates into glycolytic/gluconeogenic intermediates.

Located exclusively in the cytoplasm, the pathway is one of the three main ways the body creates molecules with reducing power, accounting for approximately 60% of NADPH production in humans. One of the uses of NADPH in the cell is to prevent oxidative stress. It reduces the coenzyme glutathione, which converts reactive $H_2O_2$ into $H_2O$. If absent, the $H_2O_2$ would be converted to hydroxyl free radicals, which can attack the cell.

## PHASES

### Oxidative Phase

In this phase, two molecules of $NADP^+$ are reduced to NADPH, utilizing the energy from the conversion of glucose-6-phosphate into ribulose 5-phosphate.

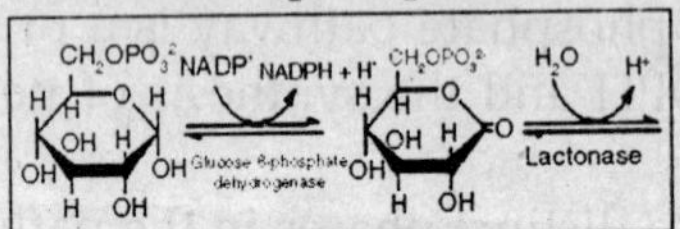


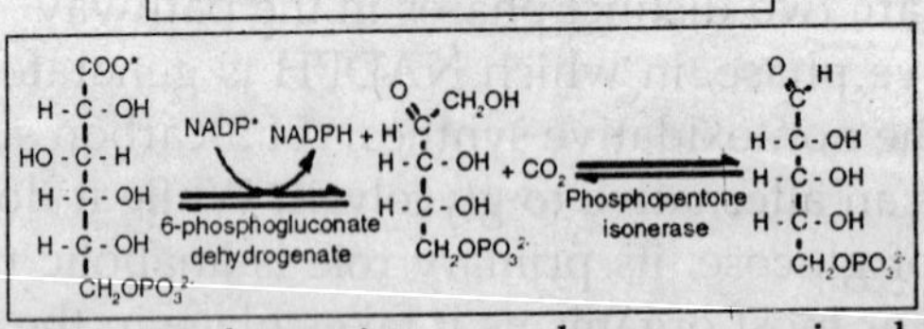


The entire set of reactions can be summarized as follows:

| *Reactants* | *Products* | *Enzyme* | *Description* |
|---|---|---|---|
| Glucose 6-phosphate + NADP+ | → 6-phosphoglucono-δ-lactone + NADPH | glucose 6-phosphate dehydrogenase | Dehydrogenation. The hemiacetal hydroxyl group located on carbon 1 of glucose 6-phosphate is converted into a carbonyl group, generating a lactone, and, in the process, NADPH is generated. |
| 6-phospho-glucono-δ-lactone + $H_2O$ | → 6-phosphogluconate + $H^+$ | 6-phosphoglu-conolactonase | Hydrolysis |
| 6-phos-phogluconate + $NADP^+$ | → ribulose 5-phosphate + NADPH + $CO_2$ | 6-phosphogluconate dehydrogenase | Oxidative decarboxylation. $NADP^+$ is the electron acceptor, generating another molecule of NADPH, a $CO_2$, and ribulose 5-phosphate. |
| ribulose 5-phosphate | ribose 5-phosphate | Phosphopentose isomerase | Isomerization. (Can also be considered part of nonoxidative phase) |

The overall reaction for this process is:
Glucose 6-phosphate + 2NADP$^+$ + $H_2O$ →
ribulose 5-phosphate + 2 NADPH + 2 H$^+$ + $CO_2$

## Non-oxidative Phase

| *Reactants* | *Products* | *Enzymes* |
|---|---|---|
| ribulose 5-phosphate | → ribose 5-phosphate | phosphopentose isomerase |
| ribose 5-phosphate | → xylulose 5-phosphate | phosphopentose epimerase |
| xylulose 5-phosphate + ribose 5-phosphate | → glyceraldehyde 3-phosphate + sedoheptulose 7-phosphate | transketolase |
| sedoheptulose 7-phosphate + glyceraldehyde 3-phosphate | → erythrose 4-phosphate + fructose 6-phosphate | transaldolase |
| xylulose 5-phosphate + erythrose 4-phosphate | → glyceraldehyde 3-phosphate + fructose 6-phosphate | transketolase |

## Regulation

Glucose-6-phosphate dehydrogenase is the rate-controlling enzyme of this pathway. It is allosterically stimulated by NADP$^+$. The ratio of NADPH:NADP$^+$ is normally about 100 : 1 in liver cytosol. This makes the cytosol a highly-reducing environment. Formation of NADP$^+$ by a NADPH-utilizing pathway, thus, stimulates production of more NADPH.

## PENTOSE PHOSPHATE PATHWAY

The pentose phosphate pathway is a process that serves to generate NADPH and the synthesis of pentose (5-carbon) sugars. There are two distinct phases in the pathway.

The first is the oxidative phase, in which NADPH is generated, and the second is the non-oxidative synthesis of 5-carbon sugars.

This pathway is an alternative to glycolysis. While it does involve oxidation of glucose, its primary role is anabolic rather than catabolic. For most organisms it takes place in the cytosol; in plants most steps take place in plastids.

### Uses

The primary functions of the pathway are:

- To generate reducing equivalents, in the form of

NADPH, for reductive biosynthesis reactions within cells.

- To provide the cell with ribose-5-phosphate (R5P) for the synthesis of the nucleotides and nucleic acids.
- Although not a significant function of the PPP, it can operate to metabolize dietary pentose sugars derived from the digestion of nucleic acids as well as to rearrange the carbon skeletons of dietary carbohydrates into glycolytic/gluconeogenic intermediates.

Located exclusively in the cytoplasm, the pathway is one of the three main ways the body creates molecules with reducing power, accounting for approximately 60% of NADPH production in humans.

One of the uses of NADPH in the cell is to prevent oxidative stress. It reduces the coenzyme glutathione, which converts reactive $H_2O_2$ into $H_2O$. If absent, the $H_2O_2$ would be converted to hydroxyl free radicals, which can attack the cell. Significantly, erythrocytes utilize the reactions of the PPP to generate large amounts of NADPH used in the reduction of glutathione. It is also used to generate hydrogen peroxide for phagocytes.

## PHASES

### Oxidative Phase

In this phase, two molecules of $NADP^+$ are reduced to NADPH, utilizing the energy from the conversion of glucose-6-phosphate into ribulose 5-phosphate.

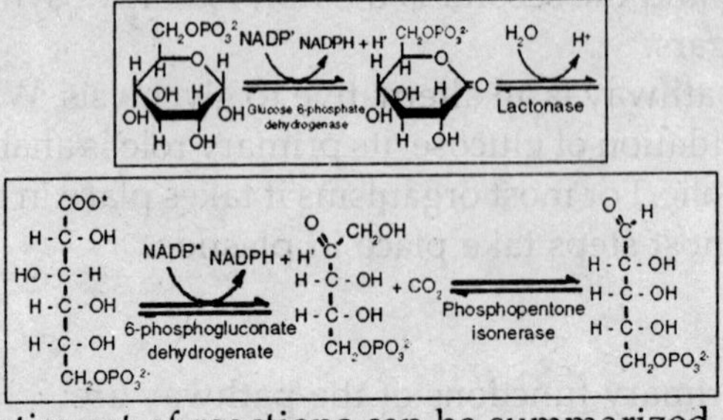


The entire set of reactions can be summarized as follows:

| *Reactants* | *Products* | *Enzyme* | *Description* |
|---|---|---|---|
| Glucose 6-phosphate + NADP+ | → 6-phosphoglucono-δ-lactone + NADPH | glucose 6-phosphate dehydrogenase | Dehydrogenation. The hemiacetal hydroxyl group located on carbon 1 of glucose 6-phosphate is converted into a carbonyl group, generating a lactone, and, in the process, NADPH is generated. |
| 6-phospho-glucono-δ-lactone + $H_2O$ | → 6-phosphogluconate + $H^+$ | 6-phosphoglu-conolactonase | Hydrolysis |
| 6-phos-phogluconate + $NADP^+$ | → ribulose 5-phosphate + NADPH + $CO_2$ | 6-phosphogluconate dehydrogenase | Oxidative decarboxylation. $NADP^+$ is the electron acceptor, generating another molecule of NADPH, a $CO_2$, and ribulose 5-phosphate. |
| ribulose 5-phosphate | ribose 5-phosphate | Phosphopentose isomerase | Isomerization. (Can also be considered part of nonoxidative phase) |

The overall reaction for this process is:

Glucose 6-phosphate + $2NADP^+$ + $H_2O$ →
ribulose 5-phosphate + 2 NADPH + 2 $H^+$ + $CO_2$

## Non-oxidative Phase

| *Reactants* | *Products* | *Enzymes* |
|---|---|---|
| ribulose 5-phosphate | → ribose 5-phosphate | phosphopentose isomerase |
| ribose 5-phosphate | → xylulose 5-phosphate | phosphopentose epimerase |
| xylulose 5-phosphate + ribose 5-phosphate | → glyceraldehyde 3-phosphate + sedoheptulose 7-phosphate | transketolase |
| sedoheptulose 7-phosphate + glyceraldehyde 3-phosphate | → erythrose 4-phosphate + fructose 6-phosphate | transaldolase |
| xylulose 5-phosphate + erythrose 4-phosphate | → glyceraldehyde 3-phosphate + fructose 6-phosphate | transketolase |

## Regulation

Glucose-6-phosphate dehydrogenase is the rate-controlling enzyme of this pathway. It is allosterically stimulated by $NADP^+$. The ratio of NADPH:$NADP^+$ is normally about 100 : 1 in liver cytosol. This makes the cytosol a highly-reducing environment. Formation of $NADP^+$ by a NADPH-utilizing pathway, thus, stimulates production of more NADPH.

## THE GLYOXYLATE CYCLE

It was stated above that fatty acids cannot be utilized for gluconeogenesis. Also, it was mentioned earlier that fatty acids are degraded via acetyl-CoA. Since acetyl-CoA is degraded via the TCA and TCA intermediates can be utilized for gluconeogenesis, why is it not possible to turn fatty acids into glucose? This is because of the different roles of the TCA intermediates in the two processes:

- In the degradation of acetyl-CoA, they have a catalytic role – they are required for the cycle to run, but there is no net gain or loss of intermediates. Acetyl-CoA fed into the TCA is consumed without increasing the pool of TCA intermediates.
- In gluconeogenesis, oxaloacetate is drained from the pool of TCA intermediates. Thus, the TCA can only sustain gluconeogenesis to the extent it is supplied with new intermediates from other sources.

To turn fatty acids into glucose, we would therefore need a means to use acetyl-CoA for increasing the pool of intermediates. Such a pathway does not exist in mammals, but it does exist in plants; it is known as the glyoxylate cycle. It is a 'side-road' to the TCA that allows them to use two molecules of acetyl-CoA per cycle to bring about the net synthesis of one $C_4$-intermediate. Two reactions are required for this cycle:

- Isocitrate is split into succinate and glyoxylate by isocitrate lyase. Since the isocitrate dehydrogenase and the á-ketoglutarate dehydrogenase reactions are bypassed, the loss of two carbons as $CO_2$ is avoided; these carbons are retained in the form of glyoxylate.

- Glyoxylate combines with an 'extra' acetyl-CoA to form one molecule of malate. This reaction is catalysed by malate synthase and like the citrate synthase reaction it is pushed forward by the concomitant hydrolysis of coenzyme A.

We know that many plant seeds are very rich in oil (=fat). The glyoxylate cycle enables plant seeds to store metabolic energy and carbon as fat and to use it for the synthesis of glucose and other carbohydrates during germination. Fat is water-free and has approximately twice the content of metabolic energy per gram of carbohydrates; it is therefore both more compact and more resistant to microbial degradation than starch.

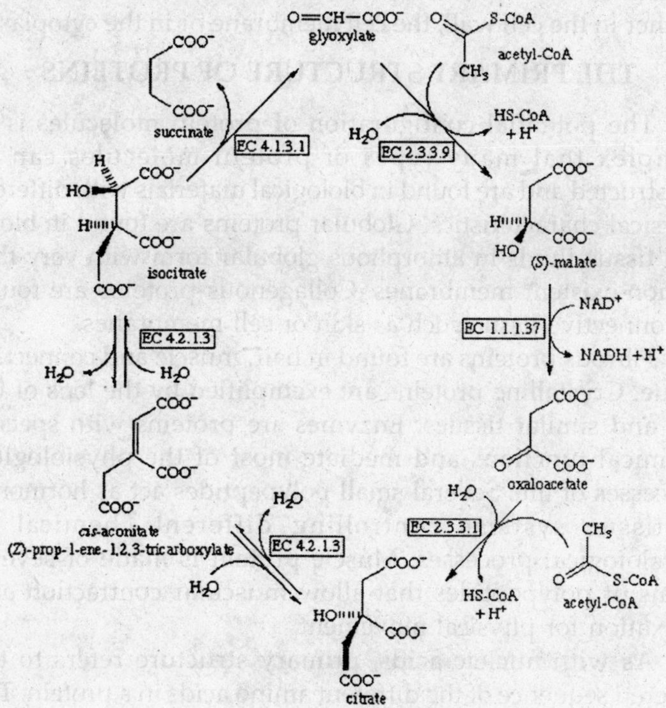


Bold bonds show the carboxymethyl group originating from the acetyl-CoA

Fig. Glyoxylate Cycle

# Chapter 3

# Proteins and Amino Acids

Proteins and peptides (small proteins) are essential to the cell and serve two major functions. Many proteins are enzymes that catalyze almost all biological reactions in a living organism. Other proteins perform a structural role for the cell - either in the cell wall, the cell membrane or in the cytoplasm.

## THE PRIMARY STRUCTURE OF PROTEINS

The potential configuration of protein molecules is so complex that many types of protein molecules can be constructed and are found in biological materials with different physical characteristics. Globular proteins are found in blood and tissue fluids in amorphous globular form with very thin or non-existent membranes. Collagenous proteins are found in connective tissue such as skin or cell membranes.

Fibrous proteins are found in hair, muscle and connective tissue. Crystalline proteins are exemplified by the lens of the eye and similar tissues. Enzymes are proteins with specific chemical functions and mediate most of the physiological processes of life. Several small polypeptides act as hormones in tissue systems controlling different chemical or physiological processes. Muscle protein is made of several forms of polypeptides that allow muscular contraction and relaxation for physical movement.

As with nucleic acids, primary structure refers to the ordered sequence of the different amino acids in a protein. The carboxyl group and the amino group of amino acids are reactive. Cells synthesize proteins by attaching the carboxyl group of one amino acid to the amine group the next, with polymerization taking place at the ribosome. This is termed a peptide bond. Since each amino acid has a carboxyl group and an amino group, hundreds of amino acids can be linked together.

Peptide bonds can form between all 20 amino acids and consist of a link between the carboxyl of one a amino acid and the amine group of the next.

There are 20 common amino acids found in proteins and these amino acids can be roughly classified into 3 groups: polar, non-polar and charged. Polar and charged amino acids are hydrophilic and are often found on the surface of a protein, interacting with the surrounding water. In contrast, non-polar (or hydrophobic) amino acids will avoid water. While this simple categorization is adequate for most purposes, you should recognize that it is a bit simplistic. For example, arginine *does* have a charged hydrophilic group at one end, but the $-CH_2-$ backbone that makes up most of the amino acid is actually quite hydrophobic.

The amino acids are organized into their chemical type and characteristics. Acidic and basic amino acids carry a charge, amino acids with a sulfur or hydroxyl comonent are polar and aliphatic and aromatic amino acids are non-polar.

## SECONDARY STRUCTURE

Peptides and proteins are formed when a ribosome and the rest of the translation machinery link amino acids together in polymers that range from 10 to 10,000 residues in length. During and after protein synthesis, the residues of the primary sequence dictate how the protein folds. The simplest aspect of protein folding is termed its secondary structure, which refers to the geometry of the local polypeptide chain with respect to their immediate neighbors. How a protein folds is dictated by the primary sequence of amino acids, but predicting the overall structure from the primary sequence remains one of the most important unsolved problems in biology.

Nevertheless, it is clear that the major determinants of this final structure are hydrophobic interactions. During protein folding, hydrophobic amino acids must be hidden from the water interface by being buried in the interior of the protein. This burying defines the protein core which then influences the immediate structure around it and greatly affects the protein's overall structure.

## COMMON SECONDARY STRUCTURES

Peptide bonds between adjacent amino acids can rotate and twist to allow a large number of interactions, but two local organization schemes, the á helix and the â sheet, are found in many proteins. Their prevalence is certainly because they happen to form particularly energetically favorable structures. Formation of these structures is driven by favorable hydrogen bonding and hydrophobic interactions between nearby amino acids in the protein. The á helix resembles a ribbon of adjacent amino acids wrapped around a tube to form a staircase-like structure. This structure is very stable, yet flexible and is often seen in parts of a protein that may need to bend or move.

In the â sheet, two protein chains (perhaps different segments of the same protein) align themselves in a planar structure such that hydrogen bonds can form between facing amino acids in each sheet. The â sheet is different from the á helix in that it can involve amino acids from different sections of the protein, which come together to form this structure. Also, the structure tends to be rigid and less flexible than the á helix.

## TERTIARY STRUCTURE

The most important stabilizing force in proteins is burying hydrophobic residues from the surrounding water, but there are other chemical features that are important in creating and stabilizing tertiary structure. These include hydrogen bonds, ionic interactions and sulfhydryl bonds. Ionic interactions are attractions between groups of opposite charge. There are amino acids with negatively charged side groups (aspartate, glutamate) and amino acids with positively charged side groups (typically lysine and arginine). If opposite charges are brought close enough together, there will be an attractive force that can contribute to protein structure. It also follows that like charges will repel each other and this can also dictate protein structure. Sulfhydryl linkages are covalent bonds between cysteine groups. Cysteine is a unique amino acid in that it has a sulfur group at the end of its variable side group that is available for binding to other groups. Often in proteins, nearby sulfhydryl groups on cysteines will form a covalent bond and these are often crucial for the stabilization of the mature protein or for it to perform its function.

Several views of sulfhydryl linkages.

- The chemical formula for a sulfhydryl linkage showing the covalent bond formed.
- A cartoon showing a peptide and a sulfhydryl bond between two cysteines.
- A space filling model of a sulfhydryl linkage in a protein.

one of the most abundant enzymes on this planet. This enzyme takes energy, obtained most often from the sun and uses it to convert carbon dioxide into carbohydrate. Its abundance reflects the fact that virtually all photosynthetic organisms, from bacteria to plants, use this enzyme to produce much of the cell's carbon. Note the á helices throughout the protein and the â sheet near the bottom. These structures help to hold the protein in its proper conformation so that it can carry out its enzymatic activity.

Now we have talked about certain secondary structures that recur in many proteins, but it also happens that there are conserved tertiary structures as well. In other words, there are regions of proteins, which are sometimes termed motifs or domains, that are structurally identical in two or more different proteins. By "structurally identicalΔ, we mean that they have similarly sized á helices and â sheets that are organized in a similar overall way. Interestingly, these domains typically will not have identical primary structures. Instead, it appears that a variety of different sequences can fold to give the same overall tertiary form. It is usually clear from the sequence that structurally similar domains from different proteins are evolutionarily related.

That is, at one time there was a single protein with the

domain and through evolution, an extra copy of the gene was created. Subsequent random mutations in each gene altered the primary sequences of the proteins, but the only changes that were tolerated were those that maintained the structure and therefore the function, of each copy of the domain. This implies that these motifs often have a specific function. For example, there is a very common motif that proteins use for binding ATP. A second class of examples would be the set of motifs that are used to bind DNA.

Through genetic changes, these motifs have been grafted onto other protein regions, so that the same ATP-binding (or the DNA-binding) domain performs a similar function for proteins with completely different overall activities. Of course the unique functions are defined by those portions of the protein that are not in common.

The existence of these conserved motifs was revealed by the analysis of the crystal structures of many different proteins. One saw similar structural regions performing similar functions in proteins that otherwise were quite dissimilar. Not surprisingly, since there are some residues that need to be conserved to form these structures, we can now predict some of these motifs based on the sequence of the proteins as well. The explosion of DNA sequence information and tools that can find these motifs has allowed the prediction of protein functions solely from sequences to be done with ever greater accuracy.

The enzyme pyruvate kinase, which generates ATP by removing it from phosphoenol pyruvate, has three domains, termed A, B and C. The A domain is responsible for binding substrate (phosphenolpyruvate and ADP) while the C domain serves a regulatory role, binding fructose-6-bisphosphate and stimulating activity.

Up to this point, we have talked as if proteins are static rigid structures, but this is certainly wrong. Instead, most proteins are flexible, dynamic structures that respond to the conditions around them. Movement is typically essential to their function, especially for enzymes that need to grab a substrate, react with it and release product. In some cases, the flexibility is over very small distances; moving the residues of an active site just a few tenths of an angstrom ($10^{-10}$ meters) would have a major effect on binding and reacting with a substrate that is itself only a few Angstroms in size. However, some protein movements are huge, where one domain changes its position relative to that of the rest of the protein. When we talk about tertiary structures as found in crystallography, we are then usually talking about only one of several possible structures that a protein can exist in while performing its biological function. The structure through crystallography is simply the form of the protein that happened to form crystals and whose structure was then solved.

This flexibility has many important implications for a protein. At too high a temperature the protein might become too dynamic. This might overwhelm the stabilizing energy that maintains the protein, so that it unfolds or denatures. At too low a temperature, a protein might be so stable that it locks into a single structure and lacks the flexibility necessary to carry out its function. Interestingly, thermophilic microbes have been found to have more rigid proteins that can tolerate the turbulent environment of higher temperatures.

## QUATERNARY STRUCTURE

Many proteins are actually complexes of several polypeptides. The arrangement of more than a single polypeptide into one protein is termed the quaternary structure of that protein. Protein complexes might contain two or more copies of the same protein or they may consist of any number of different polypeptides in various ratios. Such complexes are certainly not random, but reflect precise interactions among the protein subunits based on the same sorts of interactions (e.g. hydrophobic, hydrogen bonds, etc)

An example of a protein containing identical subunits. In this case is has two subunits, so it is termed a dimer. Each subunit has a cyclic adenosine monophosphate molecule bound that activates the protein. In the active state this protein will bind DNA and activate various genes in the cell.

Dinitrogenase is an example of a protein containing non-identical subunits.It has two copies of one protein (termed $\alpha$) and two copies of a second protein (termed $\beta$) and therefore is also referred to as a $\alpha_2\beta_2$ tetramer. This protein is responsible for the reduction of $N_2$ gas to ammonia and is critical to global nitrogen cycling. In contrast to the case of hemoglobin, the importance of the quaternary structure of dinitrogenase is only partially clear. The á subunits contain the metal cluster where $N_2$ is actually reduced (or "Fixed") and the $\beta$ subunit helps form another metal cenre that helps transfer electrons to that active site. We can therefore understand why there might need to be an $\alpha\beta$ dimer.

However, it is not clear why there are two such $\alpha\beta$ dimers hooked together in the tetramer found in nature. There is, for example, no apparent communication between the two active sites. Quite possibly, the fact that dinitrogenase is a tetramer is simply a relic of evolution; perhaps a precursor to dinitrogenase really did have a need to be a tetramer and there is simply no disadvantage to the protein retaining that organization. Glycoproteins are proteins with attached sugars that are typically important for the proper function of the protein. In most cases the sugars are added to the protein after it has been translated. Proteins that encounter the outside environment of the cell will sometimes be glycosylated to stabilize the protein to attack from degradative enzymes and destructive physical forces. glycolipids are combinations of lipids and polysaccharides. In these cases the third glycerol hydroxyl attaches to a polysaccharide instead of a small polar molecule. These polysaccharides can be quite long, extending several microns into the outside environment.

Lipopolysaccharides (one type of glycolipid and usually abbreviated LPS) impart several unique properties to the surface of gram-negative bacteria and we will have more to say about them when examining the cell wall.

## CLASSIFICATION

Proteins can be classified as:

- *Simple proteins:* On hydrolysis they yield only the

amino acids and occasional small carbohydrate compounds. Examples are: albumins, globulins, glutelins, albuminoids, histones and protamines.

- *Conjugated proteins:* These are simple proteins combined with some non-protein material in the body. Examples are: nucleoproteins, glycoproteins, phosphoproteins, haemoglobins and lecithoproteins.
- *Derived proteins:* These are proteins derived from simple or conjugated proteins by physical or chemical means. Examples are: denatured proteins and peptides.

### Properties

Proteins can also be characterized by their chemical reactions. Most proteins are soluble in water, in alcohol, in dilute base or in various concentrations of salt solutions. Proteins have the characteristic coiled structure which is determined by the sequence of amino acids in the primary polypeptide chain and the stereo configuration of the radical groups attached to the alpha carbon of each amino acid.

Proteins are heat labile exhibiting various degrees of lability depending upon type of protein, solution and temperature profile. Proteins can be reversible or irreversible, denatured by heating, by salt concentration, by freezing, by ultrasonic stress or by aging.

Proteins undergo characteristic bonding with other proteins in the so-called plastein reaction and will combine with free aldyhyde and hydroxy groups of carbohydrates to form Maillard type compounds.

### Chemical Determination

The nitrogen content of most proteins found in animal, nut and grain tissue is about 16 percent; therefore, protein content is commonly expressed as nitrogen content × 6.25.

## PROTEIN DIGESTION AND METABOLISM

Ingested proteins are first split into smaller fragments by pepsin in the stomach or by trypsin or chymotrypsin from the pancreas. These peptides are then further reduced by the action of carboxypeptidase which hydrolyzes off one amino acid at a time beginning at the free carboxyl end of the molecule or by aminopeptidase which splits off one amino acid at a time beginning at the free amino end of the polypeptide chain.

The free amino acids released into the digestive system are then absorbed through the walls of the gastro intestinal tract into the blood stream where they are then resynthesized into new tissue proteins or are catabolyzed for energy or for fragments for further tissue metabolism.

## GROSS PROTEIN REQUIREMENTS

Gross protein requirements have been determined for a few species of fish. Simulated whole egg protein component of test diets contains an excess of indispensable amino acids. These diets were kept approximately isocaloric by adjusting total protein plus digestible carbohydrate components to a fixed amount as the protein diet treatments were varied over the ranges tested.

Tests in feeding fry, fingerling and yearling fish have shown that gross protein requirements are highest in initial feeding fry and that they decrease as fish size increases. To grow at the maximum rate, fry must have a diet in which nearly half of the digestible ingredients consist of balanced protein; at 6-8 weeks this requirement is decreased to about 40 percent of the diet for salmon and trout and to about 35 percent of the diet for yearling salmonids raised at standard environmental temperature (SET).

Initially feeding fry require that about 50 percent of the digestible components of the ration be protein and the requirement decreases with size. Some feeding trials with salmon have indicated direct relationships between changes in the protein requirements of young fish and changes in water temperature.

Chinook salmon in 7 C water require about 40 percent whole egg protein for maximum growth; the same fish in 15 C water require about 50 percent protein.

Salmon, trout and catfish can use more protein than required for maximum growth because of efficiency in eliminating nitrogenous wastes in the form of soluble ammonia compounds through the gill tissue directly into the water environment. This system for eliminating nitrogen is more efficient than that available to fowl and mammals. Fowl and mammals consume energy to synthesize urea, uric acid, or other nitrogen compounds which are excreted through the kidney tissue and expelled in urine.

Digestible carbohydrate and fat will spare excess protein in the diet as long as the protein requirement for maximum growth is met

***Table:** Estimated Dietary Protein Requirement of Certain Fish*

| *Species* | *Crude protein level in diet for optimal growth (g/kg)* |
|---|---|
| Rainbow trout (Salmo gairdneri) | 400-460 |
| Carp (Cyprinus carpio) | 380 |
| Chinook salmon (Oncorhynchus tshawytscha) | 400 |
| Eel (Anguilla japonica) | 445 |
| Plaice (Pleuronectes platessa) | 500 |
| Gilthead bream (Chrysophrys aurata) | 400 |
| Grass carp (Ctenopharyngodon idella) | 410-430 |
| Brycon sp. | 356 |
| Red sea bream (Chrysophrys major) | 550 |
| Yellowtail (Seriola quinqueradiata) | 550 |

Basically the fish must be given a diet containing graded levels of high quality protein and energy and adequate balances of essential fatty acids, vitamins and minerals over a prolonged period.

From the resulting dose/response curve the protein requirement is usually obtained by an Almquist plot. These differences in apparent protein requirement are thought to be due to differences in culture techniques and diet composition.

The relatively high dietary protein levels required for maximal growth of certain fish such as grass carp,

Ctenopharyngodon idella and Brycon spp. are surprising as these fish are omnivorous. Brycon spp. are grown on unwanted fruit and other plant material of low protein content and under these conditions there is presumably a substantial contribution to their protein intake from a natural food chain.

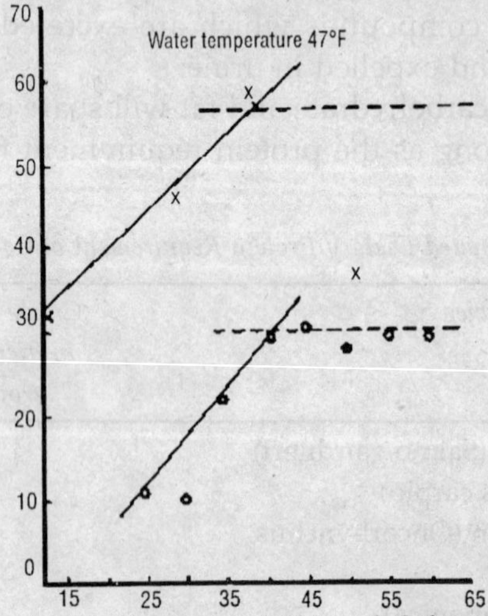


**Fig.** Protein Requirement of Chinook Salmon at 47°F.
Top Curve: Initial Individual Average Weight of Fish, 1.5g.
Bottom Curve: Initial Individual Average Weight of Fish, 5.6g.

Protein requirement of eurythaline fish such as the rainbow trout, Salmo gairdneri and the coho salmon, Oncorhynchus kisutch, reared in water of salinity 20 ppt is about the same as the requirement in freshwater. No data are available for the protein requirement of these species in full strength sea water..

## PROCEDURE OF PROTEIN SYNTHESIS

### Initiation

Peptide synthesis always starts from methionine (Met). Therefore, the initial aminoacyl-tRNA is Met-tRNAiMet, where the subscript "*i*Δ specifies "initiation". In bacteria, the methionine of the initial aminoacyl-tRNA has been modified by the addition of a formyl group (HCO) to its amino group.

The modified methionine is called formylmethionine (fMet), which is unique for bacteria. Thus, fMet is an obvious foreign substance in eukaryotes. It can elicit a strong immune response. In humans, the immune response elicited by the peptide "fMet-Leu-Phe" is about a thousand times greater than "Met-Leu-Phe".

## Elongation

A ribosome contains two major tRNA-binding sites: A site and P site. After the large subunit joins the initiation complex, the initial Met-tRNAiMet enters the P site and the newly arrived aminoacyl-tRNA is always placed at the A site ("A" for "aminoacyl"). Then, methionine is transferred to the new aminoacyl-tRNA, forming a "peptidyl-tRNA" where a peptide is attached to the tRNA. Subsequently, the empty tRNA at the P site is ejected from the ribosome and the peptidyl-tRNA jumps to the P site ("P" for "peptidyl"). During this translocation step, the ribosome also moves one codon down the mRNA chain. Similar steps are repeated in the next cycles of elongation.

## Termination

Protein synthesis will terminate when the ribosome arrives at one of three stop codons. The termination process is assisted by special proteins called termination factors which recognize the stop codons.

Their association stimulates the release of the peptidyl-tRNA from the ribosome. Subsequently, the released peptidyl-tRNA divides into tRNA and a newly synthesized peptide chain. The ribosome also divides into the large and small subunits, ready for synthesizing another peptide.

(a) Initiation

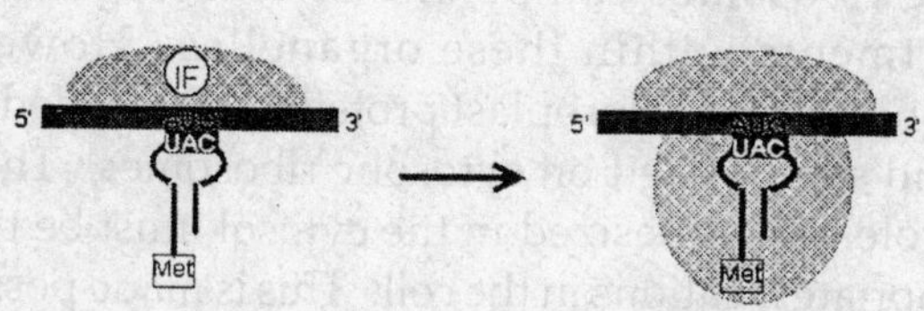


**Fig.** The Steps Involved in Protein Synthesis

- In the absence of mRNA, the large and small subunits of a ribosome are separated. At the beginning of peptide synthesis, initiation factors (IF) first assist the assembly of the small subunit, mRNA and the initial aminoacyl-tRNA. Then, the large subunit is recruited to join the complex.
- In the elongation process, one cycle involves the following steps:
  - *New entry.* A new aminoacyl-tRNA with a correct anticodon is brought to the A site. This step is catalyzed by elongation factors Tu and Ts in prokaryotes and by elongation factors EF1 and EF1b in eukaryotes..
  - *Peptide synthesis.* The peptide attached to the peptidyl-tRNA at the P site is transferred to the new aminoacyl-tRNA at the A site, generating a peptidyl-tRNA with a longer peptide. This step is catalyzed by peptidyl transferase.
  - *Translocation.* The empty tRNA at the P site is ejected from the ribosome and the peptidyl-tRNA generated at the A site takes over the vacant P site. In the mean time, the ribosome moves one codon down the mRNA chain. The A to P switch is catalyzed by the elongation factor G in bacteria and by the elongation factor EF2 in eukaryotes.

## SORTING OF PROTEINS

Proteins are synthesized on ribosomes which are located mainly in the cytosol. Only a small number of ribosomes are located in mitochondria and chloroplasts. Proteins synthesized on these ribosomes can be directly incorporated into the compartments within these organelles. However, most mitochondrial and chloroplast proteins are encoded by nuclear DNA and synthesized on cytosolic ribosomes. These and all other proteins synthesized in the cytosol must be transported to appropriate locations in the cell. This is made possible by the specific signal sequence in the newly synthesized peptide.

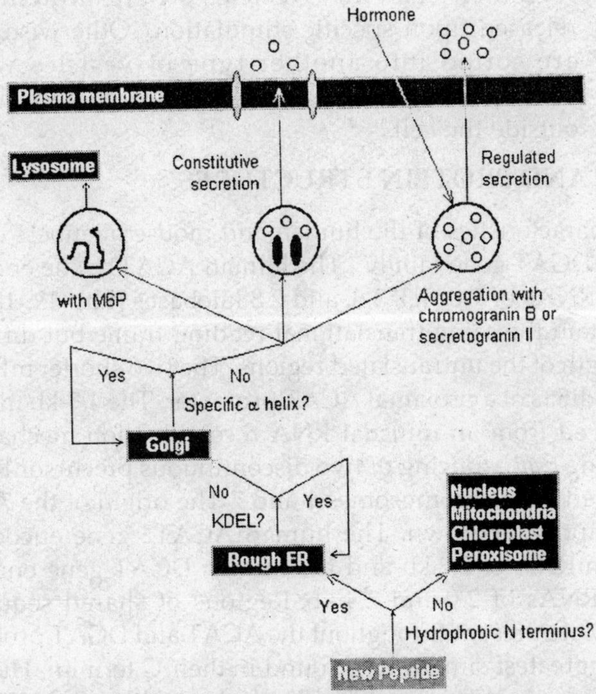


**Fig.** Protein Sorting

- If the N terminus of the new peptide contains a stretch of hydrophobic residues, it is sent to the rough endoplasmic reticulum (ER) for further sorting. Otherwise, it goes to non-ER pathways.
- The new peptide is retained in the rough ER if its C-terminus contains the sequence "Lys-Asp-Glu-Leu" (KDEL in one-letter code). Otherwise, it will move to the Golgi apparatus.
- Proteins containing a specific transmembrane a helix will be localized in the Golgi apparatus.
- After glycosylation at the Golgi apparatus, the modified protein containing a mannose 6-phosphate (M6P) will be delivered to the lysosome.
- Proteins that can aggregate with chromogranin B (secretogranin I) or secretogranin II will be sorted into

regulated secretory vesicles where proteins are released upon specific stimulation. Otherwise, they are sorted into another type of vesicles which continuously move to the plasma membrane or outside the cell.

## GENE AND PROTEIN STRUCTURES

Characteristics of the human and mouse members of the ACAT/DGAT gene family . The human ACAT1 gene encodes four mRNAs of 7.0, 4.3, 3.6, and 2.8 kilobases (kb) (12, 18-20); all contain the same translational reading frame but differ in the length of the untranslated regions. The two shorter mRNAs are products of a proximal ACAT promoter. The 4.3-kb mRNA is derived from an unusual RNA recombination mechanism involving *trans*-splicing of two discontinuous precursor RNAs produced from chromosomes 1 and 7 The origin of the 7.0-kb transcript is unknown. The human ACAT2 gene encodes a single mRNA of 2.2 kb and the human DGAT gene encodes two mRNAs of 2.0 and 2.4 kb. Regions of shared sequence identity are present throughout the ACAT and DGAT proteins, but the greatest similarity is found in their C termini. Human and mouse ACAT2 are ~40% identical to human ACAT1 and DGAT is ~20% identical to ACAT1.

Little is known about the relationship between protein structure and function in ACAT/DGAT family members. One motif, FYXDWWN (amino acids 403-409 of human ACAT1), is highly conserved in all family members and may be involved in fatty acyl-CoA binding. Another motif, MKXXSF (amino acids 265-270 of human ACAT1), is conserved in ACAT family members. The serine of this motif is required for ACAT activity. All three family members have a potential tyrosine phosphorylation motif and at least one *N*-linked glycosylation site, both of unknown significance. ACAT1 and ACAT2 contain an N-terminal leucine zipper motif. Although ACAT1 functions as a homotetramer it is unknown whether this leucine zipper motif participates in tetramer formation or whether ACAT2 and DGAT act as multimers. ACAT1 and ACAT2 do not form hetero-oligomeric complexes.

All ACAT/DGAT proteins possess multiple hydrophobic regions predicted to serve as transmembrane domains, although there is not yet a consensus structural model for the proteins. Investigations of human ACAT1 topology have yielded different results.

One study, performed by inserting epitopes at different positions in the expressed protein found seven transmembrane domains. Another study, performed by expressing truncated forms of the protein found five transmembrane domains. In both studies, the N terminus of the protein localized to the cytoplasm and the C terminus to the ER lumen.

Topologic studies of human ACAT2 performed with the truncation mutant method also found five transmembrane domains with the N terminus in the cytoplasm and the C terminus in the ER lumen. The latter study found that a serine residue essential for ACAT activity was located in the cytoplasm for ACAT1 and in the ER lumen for ACAT2, suggesting that the active sites of these enzymes reside on opposite sides of the ER membrane. Several putative transmembrane sequences are conserved in ACAT1 and ACAT2 but not in DGAT suggesting that these regions are involved in cholesterol binding. The topology of DGAT has not been reported, but a putative diacylglycerol binding motif has been identified

## TISSUE AND CELLULAR DISTRIBUTIONS

The tissue distributions of ACAT1 and ACAT2 are largely complementary. ACAT1 mRNA is present in many tissues, with the highest levels in macrophages and in adrenal and sebaceous glands all of which store cholesterol esters in cytoplasmic droplets. ACAT1 is also highly expressed in human atherosclerotic lesions, particularly in macrophage foam cells. ACAT2 is expressed predominantly in the liver and small intestine. In humans, nonhuman primates and mice, ACAT2 appears to be the major ACAT in the small intestine and the predominant isozyme in hepatocytes of nonhuman primates and mice.

Whether ACAT1 or ACAT2 is the predominant enzyme

in human hepatocytes is controversial. Both are expressed in human hepatocytes and HepG2 cells. However, immunodepletion experiments found that ACAT1 accounts for ~90% of ACAT activity in human liver, whereas ACAT2 accounts for most of the ACAT activity in adult small intestine and fetal liver but only 10-20% of activity in adult liver.

It is unclear whether the hepatic ACAT1 activity resides in hepatocytes or macrophage-derived Kupffer cells. ACAT1 expression was detected immunohistochemically in both cell types, with stronger staining in Kupffer cells. In nonhuman primates, ACAT1 was expressed only in Kupffer cells in the liver. A significant proportion of the ACAT1 activity in human liver thus may be attributable to Kupffer cells. Another study found little evidence of ACAT2 expression in human hepatocytes.

DGAT mRNA and activity are ubiquitous in mouse and human tissues, with the highest levels in liver, small intestine and adipose tissue and somewhat lower levels in testis and adrenal gland.

This wide range of tissue expression is consistent with the involvement of DGAT in the glycerol phosphate pathway of triglyceride synthesis, which is common to most cells. DGAT protein expression in tissues has not been studied because of the lack of suitable antibodies.

The intracellular localization of enzymes in this class has been examined only for ACAT1. In human melanoma cells and fibroblasts and in mouse macrophages, ACAT1 is located primarily in the ER. In mouse macrophages, a small portion of ACAT1 immunoreactivity localizes to a region near the *trans*-Golgi network. In macrophages, ACAT1 localization may change with different conditions

## BIOCHEMICAL ACTIVITY AND REGULATION

ACAT and DGAT substrates have been identified by expressing the proteins in insect cells or yeast devoid of sterol esterification activity. In insect cells, mouse and human ACAT1 and ACAT2 utilize a variety of oxysterols in addition to cholesterol as substrates. Ergosterol and plant sterols are poor substrates. Both enzymes also utilize a wide variety of long chain fatty acyl-CoAs as substrates although arachidonyl-CoA appears to be a better substrate for ACAT1 than ACAT2. In insect cells, DGAT utilizes only diacylglycerol as the acyl acceptor. DGAT has a broad fatty acyl-CoA specificity. ACAT1 is regulated primarily by post-translational mechanisms. In cholesterol-loaded cells, esterification increases without changes in the expression level of ACAT1 mRNA or protein. Purified ACAT1 is activated allosterically by cholesterol or oxysterols. ACAT1 activity may also be regulated by its intracellular localization.

For example, a portion of ACAT1 immunoreactivity in macrophages is found near the *trans*-Golgi network and the endocytic recycling compartment. These cholesterol-rich organelles are involved in membrane traffic with the plasma membrane and the proximity of ACAT1 to these organelles may be important for cholesterol ester synthesis during the formation of macrophage foam cells.

The observation that 30-40% of ACAT1 redistributed from the ER to small vesicles during foam cell formation is consistent with this possibility. Regulation by phosphorylation has not been demonstrated.

ACAT1 is also transcriptionally regulated; however, only the expression levels of the shorter ACAT1 mRNAs (2.8 and 3.6 kb) change. In mice and rabbits a fat and cholesterol-rich diet increases ACAT1 mRNA levels 2-3-fold in the liver. In human HepG2 cells, free fatty acids, but not 25-hydroxycholesterol, increase ACAT1 mRNA levels 1.5-2-fold. ACAT1 mRNA and protein expression levels also increase during the differentiation of monocytes into macrophages.

Two factors that induce gene expression during monocyte-macrophage differentiation1,25-dihydroxyvitamin $D_3$ and 9-*cis*-retinoic acid, also increased ACAT1 mRNA, protein and cholesterol esterification activity in THP-1 cells.

Little is known about the regulation of ACAT2 and DGAT. Levels of DGAT mRNA increase markedly in parallel with enzyme activity in NIH 3T3-L1 cells during their differentiation into adipocytes. This finding suggests that transcription factors involved in adipogenesis activate DGAT expression. Like other family members, DGAT possesses a putative tyrosine phosphoryla-tion site. The use of this site in DGAT regulation has not been demonstrated, although two reports have indicated that the enzyme may be posttranslationally regulated by a tyrosine kinase.

## FUNCTIONAL ANALYSIS OF ACAT1 AND ACAT2

Gene knockout experiments in mice have provided insights into the biological roles of ACAT1 and ACAT2. In ACAT1-deficient (ACAT1) mice, plasma cholesterol levels, hepatic cholesterol esterification activity and intestinal cholesterol absorption are normal, consistent with the presence of a second ACAT enzyme in the small intestine and liver of mice. ACAT1 deficiency has its greatest effect in the adrenal glands and macrophages.

Cholesterol esters and cholesterol esterification activity are virtually undetectable in the adrenal cortex of ACAT1 mice, although the depletion of cholesterol esters does not affect corticosterone synthesis. In addition, peritoneal macrophages from ACAT1 mice do not accumulate cholesterol esters when incubated with cholesterol-containing lipoproteins.

This finding provides a unique tool for testing the role of ACAT1 in macrophage foam cell formation and atherosclerosis. Another ACAT1 gene disruption study in mice confirmed general aspects of the ACAT1 phenotype and showed that ACAT1 deficiency reduces cholesterol esterification activity by > 95% in testis and ovary and results in atrophy of the meibomian glands of the eyelids.

A naturally occurring mutation in the mouse ACAT1 gene is present in the AKR strain of inbred mice which are homozygous for the *adrenocortical lipid depletion* (*ald*) allele.

The analysis of ACAT2-deficient mice (ACAT2) complemented findings from ACAT1$^{/}$ mice. ACAT2 mice have nearly undetectable ACAT activity and lack cholesterol esters in the liver and small intestine. Because they cannot synthesize cholesterol esters in the small intestine, ACAT2 mice have a reduced capacity to obtain cholesterol from the diet. As a result, they do not develop hypercholesterolemia or form cholesterol gallstones when fed a diet high in fat, cholesterol and cholic acid. ACAT2 deficiency markedly decreases the cholesterol ester content of apolipoprotein (apo) B-containing lipoproteins in mice fed either a chow diet or the above diet. Thus, ACAT2 is the cholesterol esterification enzyme in mouse liver and small intestine and plays a major role in mediating the response to dietary cholesterol.

Gene disruption studies suggest that both ACAT enzymes synthesize cholesterol esters: ACAT1 for storage in cytosolic droplets and ACAT2 for secretion in apoB-containing lipoproteins. However, these results may reflect their different tissue distributions rather than inherent biochemical differences. Both ACAT1 and ACAT2, when expressed in a cell line deficient in cholesterol esterification activity (AC29 cells), synthesize cholesterol esters that are stored in cytoplasmic droplets Also, overexpression of ACAT1 in mouse liver causes both cholesterol ester storage and secretion.

## PROTEIN CLASSIFICATION

Proteins are linear seqeuences of amino acids that perform essential functions in organisms. To determine the function of a protein, expensive experimental procedures are required. With better sequencing techniques, it becomes evident that computational prediction/classfication of an unknown protein's function will help drive the labor and cost down.

Protein remote homology detection involves detecting functionally related proteins with very little sequence identity. The task is challanging due to two major reasons: first, the length of the sequences vary and second, remotely homologous proteins need not share great sequence identity. Some share sequence identity as little as 30%.

### Structural Classification of Proteins

The Structural Classification of Proteins (SCOP) database is a largely manual classification of protein structural domains based on similarities of their amino acid sequences and three-dimensional structures.

## Comparison to other Classification Systems

This classification is more significantly based on the human expertise than semi-automatic CATH, its chief rival. It is usually accepted that SCOP provides a better justified classification.

Human expertise is needed to decide whether certain proteins are evolutionary related and therefore should be assigned to the same superfamily, or their similarity is a result of structural constraints and therefore they belong to the same fold.

Another database, FSSP, is purely automatically generated (including regular automatic updates) but offers no classification, allowing the user to draw their own conclusion as to the significance of structural relationships based on the pairwise comparisons of individual protein structures.

## Protein Functions

Consumed proteins are digested into amino acids, and these amino acids join other amino acids produced by the body to constitute the amino acid pool. The tissues take the amino acids from this pool to synthesize the specific proteins the body needs (muscle, hair, nails, hormones, enzymes, and so on). This amino acid pool is also available for use as energy (via a deamination process) to be burned if other fuels (carbohydrate and fat) cannot satisfy energy needs.

The major functions of protein include the following:

1. Protein provides a source of carbon for energy-yielding reactions. Certain amino acids can be converted to glucose and metabolized to provide ATP, while others can be stored as fat that can subsequently be catabolized to provide ATP.
2. Protein is a critical compound in controlling fluid volume and osmolarity in the blood and body tissues. This function is a major controlling factor in maintaining water balance.
3. Proteins are amphoteric, with the capacity to buffer both acid and alkaline environments to maintain an optimal blood *pH*.

4. Antibodies are protein-based substances critical for maintaining health.
5. Proteins form enzymes that are involved in digestive and other cellular processes that create needed chemical end products.
6. Proteins are a critical component of body tissues, including the organs (heart, liver, pancreas, and so on), muscles, and bones.
7. Proteins are "smart" transporters of substances in the blood, moving substances to the correct receptor sites. For instance, transferrin is the protein transporter for iron.
8. Proteins are synthesized into specific hormones (such as insulin) and neurotransmitters (such as serotonin) that control body function.

## Protein Metabolism

Protein is composed of carbon, hydrogen, oxygen, nitrogen, and, in some cases, sulfur. Protein is the only nutrient that contains nitrogen, a fact that makes it both essential and potentially toxic.

Amino acid building blocks constitute the larger molecular structures of protein. Some of these amino acids can be synthesized from other amino acids (referred to as nonessential amino acids), while some must be obtained from the foods we eat (referred to as essential amino acids).

There is widespread confusion about essential and nonessential amino acids because the word *nonessential* suggests that, although they are present, they are not really needed. This, however, could not be further from the truth. Both the essential and nonessential amino acids are equally important (and therefore equally essential) in human metabolic processes.

Amino acids combine together to make much larger proteins. The sequence of the amino acids and the secondary and tertiary structures of the protein determine its function. When dietary protein is consumed, it is digested into polypeptides (small protein molecules) and eventually into individual amino acids. The amino acids are absorbed into the blood and transported to different tissues, where they are manufactured into necessary proteins. Tissues manufacture the proteins that are needed through the amino acids available to them.

**Table. Amino Acid and Protein Functions**

| | |
|---|---|
| Working proteins (produces hemoglobin, enzymes, and hormones; maintain normal blood osmolarity; form antibodies; used as energy source) | Enzyumes<br>Antibodies<br>Transport proteins<br>Homones |
| Structural proteins (censtitute cell structure; help develop, repair, and maintain tissues) | Muscles, tendons, ligaments<br>Skin<br>Bone and teeth cores<br>Hair and nails |

To ensure that the tissues are capable of manufacturing needed proteins, the essential amino acids must be present for protein synthesis to take place. Some people believe that if you want healthy looking hair and nails, you should consume the primary protein that hair and nails are made of (gelatin).

Gelatin is a low-quality protein with a poor distribution of the essential amino acids, so it would not encourage an optimal protein synthesis.

Put simply, eating hair and nails does not encourage optimal synthesis of hair and nails. The best way to ensure an optimal synthesis of needed proteins is to make all the essential amino acids available to cells so they are capable of making whatever they need.

Several amino acids have specific effects on the central nervous system. Because of these known effects, single amino acids have been sold with the purpose of imparting these known outcomes.

Tryptophan, for instance, has been sold as an agent that causes relaxation or sleepiness. The dangers, however, of pushing high doses of single amino acids into the system are sufficiently large that this should not be done unless under the careful supervision of a physician.

The best strategy is to supply the widest possible array of

essential amino acids from food alone, which allows the individual tissues to synthesize the amino acids needed for optimal body function.

The liver is the central processing unit for protein synthesis, continually monitoring protein needs and synthesizing amino acids and proteins to satisfy a variety of needs.

This protein synthesis is accomplished through transamination and deamination reactions. In transamination, the nitrogen from one amino acid is used for the manufacture of another amino acid; in deamination the amino group is removed from an amino acid and converted to ammonia. The remaining carbon structure is either reconstructed as fat and stored, converted to glucose (as happens with the amino acids alanine and glutamine), or burned for energy.

Once all protein needs are met, the fate of all remaining amino acids is deamination. The ammonia created during deamination is toxic to the body, but enzymes in the liver convert the ammonia to urea.

Urea is excreted from the body in the urine. Therefore, the greater the extra protein consumed, the greater the production of ammonia that must be removed from the system (as urea).

The majority of the remaining deaminated carbon chains are typically stored as fat.

**Table. Amono Acids With Neurotransmitter Products**

| *Amino acid* | *Product* | *Function* |
|---|---|---|
| Tryptophan | Serotonin<br>Melatonin | Mood, pain, food intake, arousal |
| Tyrosine Phenylalanine | Dopamine<br>Norepinephrine<br>Epinephrine | Motor function, mood, arousal, attention, anxiety |
| Histidine | Histamine | Fod intake, arousal, thermoregulation |
| Arginine | Nitric oxide | Arousal, anxiety, memory |
| Threonine | Glycine | Motor function |

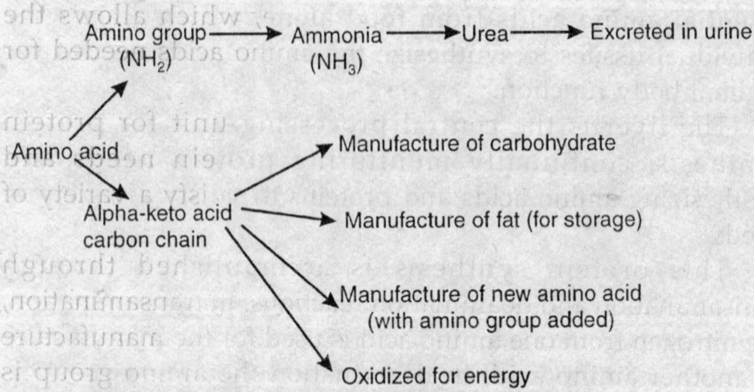


**Fig.** Deaminatioin and the Reconstruction of th.e Remaining Carbon Structure

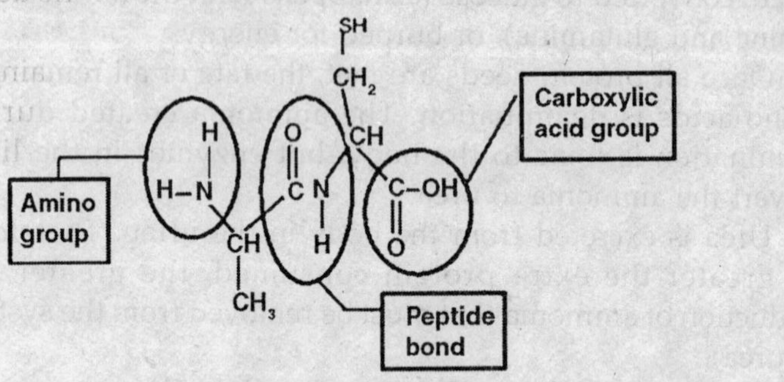


**Example:** Chemical structure of alanine (Note: Two alanine molecules can be combined with the nitrogen removed to make glucose in the liver via the glucose-alanine cycle. The process of making new glucose is referred to as gluconeogenesis).

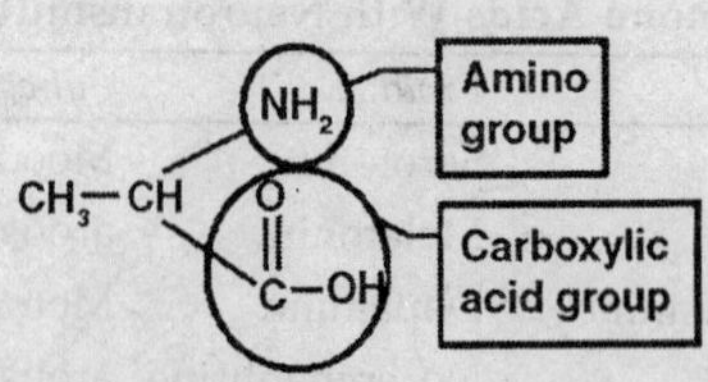


## AMINO ACIDS

The amino acids are the building blocks of proteins; about 23 amino acids have been isolated from natural proteins. Ten of these are indispensable for fish. The animal is incapable of synthesizing indispensable amino acids and must therefore obtain these from the diet.

## ESSENTIAL AND NON-ESSENTIAL AMINO ACIDS

Salmon, trout and channel catfish fed diets devoid of arginine, histidine, isoleucine, leucine, lysine, methionine, phenylalanine, threonine, tryptophan or valine failed to grow. These same fish fed diets devoid of other L-amino acids grew as well as fish receiving all 18 amino acids tested.

The nitrogen component in the test diets was made up of 18 L-amino acids in the pattern found in whole egg protein. All fish on test recovered rapidly when the missing amino acid was replaced in the diet. The slope of the growth curve of the recovery group was identical with that of fish receiving the complete amino acid test diet. Dispensable amino acids tested were alanine, aspartic acid, cystine, glutamic acid, glycine, proline, serine and tyrosine.

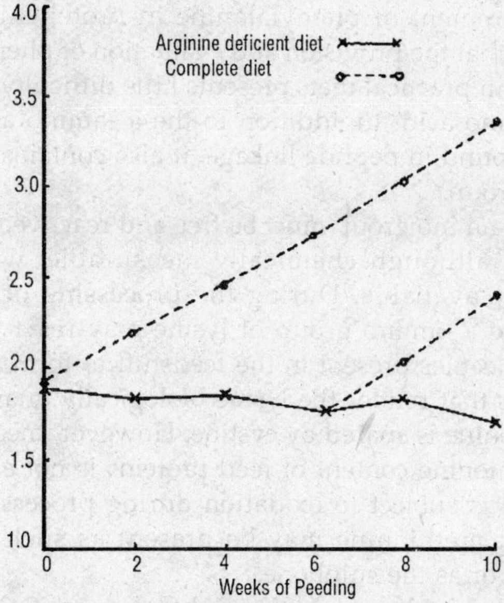


**Fig.** Growth of Arginine Deficient Fish. The Deficient Group was Divided After Six Weeks on the Deficient Diet and the Missing Amino Acid was Replaced in One of the Two Sub-lots

These amino acids were found to be not essential for the growth of salmon, trout and channel catfish. Quantitative studies on the requirements of the 10 indispensable amino acids used a casein-gelatin mixture supplemented with crystalline L-amino acids.

The test diet had an amino acid pattern of 40 percent whole egg protein for the nitrogen component. Experiments conducted with carp and eel showed a similar lack of growth when an indispensable amino acid was absent from the diet.

## Essential Amino Acids and Protein Quality

If the essential amino acid requirements of fish are known, it should be possible to meet these needs in culture systems in a number of ways from different food proteins or combinations of food proteins. Phenylalanine is spared by tyrosine. It is not known to be chemically modified nor rendered unavailable by the harsh conditions to which feedstuff proteins are normally subjected during processing.

Measurement of phenylalanine in proteins is uncomplicated so that the provision and evaluation of phenylalanine in proteins in practical diets presents little difficulty. Lysine is a basic amino acid. In addition to the a -amino acid group normally bound in peptide linkage, it also contains a second, a -amino group.

This a -amino group must be free and reactive, otherwise the lysine, although chemically measurable, will not be biologically available. During the processing of feedstuff proteins the a -amino group of lysine may react with non-protein molecules present in the feedstuff to form additional compounds that render the lysine biologically unavailable.

Methionine is spared by cystine. However, measurement of the methionine content of feed proteins is not easy as the amino acid is subject to oxidation during processing. After processing, methionine may be present as such or as the sulphoxide or as the sulphone.

The sulphoxide may be formed from methionine during acid hydrolysis of the feed protein prior to measurement of its any-no acid composition. Acid hydrolysis of proteins before analysis disturbs the original equilibrium between the two compounds so that the composition of the hydrolysate no longer reflects that of the protein.

In determining the methionine content of pure proteins, oxidation of the amino acid to methionine sulphone is normally quantitative. In the case of feed proteins, however, this will not reveal how much methionine or methionine sulphoxide was present in the protein prior to performate oxidation and hydrolysis.

Methionine sulphoxide may have some biological value for fish which may have some capability of reconverting it to methionine and thus partially make up for some of the methionine oxidized during processing. Methods have recently been reported for measurement of methionine in proteins using an iodoplatinate reagent before and after reduction with titanium trichloride, to give values for both methionine and the sulphoxide in the original protein. A method for measuring methionine specifically by cyanogen bromide cleavage has also been described.

Both methods remain to be independently assessed. Microbiological assay of methionine in feed proteins is a valuable tool although there is the danger that oxides of methionine may differ in their activity for micro-organisms and misrepresent values.

## QUANTITATIVE REQUIREMENTS OF AMINO ACID

Quantitative requirements by salmonids for the ten indispensable amino acids were determined by feeding linear increments of one amino acid at a time in a test diet containing an amino acid profile identical with whole egg protein except for the amino acid tested.

Replicate groups of fish were fed the diet treatments until gross differences appeared in the growth of test lots. An Almquist plot of growth response indicated the level of amino acids required for maximum growth under those specific test conditions.

Diets were designed to contain protein at or slightly below the optimum protein requirement for that species and test condition to assure maximum utilization of the limiting amino acid. A comparison of the requirements for the ten indispensable amino acids.

A recent innovation has been the use in test diets of proteins relatively deficient in a given essential amino acid. Thus combinations of fishmeal and zein have been used in test diets to define the requirement of rainbow trout for arginine. Diets containing different relative amounts of casein and gelatin showed that an increase in the level of protein-bound arginine from 11 to 17 g/kg resulted in a significant increase in the growth of channel catfish.

**Table:** *Amino Acid Requirements of Seven Animals*

| *Amino acid* | *Eel fingerling* | *Carp fry* | *Channel catfish* | *Chinook salmon fingerling* | *Chick* | *Young Pig* | *Rat* |
|---|---|---|---|---|---|---|---|
| Arginine | 3.9 (1.7/42) | 4.3 (1.65/38.5) | | 6.0 (2.4/40) | 6.1 (1.1/18) | 1.5 (0.2/13) | 1.0 (0.2/19) |
| Histidine | 1.9 (0.8/42) | | | 1.8 (0.7/40) | 1.7 (0.3/18) | 1.5 (0.2/13) | 2.1 (0.4/19) |
| Isoleucine | 3.6 (1.5/42) | 2.6 (1.0/38.5) | | 2.2 (0.9/41) | 4.4 (0.8/18) | 4.6 (0.6/13) | 3.9 (0.5/13) |
| Leucine | 4.1 (1.7/42) | 3.9 (1.5/38.5) | | 3.9 (1.6/41) | 6.7 (1.2/18) | 4.6 (0.6/13) | 4.5 (0.9/19) |
| Lysine | 4.8 (2.0/42) | | 5.1 (1.23/24.0) | 5.0 (2.0/40) | 6.1 (1.1/18) | 4.7 (0.65/13) | 5.4 (1.0/19) |
| Methionine [2] | 4.5 (2.1/42) [3] | 3.1 (1.2/38.5) | 2.3 (0.56/24.0) | 4.0 (1.6/40) [3] | 4.4 (0.8/18) | 3.0 (0.6/20) | 3.0 (0.6/20) |
| Phenylalanine [4] | | | | 5.1 (2.1/41) [5] | 7.2 (1.3/18) | 3.6 (0.45/13) | 5.3 (0.9/17) |
| Threonine | 3.6 (1.5/42) | | | 2.2 (0.9/40) | 3.3 (0.6/18) | 3.0 (0.4/13) | 3.1 (0.2/19) |
| Tryptophan | 1.0 (0.4/42) | | | 0.5 (0.2/40) | 1.1 (0.2/18) | 0.8 (0.2/25) | 1.0 (0.2/19) |
| Valine | 3.6 (1.5/42) | | | 3.2 (1.3/40) | 4.4 (0.8/18) | 3.1 (0.4/13) | 3.1 (0.4/13) |

1. Expressed as percent of dietary protein. In parentheses, the numerators are requirements as percent of dry diet and the denominators are percent total protein in the diet
2. In the absence of cystine
3. Methionine plus cystine
4. In the absence of tyro sine
5. Phenylalanine plus tyrosine

Arginine requirement of rainbow trout has been determined from a conventional dose/response (growth) curve and also by measuring the tissue (blood and muscle) levels of free arginine in groups of trout given increasing amounts of dietary arginine.

After the dietary requirement of the trout for arginine has been met, any further increase in arginine intake led to an increase in the concentration of free arginine in blood and muscle. Good agreement was obtained between the two methods. This leads to difficulties in formulating the protein component of practical diets for those species whose amino acid requirements are not yet known.

A possible solution is to use, for each amino acid, the highest level required by any of those species for which data is available. The need for further quantitative data on the amino acid requirements of fish, especially those actually or potentially useful as farm animals, is obvious.

## SUPPLEMENTING DIETS WITH AMINO ACIDS

One solution to the use of proteins that are relatively deficient in one or more amino acids is to supplement the protein with appropriate amounts of the amino acid needed in practical diets. Fish appear to utilize free amino acids at various degrees of efficiency.

Young carp, Cyprinus carpio, were shown to be unable to grow on diets in which the protein component (casein, gelatin) was replaced by a mixture of amino acids similar in overall composition. A trypsin hydrolyzate of casein was equally ineffective.

However, if a diet containing free amino acids as the protein component is carefully neutralized with NaOH to pH 6.5-6.7 then some growth of young carp does occur. This growth was markedly inferior to that occurring on a comparable casein diet under the same conditions.

Channel catfish are also unable to utilize free amino acids given as supplements to deficient proteins. When soybean meal was substituted isonitrogenously for menhaden meal, growth and feed efficiency of channel catfish were substantially reduced. Addition of free methionine, cystine or lysine, the most limiting amino acids, to these soy-substituted diets did not enhance weight gain.

Raising the arginine level of catfish diets from 11 to 17 g/kg by isonitrogenous substitution of gelatin for casein enhanced weight gain significantly but the addition of free arginine, cystine, tryptophan or methionine to casein had little effect on growth or food conversion. Salmonids are able to utilize free amino acids for growth. A zein-gelatin diet supplemented with lysine and trytophan was shown to be markedly superior to an unsupplemented zein-gelatin diet for rainbow trout when weight gain and protein utilization were used as criteria.

Several investigators have demonstrated the potential of supplementing amino acid deficient proteins with limiting amino acids in diets for salmonids. Casein supplemented with six amino acids produced feed conversion ratios with Atlantic salmon similar to those obtained when an isolated fish protein was used as the dietary protein source.

Soybean meal supplemented with five or more amino acids (including methionine and lysine) was a superior protein source to soybean meal alone for rainbow trout. Single additions of methionine and lysine did not, however, improve the value of soybean meal.

These results suggest that the amino acid spectrum of the isolated fish protein they used may possibly approximate the amino acid requirement of rainbow trout. The nutritional value of a soy protein isolate could be enhanced by supplementing it with the first limiting amino acid; i.e., methionine.

Diets containing, as protein component, fishmeal, meat and bone meal and yeast and soybean meal could be improved by supplementing with cystine (10 g/kg) and tryptophan (5 g/kg) together. Fishmeal can be entirely replaced without a reduction in food conversion rate in diets for rainbow trout by a mixture of poultry by-product meal and feather meal together with 17 g lysine HCL/kg, 4.8 g DL-methionine/kg and 1.44 g DL-tryptophan/kg.

## STRUCTURE OF AMINO ACIDS

Proteins are polymers of amino acids.. Amino acids, with rare exception, contain an á carbon that is connected to an amino ($NH_3$) group, a carboxyl group (COOH) and a variable side group (R). The side group gives each amino acid its distinctive properties and helps to dictate the folding of the protein.

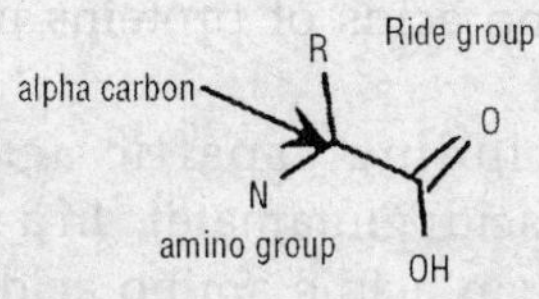


**Fig.** A General Amino Acid

Each amino acid contains a carboxyl group, an amino group and a variable side group (R). These all connect to a central carbon termed the á-carbon.

## BUILDING BLOCKS - AMINO ACIDS

An amino acid is defined as the molecule containing an amino group ($NH_2$), a carboxyl group (COOH) and an R group. It has the following general formula,

$$R\text{-}CH(NH_2)\text{-}COOH$$

The R group differs among various amino acids. In a protein, the R group is also called a sidechain.

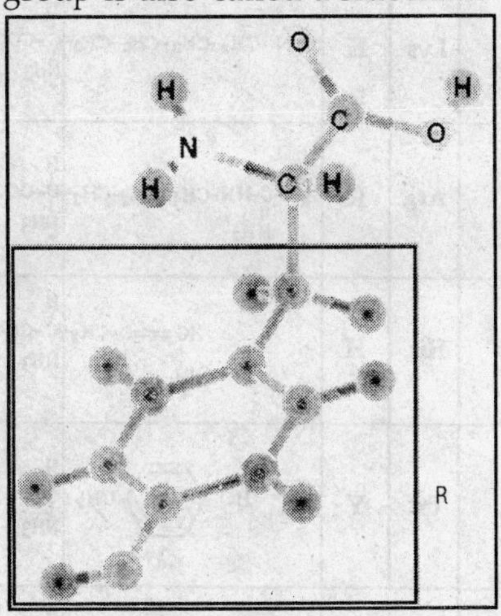


**Fig.** The Structure of an Amino Acid (Tyrosine)

There are over 300 naturally occurring amino acids on earth, but the number of different amino acids in proteins is only 20

## THE TWENTY AMINO ACIDS OF PROTEINS

Based on the physical and chemical properties of R groups,

the 20 amino acids of proteins may be classified as follows.

- *Acidic:* Including aspartic acid (aspartate) and glutamatic acid (glutamate). In a neutral solution, the R group of an acidic amino acid may lose a proton and become negatively charged.

| Name | Symbol | | R group | Hydrophobicity |
|---|---|---|---|---|
| | 3 Lett. | 1 Lett. | | |
| **Aspartate** | **Asp** | **D** | 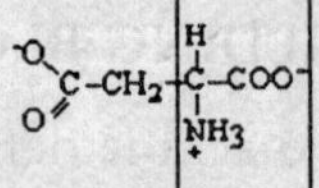 | **-3.5** |
| **Glutamate** | **Glu** | **E** | 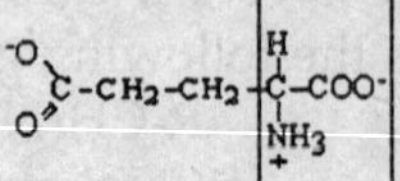 | **-3.5** |
| **Lysine** | **Lys** | **K** | 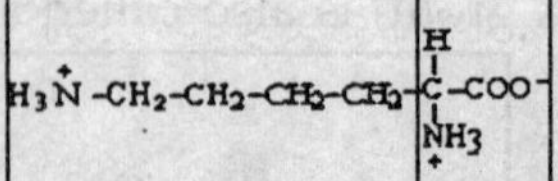 | **-3.9** |
| **Arginine** | **Arg** | **R** | 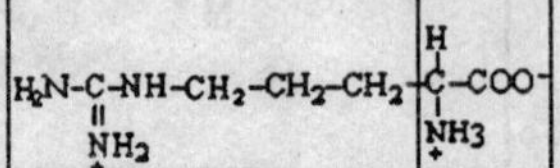 | **-4.5** |
| **Histidine** | **His** | **H** | 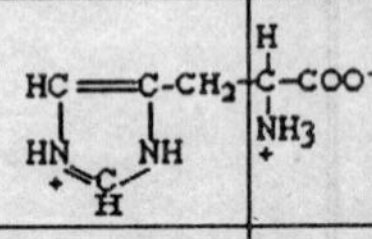 | **-3.2** |
| **Tyrosine** | **Tyr** | **Y** | 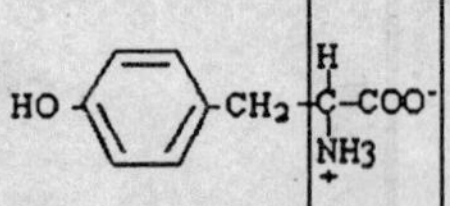 | **-1.3** |
| **Tryptophan** | **Trp** | **W** | 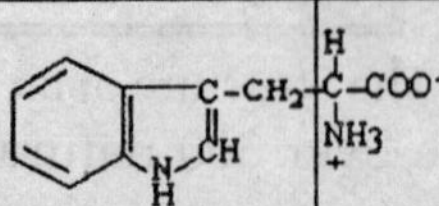 | **-0.9** |

- *Basic:* Including lysine, arginine and histidine. In a neutral solution, the R group of a basic amino acid may gain a proton and become positively charged. Interaction between positive and negative R groups may form a salt bridge, which is an important stabilizing force in proteins.

- *Aromatic:* Including tyrosine, tryptophan and phenylalanine. Their R groups contain an aromatic ring.

| | | | | |
|---|---|---|---|---|
| **Phenylalanine** | **Phe** | **F** | 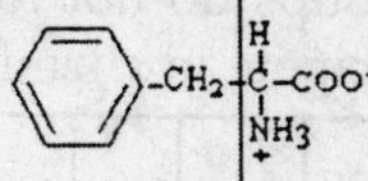 | **2.8** |
| **Cysteine** | **Cys** | **C** | 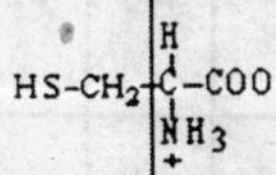 | **2.5** |
| **Methionine** | **Met** | **M** | 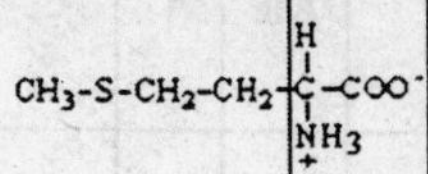 | **1.9** |
| **Serine** | **Ser** | **S** | 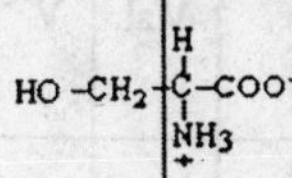 | **-0.8** |
| **Threonine** | **Thr** | **T** | 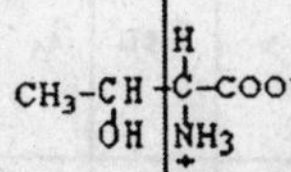 | **-0.7** |
| **Asparagine** | **Asn** | **N** | 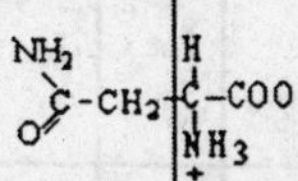 | **-3.5** |
| **Glutamine** | **Gln** | **Q** | 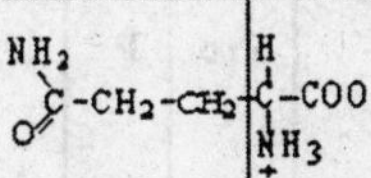 | **-3.5** |

- *Sulfur:* including cysteine and methionine. Their R groups contain a sulfur atom (S). The disulfide bond formed between two cysteine residues provides a strong force for stabilizing the globular structure. A unique feature about methionine is that the synthesis of all peptide chains starts from methionine.
- Uncharged hydrophilic: including serine, threonine, asparagine and glutamine. Their R groups are

hydrophilic and capable of forming hydrogen bonds.

- Inactive hydrophobic: including glycine, alanine, valine, leucine and isoleucine. These amino acids are more likely to be buried in the protein interior. Their R groups do not form hydrogen bonds and rarely participate in chemical reactions.

| | | | | |
|---|---|---|---|---|
| Glycine | Gly | G | 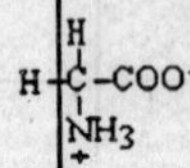 | -0.4 |
| Alanine | Ala | A | 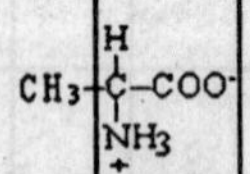 | 1.8 |
| Valine | Val | V | 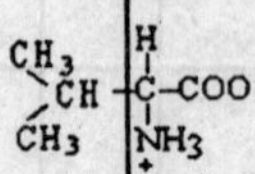 | 4.2 |
| Leucine | Leu | L | 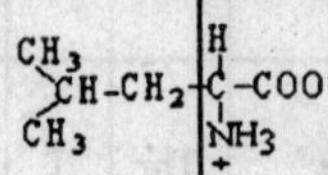 | 3.8 |
| Isoleucine | Ile | I | 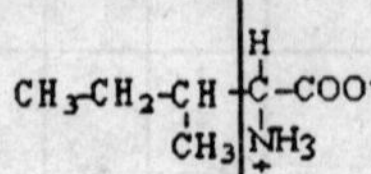 | 4.5 |
| Proline | Pro | P | 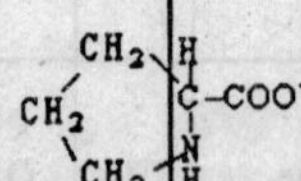 | -1.6 |

- Special structure: including proline. In most amino acids, the R group and the amino group are not directly connected. Proline is the only exception among 20 amino acids found in protein. Due to this special feature, proline is often located at the turn of a peptide chain in the three-dimensional structure of a protein.

The hydrophobicity index tells the relative hydrophobicity among amino acids. More positive value indicates stronger hydrophobicity. Hydrophilic amino acids have negative values. In a protein, hydrophobic amino acids are more likely to be located in the protein interior, whereas hydrophilic amino acids are more likely to face the aqueous environment.

## PH AND PKA

$$AH \rightleftharpoons H^+ + A^-$$

Define its equilibrium constant,

$$K_a \equiv [H^+][A^-]/[AH]$$

where [..] represents the concentration.

Taking logarithm on both sides,

$$\log K_a = \log [H^+] + \log [A^-]/[AH]$$

Define

$$pK_a \equiv -\log K_a; \quad pH \equiv -\log [H^+]$$

we obtain

$$pK_a = pH - \log [A^-]/[AH]$$

or

$$[A^-]/[AH] = 10^{(pH - pK_a)}$$

| pH | $[H^+]$ (mol/l) | |
|---|---|---|
| 0 | $10^{-0}$ | ↑ |
| 1 | $10^{-1}$ | |
| 2 | $10^{-2}$ | |
| 3 | $10^{-3}$ | Increasing acidity |
| 4 | $10^{-4}$ | |
| 5 | $10^{-5}$ | |
| 6 | $10^{-6}$ | |
| 7 | $10^{-7}$ | Neutral |
| 8 | $10^{-8}$ | |
| 9 | $10^{-9}$ | |
| 10 | $10^{-10}$ | Increasing alkalinity |
| 11 | $10^{-11}$ | |
| 12 | $10^{-12}$ | |
| 13 | $10^{-13}$ | |
| 14 | $10^{-14}$ | ↓ |

A molecule, or an atom group in a molecule, may lose or gain a proton when the molecule is placed in an aqueous solution. The exact probability that a molecule will be protonated or deprotonated depends on the $pK_a$ of the molecule and the pH of the solution. Let AH be an atom group in a molecule. AH could be neutral or charged. After AH loses a proton, it is denoted by $A^-$. The protonation/deprotonation reaction may be written as

From the last equation, we can see that half of molecules will lose protons if they are in a solution with $pH = pK_a$. The higher the pH value, the more likely a molecule will lose a proton. On the other hand, from the definition of pH, we find that the proton concentration of a solution becomes lower at higher pH, thereby capable of accepting more protons. This physical concept is consistent with the result we just obtained from the equation. z`z

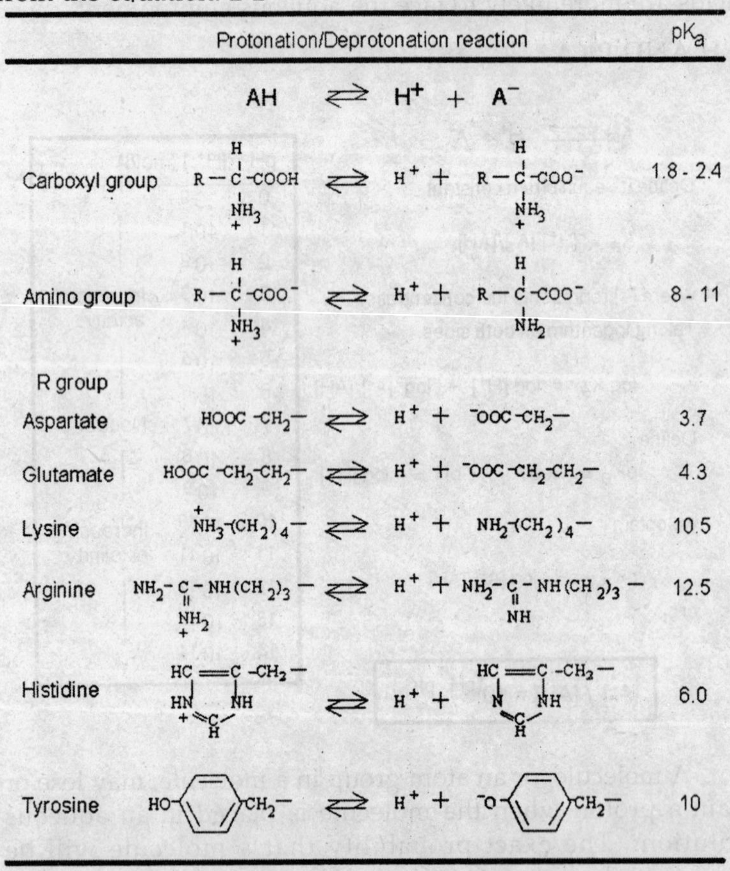


**Fig.** $pK_a$ Values of Amino Group, Carboxyl Group and a Few R Groups. This Information Tells us in a Neutral Solution:

- The carboxyl group is most likely negatively charged.
- The amino group is most likely positively charged
- The R group of aspartate and glutamate are most likely negatively charged.
- The R group of lysine and arginine are most likely positively charged.

- The R group of tyrosine at pH = 7 is most likely neutral.
- The R group of histidine has 10% probability to become positively charged at pH = 7, but the probability increases to 50% at pH = 6. Thus, histidine is very sensitive to pH change in the physiological range

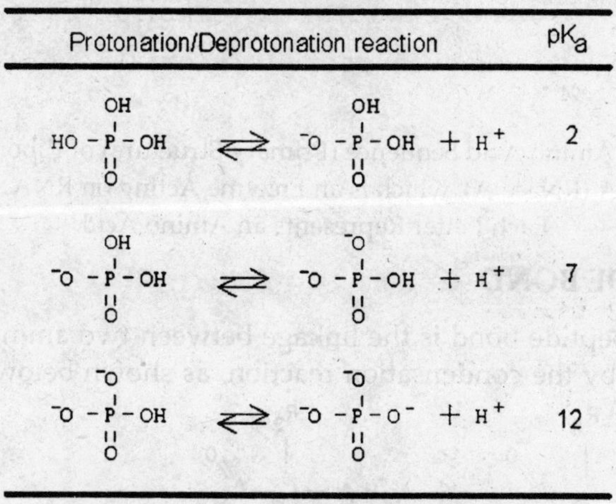


**Fig.** $pK_a$ Values of Phosphate Groups.

This information tells us that in a neutral solution $H_3PO_4$ is very likely to lose a proton and become $H_2PO4^-$.

The latter has a 50% chance of losing another proton, becoming $HPO_4^{2-}$.

The chance for $HPO_4^{2-}$ to lose an additional proton is very small. Hence, the average charge of a phosphate group is about 1.5 electron charges.

## PEPTIDES

A peptide is a chain of amino acids linked together by peptide bonds. Polypeptides usually refer to long peptides whereas oligopeptides are short peptides (< 10 amino acids). Proteins are made up of one or more polypeptides with more than 50 amino acids.

## PRIMARY STRUCTURE

The primary structure of a protein refers to its amino acid sequence. The amino acid in a peptide is also called a residue.

**Fig.** The Amino Acid Sequence (Primary Structure) of Ribonuclease A (RNase A), which is an Enzyme Acting on RNA. Each Letter Represents an Amino Acid.

## PEPTIDE BOND

A peptide bond is the linkage between two amino acids, formed by the condensation reaction, as shown below:

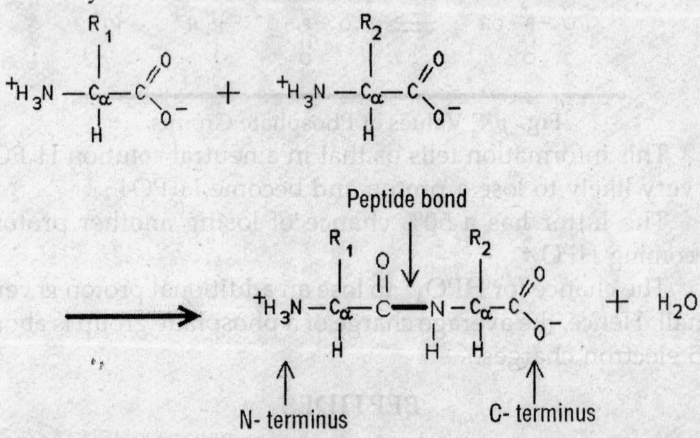


**Fig.** Formation of the Peptide Bond by Condensation Reaction

## PHI-PSI ANGLES

Due to the specific electronic structure of the peptide bond, the atoms on its two ends cannot rotate around the bond. Hence, the atoms of the group, O=C-N-H, are fixed on the same plane, known as the peptide plane. The whole plane may rotate around the N-Ca bond (f angle) or C-Ca bond (Y angle). Ca is the carbon atom connected to the R group.

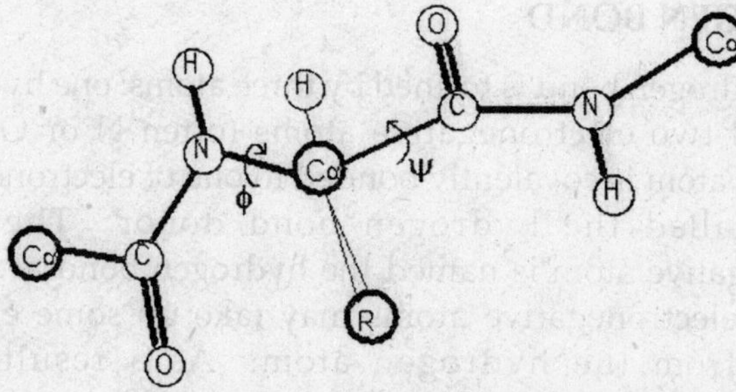


**Fig.** Illustration of the Peptide Plane (Gray Area) and f-Y Angles. The Red Line Formed by the Repeating -Ca-C-N-Ca- is the Backbone of the Peptide Chain.

Mathematically, phi (f) and psi (Y) are the dihedral angle (also known as torsional angle) which is defined as the angle between the point (e.g., Ca) at the end of a 4-point sequence and the plane (e.g., peptide plane) occupied by the other three points. In a peptide, phi-psi angles are restricted to certain ranges. A plot of their distribution is called the Ramachandran plot.

## SECONDARY STRUCTURE

In a protein, certain domains may form specific structures such as a helix and bstrand, which constitute the secondary structure of the protein.

(a)

(b)

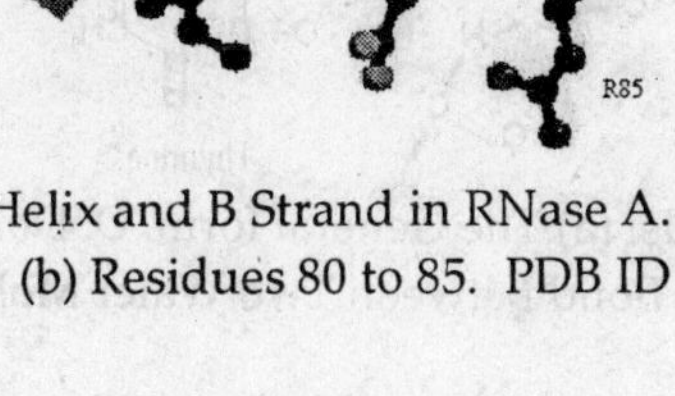


**Fig.** The a Helix and B Strand in RNase A. (a) Residues 1 to 36. (b) Residues 80 to 85. PDB ID = 1RCN.

## HYDROGEN BOND

A hydrogen bond is formed by three atoms: one hydrogen atom and two electronegative atoms (often N or O). The hydrogen atom is covalently bonded to one of electronegative atoms, called the hydrogen bond donor. The other electronegative atom is named the hydrogen bond acceptor. The two electronegative atoms may take up some electron density from the hydrogen atom. As a result, each electronegative atom carries partial negative charge and the hydrogen atom carries partial positive charge. The hydrogen atom and the hydrogen bond acceptor can then have attractive interactions. The strength of the hydrogen bond depends on the donor and acceptor as well as their environment. The bond energy usually ranges from 1 kcal/mol to 5 kcal/mol. This energy is smaller than covalent bond energy, but greater than thermal energy (0.6 kcal/mol at room temperature). Therefore, the hydrogen bond can provide a significant stabilizing force in macromolecules such as proteins and nucleic acids.

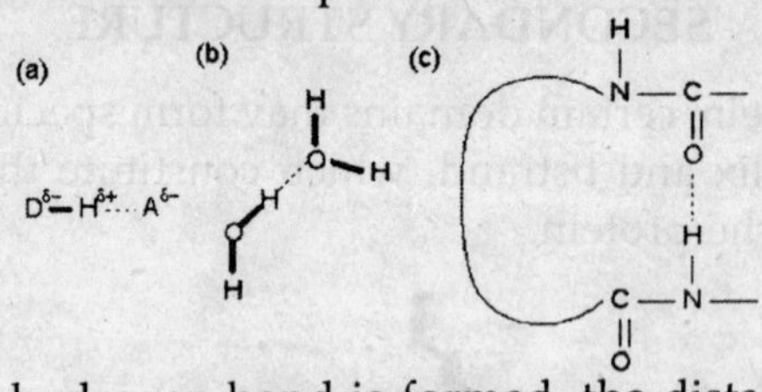


When a hydrogen bond is formed, the distance between the donor and the acceptor is about 0.28 nm (1 nm = $10^{-9}$ meter). The hydrogen atom is still covalently bonded to the donor and their distance is shorter than that between the hydrogen atom and the acceptor.

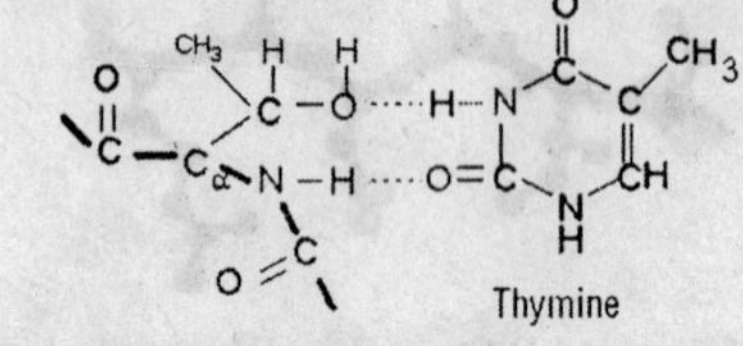


**Fig.** Hydrogen Bonds. (a) The General form. D: Donor; A: Acceptor. (b) The Hydrogen Bond between Two Water Molecules. (c) The

Hydrogen Bond in a Polypeptide, e.g., in an a Helix or between Two b Strands. (d) The Hydrogen Bond between Residue 45 (a Threonine) of RNase A and its Substrate Thymine.

## ALPHA HELIX

An a helix has the following features:

- Every 3.6 residues make one turn,
- The distance between two turns is 0.54 nm,
- The C=O (or N-H) of one turn is hydrogen bonded to N-H (or C=O) of the neighboring turn.

Hydrogen bonds play a role in stabilizing the a helix conformation. However, the size and charges of sidechains are also important factors. Alanine has a greater propensity to form a helices than proline.

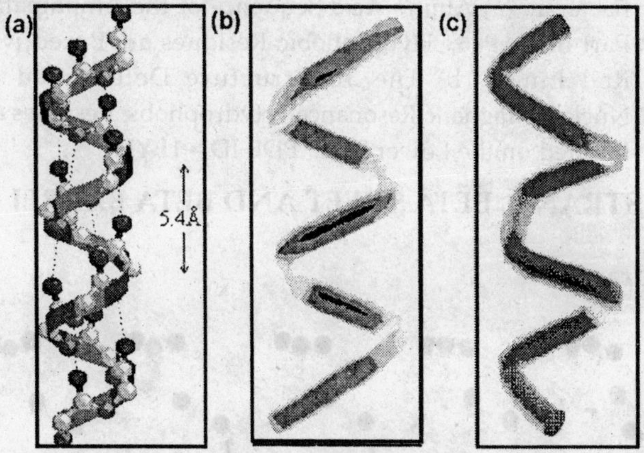


Fig. A Helix Conformations. (a) The Ideal Right-handed a Helix. C: Green; O: Red; N: Blue; H: not shown; Hydro-gen Bond: Dashed Line. (b) The Right-handed a Helix Without showing Atoms. (c) The Left-handed a Helix.

The right-handed or left-handed helix may be distinguished by stretching out your thumb and curling the other four fingers.

## AMPHIPATHIC A HELIX

In an amphipathic a helix, one side of the helix contains mainly hydrophilic amino acids and the other side contains mainly hydrophobic amino acids. The amino acid sequence of amphipathic a helix alternates between hydrophilic and hydrophobic residues every 3 to 4 residues, since the a helix makes a turn for every 3.6 residues. An example is shown below.

**Fig.** The Amphipathic a Helix Structure of CAP18, which is a Molecule Capable of Binding to the Endotoxin of Bacteria. (a) Amino Acid Sequence of the Amphipathic Part of CAP18. Hydrophobic Residues are Boxed with Red Lines. (b) The 3D Structure Determined by Nuclear Magnetic Resonance. Hydrophobic Residues are Located on the Lower Side. PDB ID = 1LYP.

## BETA STRAND, BETA SHEET AND BETA BARREL

### Beta Strand

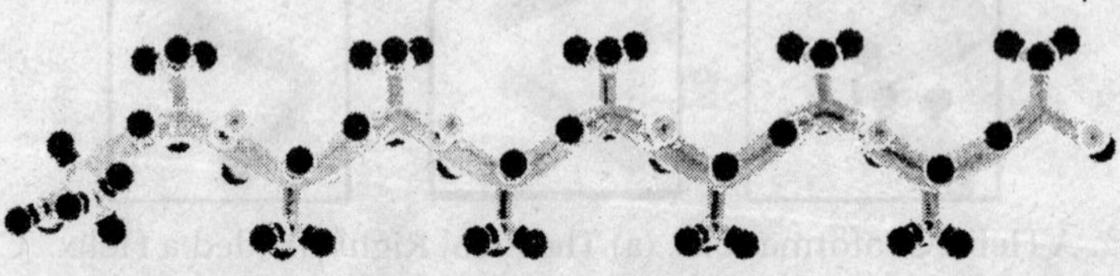

Fig. An Ideal B Strand

In a b strand, the torsion angle of N-Ca-C-N in the backbone is about 120 degrees. The conformation of an ideal b strand. sidechains of two neighboring residues project in the opposite direction from the backbone.

### Beta Sheet

A b sheet consists of two or more hydrogen bonded b strands. The two neighboring b strands may be parallel if they are aligned in the same direction from one terminus (N or C) to the other, or anti-parallel if they are aligned in the opposite direction.

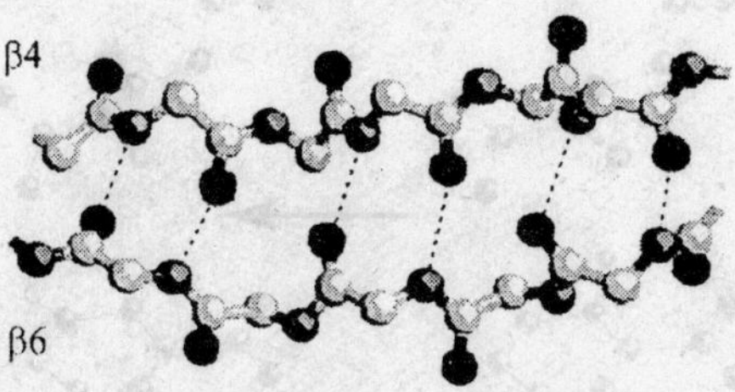


**Fig.** The b Sheet Structure Found in RNase A. This Figure shows Only the Backbone Atoms, Excluding Hydrogens. RNase A Contains a Single Peptide Chain, which Makes a Turn at the Junction between b4 and b6. Therefore, the Two Strands are Antiparallel.

## BIOSYNTHESIS OF AMINO ACIDS

### NITROGEN SOURCES

In order to synthesize amino acids, a source of nitrogen is needed. In animals glutamate and glutamine play the pivotal roles. The ±"amino group of most of the amino acids comes from the transamination reaction transferring the amino group from glutamate to an ±-ketoacid acceptor. Glutamate is synthesized from ammonia and ±-ketoglutarate by the action of glutamate dehydrogenase.

$NH_4^+$ α-ketoglutarate + NAD(P)H + $H^+ \rightleftarrows$ Glutamate + $NAD(P)^+ + H_2O$

This reaction occurs in two steps first a Schiff base is formed between the ammonia and the ketone of aketoglutarate. The Schiff base is reduced by hydride transfer from either NADH or NADPH to form glutamate.

The importance of the hydride transfer is that it establishes the stereochemistry at the ±-carbon. Glutamate dehyrogenase binds ±-ketoglutarate is such a way that the hydride is transferred to specific face to form only the L-isomer of glutamate.

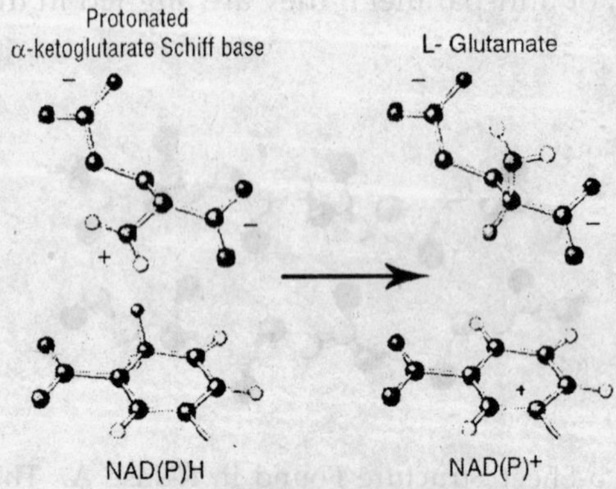


A second ammonium ion is incorporated into glutamate by the action of glutamine synthetase to form glutamine. The regulation of glutamine synthetase plays a crucial role in controlling nitrogen metabolism. The dynamic duo of glutamate dehydrogenase and glutamine synthetase are found in all living organisms.

## AMINO ACIDS ARE SYNTHESIZED FROM METABOLITES OF OTHER PATHWAYS

Plants and bacteria can synthesize all 20 of the amino acids. Where as we humans cannot synthesize of them. These nine amino acids must come from our diets and are called essential amino acids. The essential amino acids are: Histidine, Isoleucine, Leucine, Lysine, Methionine, Phenylalanine, Threonine, Tryptophan and Valine.

The other 11 amino acids are called nonessential amino acids they include: Alanine, Arginine, Asparagine, Aspartate, Cysteine, Glutamate, Glutamine, Glycine, Proline, Serine and Tyrosine. These nonessential amino acids are synthesize by simple pathways. Whereas the biosynthesis of the essential amino acids are quite complex.

The major metabolic precursors of the amino acids. The amino acids that are precursors for other amino acids are shown in yellow. The nine essential amino acids are shown in boldface. The carbon skeletons come from intermediates of glycolysis, the pentose phosphate pathway and the citric acid cycle. On the basis of the starting points the 20 amino acids can be group into 6 categories depending on the precursor: oxaloacetate, PEP, ±-ketoglutarate, pyruvate, 3-phosphoglycerate, ribose-5P.

## ASPARTATE, GLUTAMATE AND ALANINE BIOSYNTHESIS

Oxaloacetate, pyruvate and ±-ketoglutarate are all ±-ketoacids which are substrates for transamination reactions that we very familiar with by now. Asparatate aminotransferase transfers an amino group from glutamine to oxaloacetate to form aspartate and ±-ketoglutarate.

There are a number of transaminase that will transfer an amino group from an amino acid such as aspartate or alanine and transfer it to ±- ketoglutarate to form glutamate. Alanine aminotransferase transfers the amino group of glutamine to pyruvate to form alanine and ±-ketoglutarate. All of these transamination reactions are catalyzed by enzymes that require pyridoxal phosphate as a cofactor. Review the pyridoxal phosphate mechanism.

The enzyme reaction begins with the formation of the external aldimine between PLP and the amino group donor displacing the active site lysine residue.

The amino group is transferred to PLP to form the pyridixoamine phosphate intermediate releasing the ±-ketoacid remnant of the donor. The second half of the reaction is the reverse of the first. The amino group acceptor is another ±-ketoacid which forms a Schiff base with the PMP to form a ketimine intermediate.

The amino group is transferred to the acceptor to form the amino acid. The important step in the transamination reaction is the protonation of the ±-carbon of the quinonoid intermiate.

This protonation determines the stereochemistry at the ±-carbon. In the case of aspartate amino transferase (the postor child of transaminases) the chirality of the amino acid is determined by binding interactions with the substrate. Particularly, the interaction of the conserved arginine residue

(Arg-386) with the carboxylate group of the substrate. This interaction orients the substrate so that when Lys-268 protonates the ±"carbon of the quinonoid intermediate it generates an external aldimine with the L-configuration at the ±-carbon.

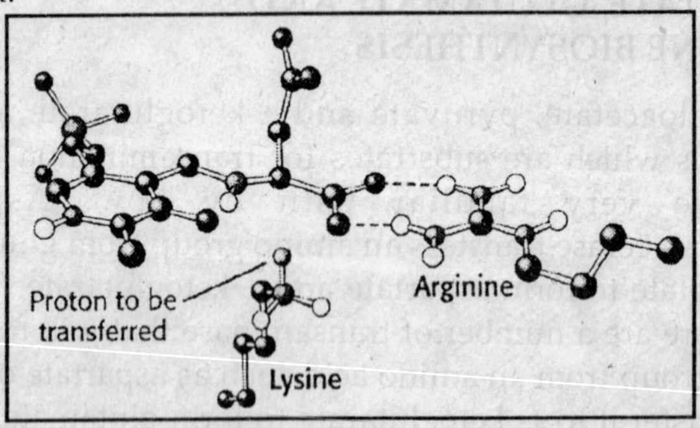


## ASPARAGINE BIOSYNTHESIS

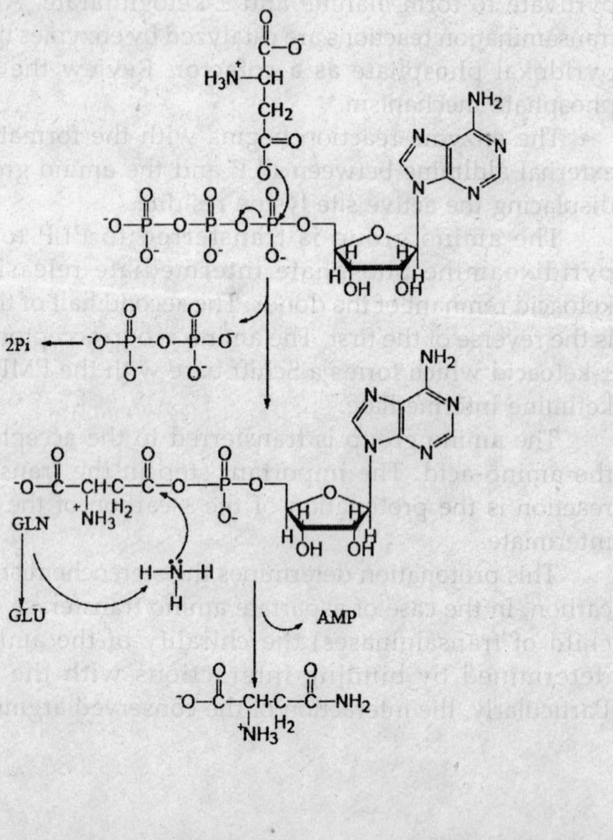

Asparagine synthetase catalyzes the formation of asparagine from aspartate. The first step is to activate the carboxylate group of aspartate by reacting with ATP to form an aspartyladenylate intermediate. The activated carbonyl of aspartyl-adenylate intermediate is attacked by an ammonia nucleophile to produce asparagines. In bacteria ammonia is the nitrogen source for this reaction. In mammals the nitrogen source for this reaction is derived from the hydolysis of glutamine into glutamate. The ammonia produced in this reaction is channeled to the aspartyl-adenylate intermediate to from asparagine.

The advantage of this pathway of using glutamine as the nitrogen source is that the ammonia is produced by the enzyme and channeled directly to the intermediate. The cells are not exposed to ammonia which is toxic to animals.

## PROLINE AND ARGININE BIOSYNTHESIS

Proline and arginine are both derived from glutamate. The first step is to form glutamate ³-semialdehyde. The ³-carboxylate group of glutamate is activated by phosphorylation with ATP to from a ³- glutamylphosphate intermediate. The mixed anhydride is then reduced with NADPH liberating the phosphate and producing glutamate ³-semialdehyde.

Glutamate ³-semialdehyde will spontaneously cyclize eliminating H2O and forming pyrroline 5- carboxylate which is then reduced by pyrolline-5-carboxylate reductase to form proline. Glutamate ³-semialdehyde is also a substrate for ornithine-′-aminotransferase which will transfer and amino group from glutamate to glutamate ³-semialdehyde to from ornithine and ±-ketoglutarate. Ornithine of course is a metabolite of the urea cycle which reacts with carbamoyl phosphate to form citrulline. Citrulline is condensed with aspartate to form argininosuccinate which is then cleaved to form arginine and fumarate.

## SERINE, CYSTEINE AND GLYCINE BIOSYNTHESIS

Serine is synthesized from 3-phosphoglycerate. Serine is a precursor for both cysteine and glycine. The first step is the oxidation of 3-phosphoglycerate into 3-phosphohydroxy-pyruvate. The enzyme that catalyzes the oxidation is 3-phosphoglycerate dehydrogenase.

This oxidation generates an ±-ketoacid which can now be transaminated by transferring an amino group from glutamate to 3-phosphoglycerate to generate 3-phosphoserine and ±-ketoglutarate.

The phosphate group is removed by phosphoserine phosphatase to produce serine. Serine is converted into glycine by serine hydroxy-methyltransferase which contains both a pyridoxal phosphate cofactor and a tetrahydrofolate cofactor. It is time to take a closer look as this remarkable cofactor and this remarkable enzyme.

## Tetrahydrofolate-Carrier of Activated One-Carbon Units

*Tetrahydrofolate:* Tetrahydrofolate is a versatile carrier of one carbon units. It consists of 3 groups. A pteridine, an amino benzoate and a chain of one or more glutamate residues. Mammals can synthesize all three components but lack the enzymes that conjugate them together.

Microorganisms do contain the necessary conjugating enzymes and can produce folate.

We must obtain folate from our diet or from the microorganism that inhabit our intestinal tract. Folic acid is member of the vitamin B complexes and is found in green vegetables, fruits, yeast and liver. Folate is reduced by dihydrofolate reductase to form THF.

This is a two step reduction requiring 2 hydride transfers from NADPH. One carbon units may be bound to THF at the N5 and N10 nitrogens. The carbons carried can exist in three oxidation state.

The most reduced state of a THF carried carbon is a methyl group, the intermediate oxidation state carries a methylene group, the most oxidized state carries a formimino group. The next higher oxidation state is the carboxylate which is carried by the biotin cofactor rather than THF.

All of these interconvertible tetrahydrofolate derivatives carry activated one carbon units which can be used for a variety of biosyntheses. THF also serves as one carbon unit acceptor in degradative reactions.

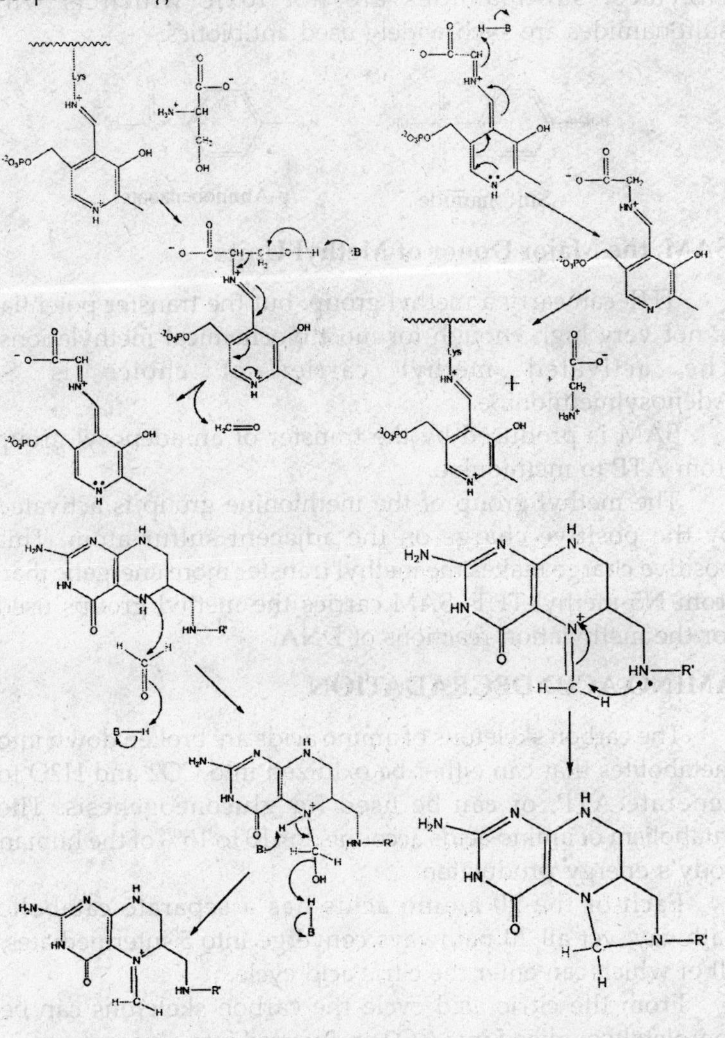

Sulfonamides such as sulfanilamide are analogs of the p-aminobenzoate substituent of folic acid. The are competitive inhibitors of the bacterial conjugating enzyme for the p-aminobenzoate binding site.

They block the synthesis of folate producing a deficiency

which inactivates all of the enzymes that require a THF cofactor. Since mammals do not possess the conjugating enzymes, sulfonamides are not toxic which is why sulfonamides are such widely used antibiotics.

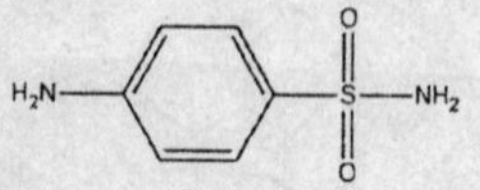

Sulfonamide

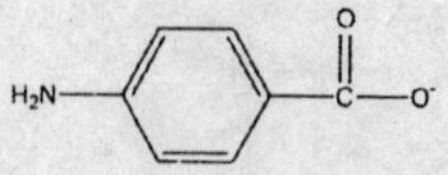

p-Aminobenzoate

## SAM, the Major Donor of Methyl Units

THF can carry a methyl group, but the transfer potential is not very high enough for most biochemical methylations. The activated methyl carrier of choice is S-Adenosylmethionine.

SAM is produced by the transfer of an adenosyl group from ATP to methionine.

The methyl group of the methionine group is activated by the positive charge on the adjacent sulfur atom. This positive charge makes the methyl transfer more energetic than from N5-methyl-THF. SAM carries the methyl groups used for the methylation reactions of DNA.

## AMINO ACID DEGRADATION

The carbon skeletons of amino acids are broken down into metabolites that can either be oxidized into CO2 and H2O to generate ATP, or can be used for gluconeogenesis. The catabolism of amino acids accounts for 10 to 15% of the human body's energy production.

Each of the 20 amino acids has a separate catabolic pathway, yet all 20 pathways converge into 5 intermediates, all of which can enter the citric acid cycle.

From the citric acid cycle the carbon skeletons can be completely oxidized into CO2 or diverted into gluconeogensis or ketogenesis.

Glucogenic amino acids are broken down into one of the following metabolites: pyruvate, ±- ketoglutarate, succinyl CoA, fumarate or oxaloacetate.

Ketogenic amino acids are broken down into acetoacetate or acetyl-CoA. Larger amino acids, tryptophan, phenylalanine, tyrosine, isoleucine and threonine are both glucogenic and ketogenic. Only 2 amino acids are purely ketogenic they are lysine and leucine.

If 2 of the amino acids are purely ketogenic and 5 amino acids are both ketogenic and glucogenic, than that leaves 13 amino acids that are purely glucogenic:

Arg, Glu, Gln, His, Pro, Val, Met, Asp, Asn, Ala, Ser, Cys and Gly.

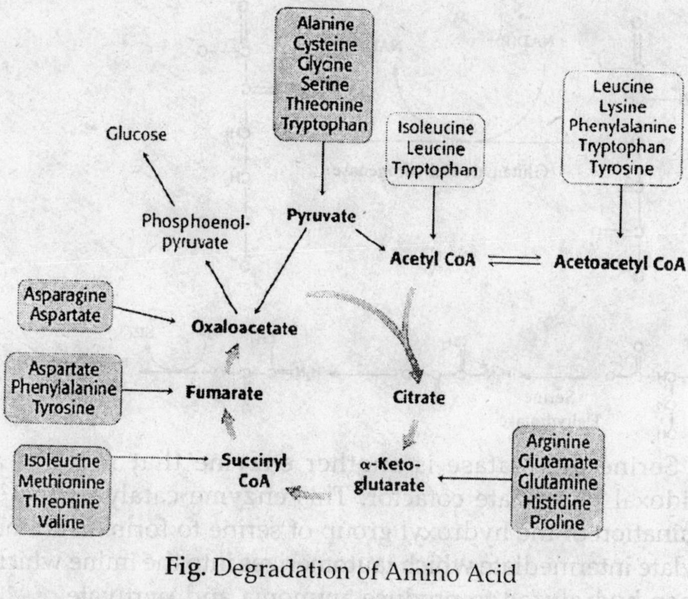


**Fig.** Degradation of Amino Acid

## Amino Acids that are Catabolized into Pyruvate

Pyruvate is the entry point for amino acids that contain 3 carbons, alanine, serine and cysteine.

Alanine transaminase reversibly transfers the amino group from alanine to ±-ketoglutarate to form pyruvate and glutamate.

Note that enzyme requires a pyridoxal phosphate cofactor. The ±-ketoglutarate is regenerated by glutamate dehydrogenase.

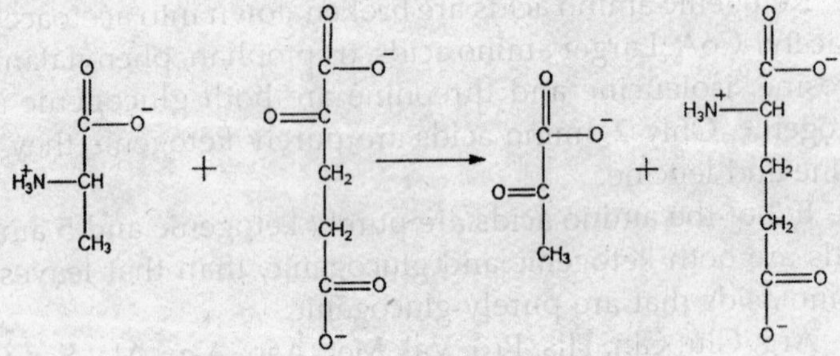

Alanine Transaminase

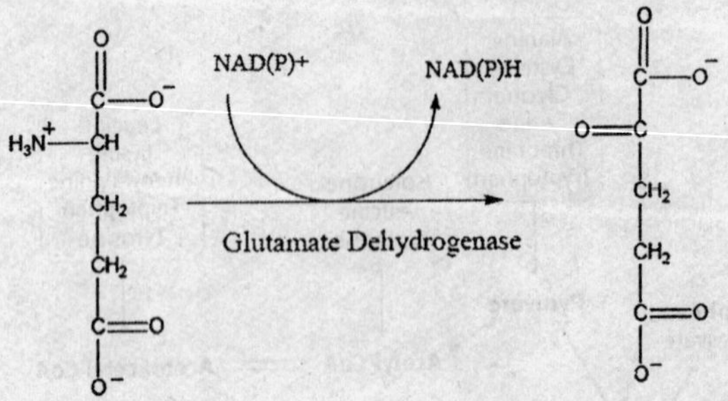


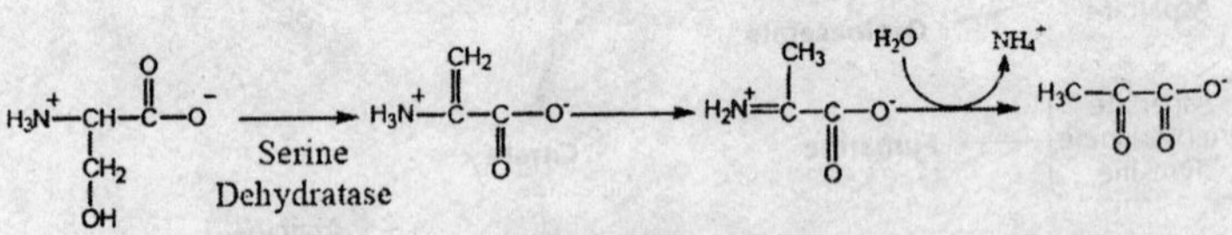


Serine dehyratase is another enzyme that requires a pyridoxal phosphate cofactor. This enzyme catalyzes the $^{2}$-elimination of the hydroxyl group of serine to form an amino acrylate intermediate which tautomerizes into the imine which is then hydrolyzed to produce ammonia and pyruvate.

Glycine is converted into pyruvate via conversion of glycine to serine by serine hydroxymethyl transferase which is an incredibly interesting enzyme. It contains a pyridoxal phosphate cofactor and a *N*5,*N*10-methylene-tetrahydrofolate which is a cofactor we have not encountered yet. The *N*5,*N*10-methylene-tetrahydrofolate is produced by the glycine cleavage system which transfers a methylene group from glycine to THF. The THF cofactor is a one carbon acceptor and donor.

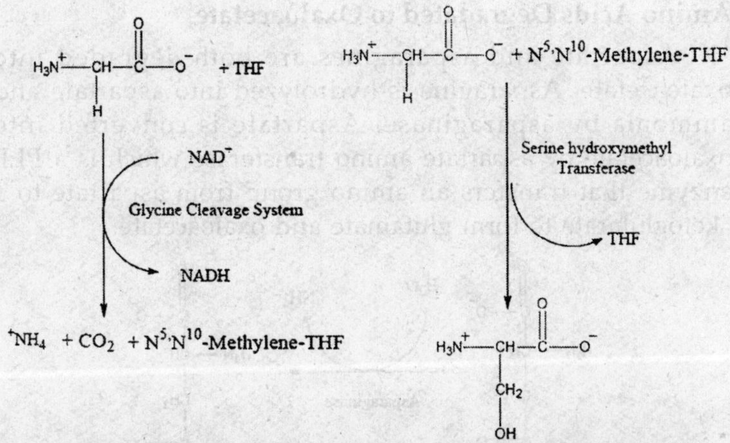


They are several pathways by which cysteine is converted into pyruvate. The three alkyl carbons of trypophan are converted into alanine which is then converted by alanine transaminase into pyruvate.

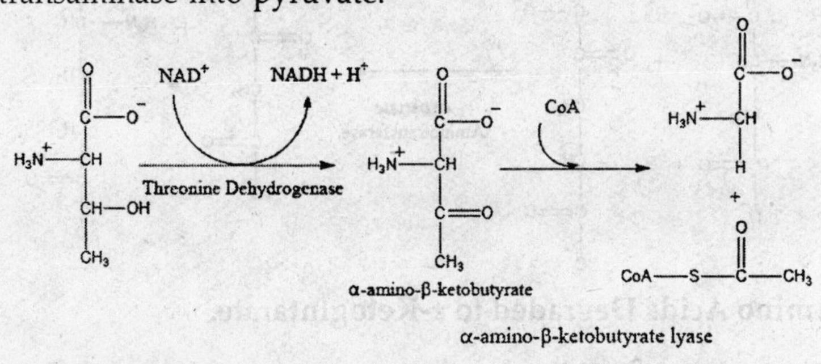


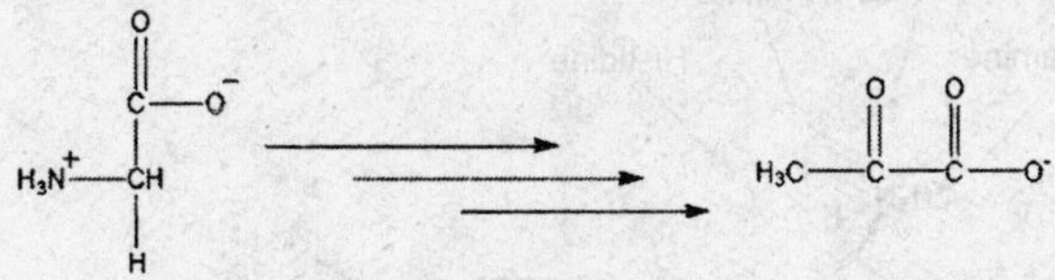

Threonine is both glucogenic and ketogenic. There are a couple of routes for the degradation of threonine. The major route is shown below. Threonine is converted into acetyl CoA and glycine. Glycine is then converted into serine by serine hydroxymethyl transferase and serine is then converted into pyruvate by serine dehydratase.

## Amino Acids Degradated to Oxaloacetate

Aspartate and asparagines are both degraded into oxaloacetate. Asparagine is hydrolyzed into aspartate and ammonia by asparaginase. Aspartate is converted into oxaloacetate by aspartate amino transferase which is a PLP enzyme that transfers an amino group from aspartate to ± ″ketoglutarate to form glutamate and oxaloacetate.

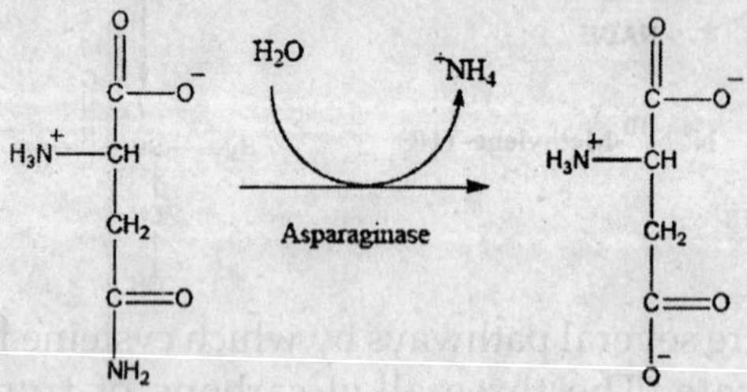


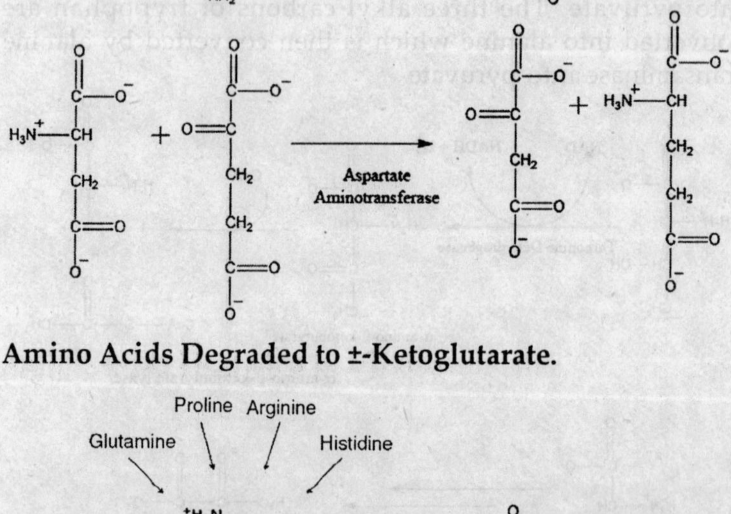


## Amino Acids Degraded to ±-Ketoglutarate.

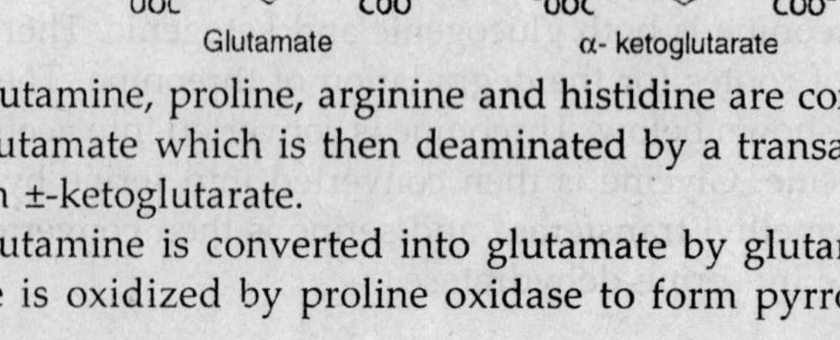


Glutamine, proline, arginine and histidine are converted into glutamate which is then deaminated by a transaminase to form ±-ketoglutarate.

Glutamine is converted into glutamate by glutaminase. Proline is oxidized by proline oxidase to form pyrroline 5-

carboxylate which spontaneously hydrolyzes to from glutamate-semialdehyde. From the urea cycle we know that arginase converts arginine into ornithine and urea. Ornithine ′-aminotransferase transfers the ′-amino group of ornithine to ±-ketoglutarate to form glutamate-semialdehyde and glutamate.

Glutamate 3-semialdehyde is oxidized to form glutamate by glutamate 5-semialdehyde dehydrogenase. Histidine is deaminated by histidine ammonia lyase which forms urocanate. Urocanate hydratase adds water to form 4-Imidazolone-5-propionate which is hydrolyzed by imidazalone propionase to form Mformiminoglutamate. Glutamate formiminotransferase transfers the formimino group to tetrahydrofolate to generate glutamate and $N^{5}$-formimino-THF.

**Amino Acids that are Broken Down in Succinyl-CoA.**

Methionine, valanine and isoleucine are broken down into propoinyl CoA. By studying oxidation of odd chain fatty acids we know that propionyl CoA is converted into D-methylmalonyl CoA by propionyl CoA carboxylase. D-methylmalonyl CoA is racemized into L-methylmalonyl CoA by methylmalonyl CoA racemase. Methylmalonyl mutase produces succinyl CoA.

The degradation of methionine requires 9 steps. One of which involves the synthesis of Sadenoylmethionine (SAM). The methyl group of SAM is highly reactive making it an important methylating reagent. SAM is a common methyl-group donor in the cell.

The first step is catalyzed by methionine adenosyl transferase which tranfers the adenosyl group of ATP to the Sulpher of methionine to form SAM. Sam methylase transfers the activated methyl group to an acceptor to form S-adenosylhomocysteine which is hydrolyzed by adenosylho-mocysteinase to form homocysteine. Cystathionine-synthase is a PLP dependent enzyme that catalyzes the condensation of a serine residue with homocysteine to form cystathionine. Cystathioniine-lyase cleaves cystathionine into cysteine and

±-ketobutyrate. ±"ketobutyrate is converted into propionyl CoA by ±-ketobutyrate dehydrogenase which catalyzes a reactioı that is analogous to pyruvate dehydrogenase and ±-ketoglutarate dehydrogenase.

## Branched Chain Amino Acids

The degradation of branched chain amino acids uses some of the enzymes we have already encountered in the citric acid cycle or ²-oxidation c.

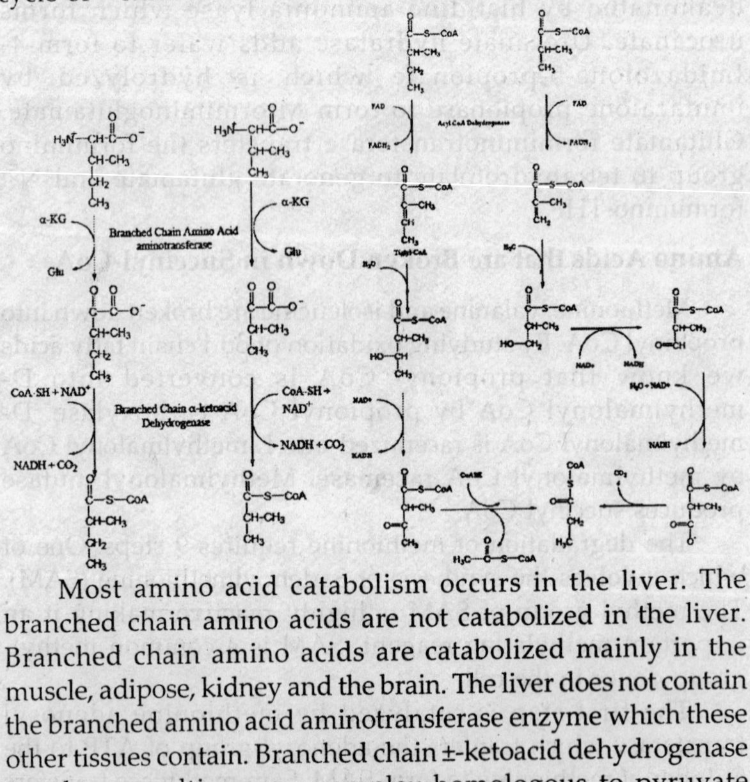


Most amino acid catabolism occurs in the liver. The branched chain amino acids are not catabolized in the liver. Branched chain amino acids are catabolized mainly in the muscle, adipose, kidney and the brain. The liver does not contain the branched amino acid aminotransferase enzyme which these other tissues contain. Branched chain ±-ketoacid dehydrogenase is a huge multienzyme complex homologous to pyruvate dehydrogenase and ±-ketoglutarate dehydrogenase.

This enzyme contains a thiamine pyrophosphate cofactor, a lipoamide cofactor, a FAD prosthetic group. The chemistry, mechanism and structure of these enzymes is very similar. Branched chain ±-ketoacid dehydrogenase is phosphorylated by a kinase which inactivates the enzyme in a similar manner that pyruvate dehydrogenase is phosphorylated and inactivated. The intake of dietary branched amino acids activates a phosphatase which activates this enzyme.

A genetic deficiency in the branched chain ±-ketoacid dehydrogenase enzyme is called maple syrup urine disease. The deficiency causes an excessive buildup of branched ±-ketoacids in the blood and the urine. The urine of these patients has the odor of maple syrup and hence the name of the disease. Maple syrup disease usually leads to mental retardation unless the patient is placed on diet that is low in valine, isoleucine and leucine early in life.

## Amino Acids that are Degraded into Acetyl CoA and Acetoacetate

There are only two amino acids that are purely ketogenic, lysine and leucine. Leucine catabolism is similar to the branched amino acids valine and isoleucine. First leucine is transaminatcdby branched amino acid aminotransferase to form ±-ketoisocaproate which is then oxidatively decarboxylated to form isovaleryl CoA by the branched chain ±-ketoacid dehydrogenase In the next step isovaleryl CoA is dehydrogenated to form β- methylcrotonyl CoA.

The enzyme that catalyzes this dehydrogenation is isovaleryl CoA dehydrogenase. β-methylcrotonyl CoA is then carboxylated by a biotin containing enzyme called methylcrotonyl CoA carboxylase to form β-methylglutaconyl CoA.

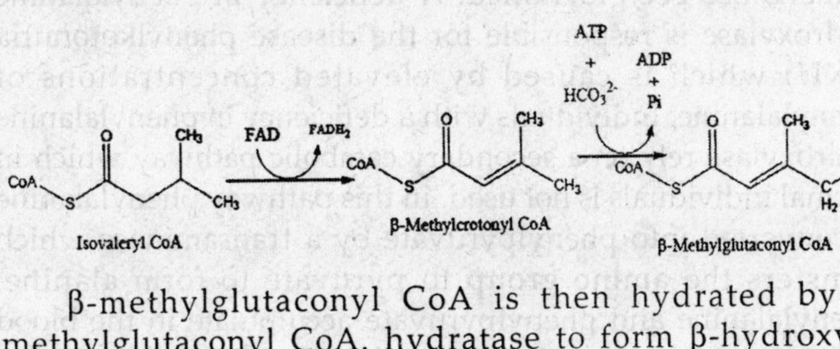


β-methylglutaconyl CoA is then hydrated by β-methylglutaconyl CoA. hydratase to form β-hydroxy-²-methylglutaryl CoA which is then cleaved into acetyl CoA and acetoacetate. The enzyme that catalyzes the last step is HMG-CoA lyase, a familiar enzyme from ketogenesis.

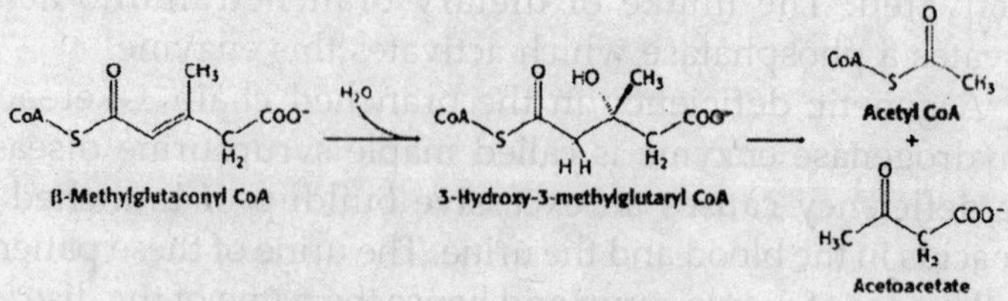


## Catabolism of Aromatic Amino Acids

- The degradation of aromatic amino acids requires molecular oxygen to break down the aromatic rings.
- The degradation of phenylalanine begins with a monooxygenase, phenylalanine hydroxylase which adds a hydroxyl group to phenylalanine to from tyrosine.
- Tyrosine aminotransferase deaminates tyrosine to form phydroxyphenylpyruvate.
- p-hydroxyphenylpyruvate dioxygenase catalyzes the formation of homogentisate.
- Homogentisate 1,2-dioxygenase catalyzes the formation of maleylacetoacetate.
- Maleylacetoacetate isomerase produces fumarylacetoacetate.
- Fumarylacetoacetase produces fumarate and acetoacetate.

Many genetic defects of phenylalanine catabolism in humans has been identified. A deficiency in phenylalanine hydroxylase is responsible for the disease phenylkotonuria (PKU) which is caused by elevated concentrations of phenylalanine. Individuals with a deficiency in phenylalanine hydroxylase rely on a secondary catabolic pathway which in normal individuals is not used. In this pathway phenylalanine is converted into phenylpyruvate by a transaminase which transfers the amino group to pyruvate to form alanine. Phenylalanine and phenylpyruvate accumulate in the blood and excreted in the urine hence the term phenylketonuria.. Some of the phenylpyruvate is decarboxylated to form phenylacetate. Some of the phenylpyruvate is reduced to phenyllactate. Phenyllactate gives the urine a distinctive odor used for diagnosis.

The high concentration of phenylalanine in the blood limits the transport of amino acids across the blood-brain barrier resulting in impairment of normal brain development, causing severe mental retardation. This can be avoided by early detection and rigid dietary control. The diet must only provide enough phenylalanine and tyrosine to meet the needs of protein synthesis. Alkapotonuria is another inheritable disease of phenylalanine catabolism. In this case the defective enzyme is homogentisate dioxygenase. This disease is less severe than PKU. Large amounts of homogentisate are excreted in the urine. The oxidation of homogentisate turns the urine black. Individuals who have this disease develop arthritis at an early age.

## REDUCTION OF ATMOSPHERIC NITROGEN

The atmosphere is a thin fragile shell that surrounds the Earth's surface. The troposphere, which is the air we breathe, extends only about 15 km above the Earth and it contains 90% of the atmosphere's mass. The stratosphere extends from 15 km to 50 km above the surface, and above it lie the mesosphere (or ionosphere) and thermosphere. Nitrogen and nitrous oxide are the chief atmospheric nitrogen compounds. Dinitrogen makes up about 78% of the atmosphere's volume, and $N_2O$ is present at about 315 ppbv (part per billion volume fraction). Atmospheric concentrations are usually stated in volume fractions, so they are insensitive to atmospheric pressure variations.

For example, the partial pressure of $N_2$ varies from 0.78 atm at sea level to 0.55 atm at 10,000 feet because the density of the atmosphere decreases with height. However, the volume fraction of $N_2$ is 0.78 at both 0 and 10,000 feet in elevation. Nitrous oxide is often called laughing gas, because of its anesthetic properties. It is a nonpolar molecule that is fat soluble, which makes it useful as a propellant for canned whipping cream.

Both $N_2$ and $N_2O$ are unreactive gases with long atmospheric lifetimes. The lifetime of an atmospheric species is roughly proportional to the total mass of the gas in the reservoir divided by the total flux of sources or sinks (at equilibrium sources = sinks, so either value can be used). It represents the average time that a molecule resides in the atmosphere, before leaving the reservoir. The long atmospheric lifetime (~$10^7$ yrs) for dinitrogen reflects the large amount in the atmosphere compared to its relatively slow rate of removal by nitrogen fixation, and its regeneration by denitrification.

Because of the low reactivity of its strong triple bond (941 kJ/mole), only the nitrogen fixing organisms and lightning or combustion (which promotes its reaction with oxygen to yield NO) remove it from the atmosphere. Gases with lifetimes longer than the mixing time (1-2 yrs for the troposphere) are evenly mixed throughout the atmosphere and their concentration is nearly constant around the Earth. Reactive gases, such as NO and $NO_2$, have short atmospheric lifetimes (< 1–8 days) and have highly variable concentrations.

Both $N_2$ and $N_2O$ are byproducts of denitrification. Nitrification produces $N_2O$ as a byproduct. The late Carl Sagan and his coworkers suggested that $N_2O$ could be used as a signature for life on other planets, since nitrous oxide is a biogenic gas with no significant geochemical sources. Nitrous oxide's long lifetime (120 - 150 yrs) can be attributed to its stable electronic configuration, which is isoelectronic to $CO_2$. Although decomposition of $N_2O$ to $N_2$ and $1/2O_2$ is thermodynamically favorable ($\Delta G^o$ = –103.59 kJ/ mole), the kinetics is unfavorable below 600°C. Unlike $CO_2$, nitrous oxide has no significant acid-base properties and this blocks aqueous ionic assimilation pathways that are available for cycling environmental carbon compounds. The only sink for nitrous oxide is photochemical destruction in the stratosphere, or its reaction with energetic oxygen atoms produced by photodissociation of stratospheric ozone.

*Atmospheric Odd Electron Nitrogen Oxides or NOx.* Atmospheric NO and $NO_2$ (called NOx) make up a very small fraction (< 50 ppbv) of the atmosphere, yet they are quite important to urban air quality. Both NO and $NO_2$ are free radicals whose Lewis dot structures are represented below.

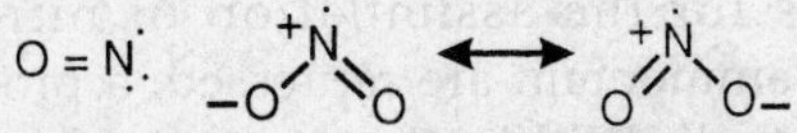

The low concentrations and short lifetime (~1-8 days) of NOx result from the rapid removal of $NO_2$ in aqueous aerosols (cloud processing and rain out) as $HNO_3$. That is why atmospheric NOx contributes significantly (~25-33%) to the acidity of rainfall in urban environments. The removal of NOx requires additional oxidation by species, such as hydrogen peroxide.

$$2NO_2 + H_2O_2 \rightarrow 2HNO_3$$

Hydrogen peroxide, which is consumed in this reaction, is an important oxidant in tropospheric aqueous aerosol (cloud) chemistry. One way that $H_2O_2$ can form is by the photolysis of tropospheric $O_3$.

$$O_3 + h\nu(\lambda < 320 \text{ nm}) \rightarrow O_2 + O$$

and then

$$O + H_2O \rightarrow 2\,OH \rightarrow H_2O_2$$

Before the NOx compounds are removed as $HNO_3$, they participate in a sequence of reactions that form ozone. Ozone is the most toxic component of urban smog. While it is beneficial in the stratosphere, its presence in polluted cities leads to respiratory distress in the sick and elderly.

## ASSIMILATION OF ATMOSPHERIC NITROGEN

The oxygenic photosynthetic bacteria, the cyanobacteria, are autotrophs that fix $CO_2$ through the reductive pentose phosphate cycle. They make an important contribution to the Earth's nitrogen cycle by incorporating nitrogen into the biosphere through assimilatory processes. Organisms of this group frequently have the capacity to assimilate as sources of nitrogen a number of different simple *N*-containing compounds including ammonium, nitrate, nitrite and urea. Many cyanobacteria can also assimilate atmospheric nitrogen, and still some strains have the capability to assimilate some amino acids, particularly arginine and glutamine. The assimilation of most of these compounds provides intracellular ammonium, which itself is a preferred nitrogen source. Thus, in the presence of ammonium, the genes encoding permeases and enzymes for the assimilation of nitrogen sources alternative to ammonium are repressed, a process known as 'nitrogen control'. Within the context of unravelling the mechanisms of nutrient incorporation in the biosphere, the aim of our research is to draw a complete picture of the mechanisms of nitrogen assimilation and its regulation in cyanobacteria.

## Mechanisms of Assimilation of Combined Nitrogen

The incorporation into the cell of nitrogen-containing compounds, which are frequently found at low concentrations in the environment, takes place through permeases that are located in the cytoplasmic membrane. Multicomponent ABC (ATP-binding cassette)-type uptake transporters have been shown to be involved in the uptake of nitrate and nitrite or urea in a number of cyanobacteria.

ABC-type permeases are also required for the transport of arginine and glutamine. These permeases use ATP to drive an active, concentrative transport of their substrates. On the other hand, a secondary transporter of the major facilitator superfamily has been identified as the nitrate–nitrite transporter in some marine cyanobacteria. The transport of ammonium is also mediated by secondary permeases, in this case of the Amt family.

Intracellular nitrate is sequentially reduced to nitrite and ammonium by nitrate reductase and nitrite reductase, which are the products of the *narB* and *nir* genes respectively. Cyanobacterial nitrite reductase is homologous with ferredoxin-dependent higher-plant nitrite reductase and contains a [4Fe-4S] cluster and sirohaem as prosthetic groups. Electrons from reduced ferredoxin are transferred to the iron–sulphur cluster and then to sirohaem, where nitrite is reduced to ammonium. Cyanobacterial nitrate reductase is homologous with Mo-containing bacterial oxidoreductases but is unique in that it uses ferredoxin as an electron donor, forming tight 1 : 1 complexes. The Mo cofactor is of the Mo-*bis*-molybdopterin guanine dinucleotide type and the enzyme also contains a [4Fe-4S] cluster. In this enzyme system, electrons flow from reduced ferredoxin to the iron–sulphur cluster and then to the Mo cofactor, where nitrate is reduced to nitrite.

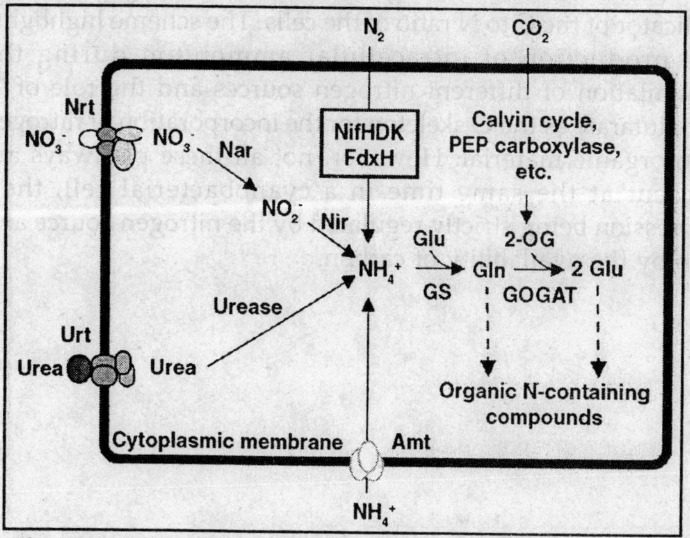


**Fig.** Main Nitrogen Assimilation Pathways in Cyanobacteria

The *narB* and *nir* genes are clustered together with the nitrate/nitrite permease-encoding genes in numerous cyanobacteria forming an operonwith the structure *nir*-permease genes-*narB*. The high conservation of this gene arrangement, in which the expression level is higher for the upstream than for the downstream genes in the operon, suggests that it ensures the production of a balanced amount of the different proteins of the pathway.

Regarding organic sources of nitrogen used by cyanobacteria, urea is degraded to ammonium and $CO_2$ by a standard bacterial $Ni^{2+}$-dependent urease, whereas arginine is catabolized by an unusual pathway that combines the urea cycle and the arginase pathway rendering ammonium and glutamate as final products.

Whatever the nitrogen source used for growth, intracellular ammonium is incorporated into carbon skeletons through the glutamine synthetase–glutamate synthase pathway. In cyanobacteria, which lack 2-oxoglutarate dehydrogenase, the main metabolic role of 2-oxoglutarate is the incorporation of nitrogen.

This metabolic arrangement makes 2-oxoglutarate an indicator of the C to N ratio of the cells. The scheme highlights the production of intracellular ammonium during the assimilation of different nitrogen sources and the role of 2-oxoglutarate as the C-skeleton for the incorporation of nitrogen into organic material. However, not all these pathways are present at the same time in a cyanobacterial cell, their expression being strictly regulated by the nitrogen source and also by the availability of carbon.

# Chapter 4

# Lipid Metabolism

## FATTY ACID SYNTHESIS

One might predict that the pathway for the synthesis of fatty acids would be the reversal of the oxidation pathway. However, this would not allow distinct regulation of the two pathways to occur even given the fact that the pathways are separated within different cellular compartments.

The pathway for fatty acid synthesis occurs in the cytoplasm, whereas, oxidation occurs in the mitochondria. The other major difference is the use of nucleotide co-factors. Oxidation of fats involves the reduction of $FADH^+$ and $NAD^+$. Synthesis of fats involves the oxidation of NADPH. However, the essential chemistry of the two processes are reversals of each other. Both oxidation and synthesis of fats utilize an activated two carbon intermediate, acetyl-CoA. However, the acetyl-CoA in fat synthesis exists temporarily bound to the enzyme complex as malonyl-CoA.

The synthesis of malonyl-CoA is the first committed step of fatty acid synthesis and the enzyme that catalyzes this reaction, acetyl-CoA carboxylase (ACC), is the major site of regulation of fatty acid synthesis. Like other enzymes that transfer $CO_2$ to substrates, ACC requires a biotin co-factor.

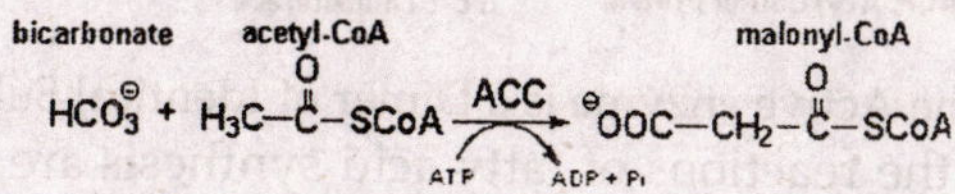


The rate of fatty acid synthesis is controlled by the equilibrium between monomeric ACC and polymeric ACC. The activity of ACC requires polymerization. This conformational change is enhanced by citrate and inhibited by long-chain fatty acids. ACC is also controlled through hormone mediated phosphorylation

The acetyl groups that are the products of fatty acid oxidation are linked to CoASH. As you should recall, CoA contains a phosphopantetheine group coupled to AMP. The carrier of acetyl groups (and elongating acyl groups) during fatty acid synthesis is also a phosphopantetheine prosthetic group, however, it is attached a serine hydroxyl in the synthetic enzyme complex. The carrier portion of the synthetic complex is called acyl carrier protein, ACP. This is somewhat of a misnomer in eukaryotic fatty acid synthesis since the ACP portion of the synthetic complex is simply one of many domains of a single polypeptide. The acetyl-CoA and malonyl-CoA are transferred to ACP by the action of acetyl-CoA transacylase and malonyl-CoA transacylase, respectively.

The synthesis of fatty acids from acetyl-CoA and malonyl-CoA is carried out by fatty acid synthase, FAS.

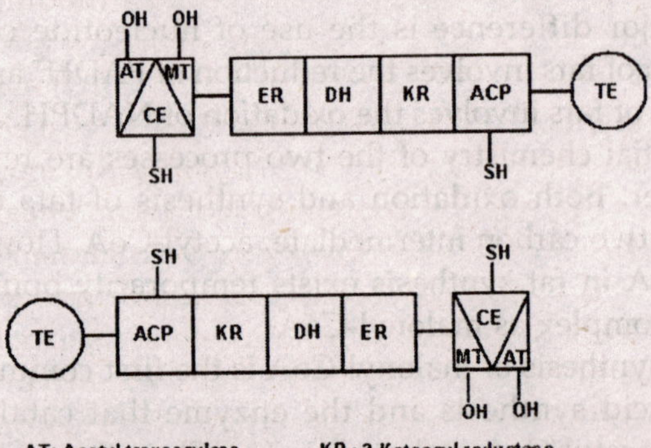


AT: Acetyl transacylase
MT: Malonyl transacylase
CE : Condensing enzyme
ACP: Acyl Carrier Protein
KR : 3-Ketoacyl reductase
DH : 3-Hydroxyacyl dehydratase
ER : Enoyl reductase
TE : Thioesterase

**Fig.** The Active enzyme is a Dimer of Identical Subunits.

All of the reactions of fatty acid synthesis are carried out by the multiple enzymatic activities of FAS. Like fat oxidation, fat synthesis involves 4 enzymatic activities. These are, b-keto-ACP synthase (condensing enzyme), b-keto-ACP reductase, 3-

OH acyl-ACP dehydratase and enoyl-ACP reductase(give a look to the FAS reaction mechanism). The two reduction reactions require NADPH oxidation to $NADP^+$.

The primary fatty acid synthesized by FAS is palmitate. Palmitate is then released from the enzyme by ***thioesterase*** reaction and can then undergo separate elongation and/or unsaturation to yield other fatty acid molecules.

## ORIGIN OF CYTOPLASMIC ACETYL

Acetyl-CoA is generated in the mitochondria primarily from two sources, the pyruvate dehydrogenase (PDH) reaction and fatty acid oxidation. In order for these acetyl units to be utilized for fatty acid synthesis they must be present in the cytoplasm. The shift from fatty acid oxidation and glycolytic oxidation occurs when the need for energy diminishes. This results in reduced oxidation of acetyl-CoA in the TCA cycle and the oxidative phosphorylation pathway. Under these conditions the mitochondrial acetyl units can be stored as fat for future energy demands.

Acetyl-CoA enters the cytoplasm in the form of citrate via the tricarboxylate transport system as diagrammed. In the cytoplasm, citrate is converted to oxaloacetate and acetyl-CoA by the ATP driven *ATP-citrate lyase* reaction. This reaction is essentially the reverse of that catalyzed by the TCA enzyme *citrate synthase* except it requires the energy of ATP hydrolysis to drive it forward. The resultant oxaloacetate is converted to malate by *malate dehydrogenase* (MDH).The malate produced by this pathway can undergo oxidative decarboxylation by *malic enzyme.* The co-enzyme for this reaction is $NADP^+$ generating NADPH. The advantage of this series of reactions for converting mitochondrial acetyl-CoA into cytoplasmic acetyl-CoA is that the NADPH produced by the *malic enzyme* reaction can be a major source of reducing co-factor for the *fatty acid synthase* activities.

## REGULATION OF FATTY ACID METABOLISM

One must consider the global organismal energy requirements in order to effectively understand how the synthesis and degradation of fats (and also carbohydrates) needs to be exquisitely regulated. The blood is the carrier of triacylglycerols in the form of VLDLs and chylomicrons, fatty acids bound to albumin, amino acids, lactate, ketone bodies and glucose. The pancreas is the primary organ involved in sensing the organisms dietary and energetic states via glucose concentrations in the blood. In response to low blood glucose, glucagon is secreted, whereas, in response to elevated blood glucose insulin is secreted.

The regulation of fat metabolism occurs via two distinct mechanisms. One is short term regulation which is regulation effected by events such as substrate availability, allosteric effectors and/or enzyme modification. ACC is the rate limiting (committed) step in fatty acid synthesis. This enzyme is activated by citrate and inhibited by palmitoyl-CoA and other long chain fatty acyl-CoAs. ACC activity also is affected by phosphorylation. Glucagon stimulated increases in PKA activity result in phosphorylation of certain serine residues in ACC leading to decreased activity of the enzyme. On the other hand insulin leads to phosphorylation of ACC at sites distinct from glucagon that result in increased ACC activity. These forms of regulation are all defined as short term regulation.

Control of a given pathways' regulatory enzymes can also occur by alteration of enzyme synthesis and turn-over rates. These changes are long term regulatory effects. Insulin stimulates ACC and FAS synthesis, whereas, starvation leads to decreased synthesis of these enzymes. Adipose tissue lipoprotein lipase levels also are increased by insulin and decreased by starvation. However, in contrast to the effects of insulin and starvation on adipose tissue, their effects on heart lipoprotein lipase are just the inverse. This allows the heart to absorb any available fatty acids in the blood in order to oxidize them for energy production. Starvation also leads to increases in the levels of fatty acid oxidation enzymes in the heart as well as a decrease in FAS and related enzymes of synthesis.

Adipose tissue contains hormone-sensitive lipase, that is activated by PKA-dependent phosphorylation leading to increased fatty acid release to the blood. This leads to increased fatty acid oxidation in other tissues such as muscle and liver. In the liver the net result (due to increased acetyl-CoA levels) is the production of ketone bodies. This would occur under conditions where insufficient carbohydrate stores and gluconeogenic precursors were available in liver for increased glucose production. The increased fatty acid availability in response to glucagon or epinephrine is assured of being completely oxidized since PKA also phosphorylates (and as a result inhibits) ACC, thus inhibiting fatty acid synthesis.

Insulin, on the other hand, has the opposite effect to glucagon and epinephrine leading to increased glycogen and triacylglyceride synthesis. One of the many effects of insulin is to lower cAMP levels which leads to increased dephosphorylation through the enhanced activity of protein phosphatases such as PP-1. With respect to fatty acid metabolism this yields dephosphorylated and inactive hormone sensitive lipase. Insulin also stimulates certain phosphorylation events. This occurs through activation of several cAMP-independent kinases. Insulin stimulated phosphorylation of ACC activates this enzyme.

Regulation of fat metabolism also occurs through malonyl-CoA induced inhibition of carnitine acyltransferase I. This functions to prevent the newly synthesized fatty acids from entering the mitochondria and being oxidized.

## ELONGATION AND DESATURATION

The fatty acid product released from FAS is palmitate (via the action of palmitoyl thioesterase) which is a 16:0 fatty acid, i.e. 16 carbons and no sites of unsaturation. Elongation and unsaturation of fatty acids occurs in both the mitochondria and endoplasmic reticulum (microsomal membranes). The predominant site of these processes is in the ER membranes. Elongation involves condensation of acyl-CoA groups with malonyl-CoA. The resultant product is two carbons longer (CO2 is released from malonyl-CoA as in the FAS reaction) which undergoes reduction, dehydration and reduction yielding a saturated fatty acid. The reduction reactions of elongation require NADPH as co-factor just as for the similar reactions catalyzed by FAS. Mitochondrial elongation involves acetyl-CoA units and is a reversal of oxidation except that the final reduction utilizes NADPH instead of FADH2 as co-factor.

Desaturation occurs in the ER membranes as well and in mammalian cells involves 4 broad specificity fatty acyl-CoA desaturases (non-heme iron containing enzymes). These enzymes introduce unsaturation at C4, C5, C6 or C9.

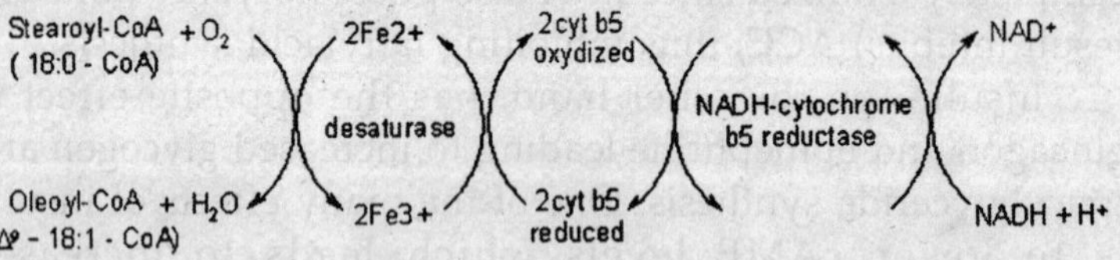


The electrons from the reduced fatty acid (saturated) and from the desaturase are transferred to molecular oxygen to form water and oxydized fatty acid (unsaturated). NADH is the donor of two electrons which keep the desaturase in the reduced form through the action of the enzyme NADH-cytochrome b5 reductase. These electrons are un-coupled from mitochondrial oxidative-phosphorylation and, therefore, do not yield ATP. Since these enzymes cannot introduce sites of unsaturation beyond C9 they cannot synthesize either linoleate (18:2 9, 12) or linolenate (18:3 9, 12, 15).

These fatty acids must be acquired from the diet and are, therefore, referred to as essential fatty acids. Linoleic is especially important in that it required for the synthesis of arachidonic acid. As we shall encounter later, arachindonate is a precursor for the eicosanoids (the prostaglandins and thromboxanes). It is this role of fatty acids in eicosanoid synthesis that leads to poor growth, wound healing and dermatitis in persons on fat free diets. Also, linoleic acid is a constituent of epidermal cell sphingolipids that function as the skins water permeability barrier.

## SYNTHESIS OF TRIGLYCERIDES

Fatty acids are stored for future use as triacylglycerols in all cells, but primarily in adipocytes of adipose tissue.

Triacylglycerols constitute molecules of glycerol to which three fatty acids have been esterified. The fatty acids present in triacylglycerols are predominantly saturated. The major building block for the synthesis of triacylglycerols, in tissues other than adipose tissue,is glycerol-3-phosphate. Adipocytes lack glycerol kinase, therefore, dihydroxyacetone phosphate (DHAP), produced during glycolysis, is the precursor for triacylglycerol synthesis in adipose tissue. This means that adipoctes must have glucose to oxidize in order to store fatty acids in the form of triacylglycerols. DHAP can also serve as a backbone precursor for triacylglycerol synthesis in tissues other than adipose, but does so to a much lesser extent than glycerol.

Phoshatidic Acid Synthesis

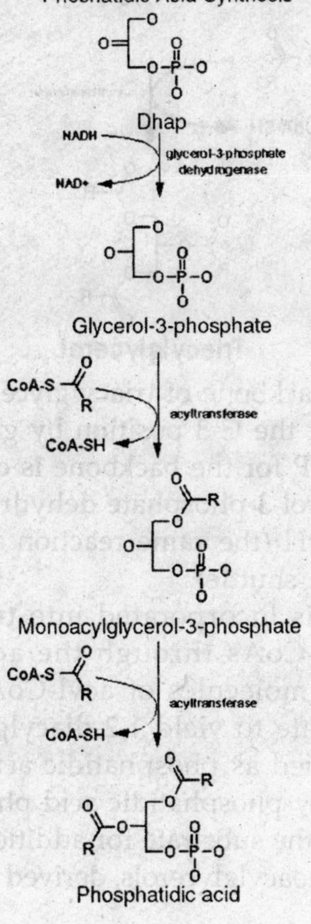

### TRIACYLGLYCEROL SYNTHESIS

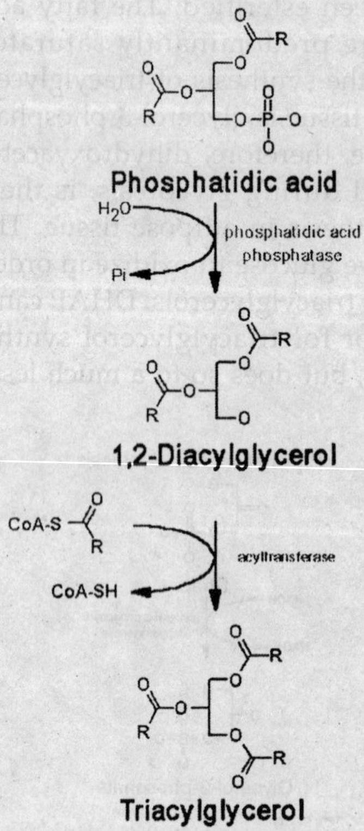


The glycerol backbone of triacylglycerols is activated by phosphorylation at the C-3 position by glycerol kinase. The utilization of DHAP for the backbone is carried out through the action of glycerol-3-phosphate dehydrogenase, a reaction that requires NADH (the same reaction as that used in the glycerol-phosphate shuttle).

The fatty acids incorporated into triacylglycerols are activated to acyl-CoAs through the action of acyl-CoA synthetases. Two molecules of acyl-CoA are esterified to glycerol-3-phosphate to yield 1,2-diacylglycerol phosphate (commonly identified as phosphatidic acid). The phosphate is then removed, by phosphatidic acid phosphatase, to yield 1,2-diacylglycerol, the substrate for addition of the third fatty acid. Intestinal monoacylglycerols, derived from the hydrolysis of dietary fats, can also serve as substrates for the synthesis of 1,2-diacylglycerols.

## PHOSPHOLIPID STRUCTURES

Phospholipids are synthesized by the addition of a basic group (predominantly a nitrogenous base) to phosphatidic acid or 1,2-diacylglycerol. Most phospholipids have a saturated fatty acid on C-1 and an unsaturated fatty acid on C-2 of the glycerol backbone.

## PHOSPHOLIPID SYNTHESIS

Phospholipids can be synthesized by two mechanisms. One utilizes a CDP-activated polar head group for attachment to the phosphate of phosphatidic acid. The other utilizes CDP-activated 1,2-diacylglycerol and an inactivated polar head group. *PC:* This class of phospholipids is also called the lecithins. At physiological pH, phosphatidylcholines are neutral zwitterions. They contain primarily palmitic or stearic acid at carbon 1 and primarily oleic, linoleic or linolenic acid at carbon 2. The lecithin dipalmitoyllecithin is a component of lung or pulmonary surfactant. It contains palmitate at both carbon 1 and 2 of glycerol and is the major (80%) phospholipid found in the extracellular lipid layer lining the pulmonary alveoli. Choline is activated first by phosphorylation and then by coupling to CDP prior to attachment to phosphatidic acid.

An other pathway to PC synthesis, involves the conversion of either PS or PE to PC. The conversion of PS to PC first requires decarboxylation of PS to yield PE; this then undergoes a series of three methylation reactions utilizing S-adenosylmethionine (SAM) as methyl group donor. *PE:* These molecules are neutral zwitterions at physiological pH. They contain primarily palmitic or stearic acid on carbon 1 and a long chain unsaturated fatty acid (e.g. 18:2, 20:4 and 22:6) on carbon 2. Synthesis of PE can occur by two pathways. The first requires that ethanolamine be activated by phosphorylation and then by coupling to CDP. The ethanolamine is then transferred from CDP-ethanolamine to phosphatidic acid to yield PE. The second involves the decarboxylation of PS.

*PS:* Phosphatidylserines will carry a net charge of -1 at physiological pH and are composed of fatty acids similar to the phosphatidylethanolamines.The pathway for PS synthesis involves an exchange reaction of serine for ethanolamine in PE and choline in PC. These exchange occur when PE and PC are in the lipid bilayer of the a membrane.

**PC,PE AND PS BIOSYNTHESIS**

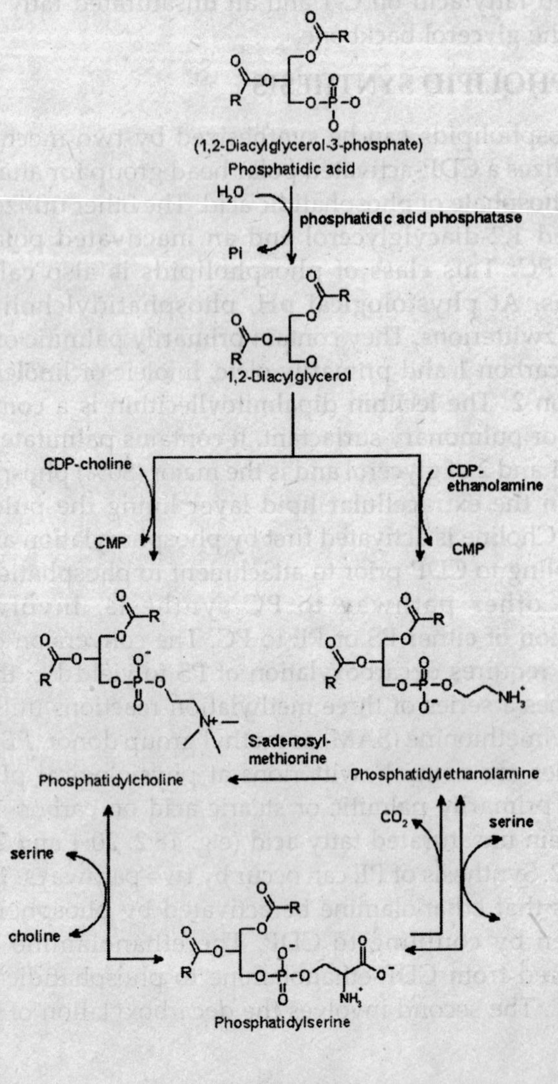

PI,PG AND CARDIOLIPIN BIOSYNTHESIS

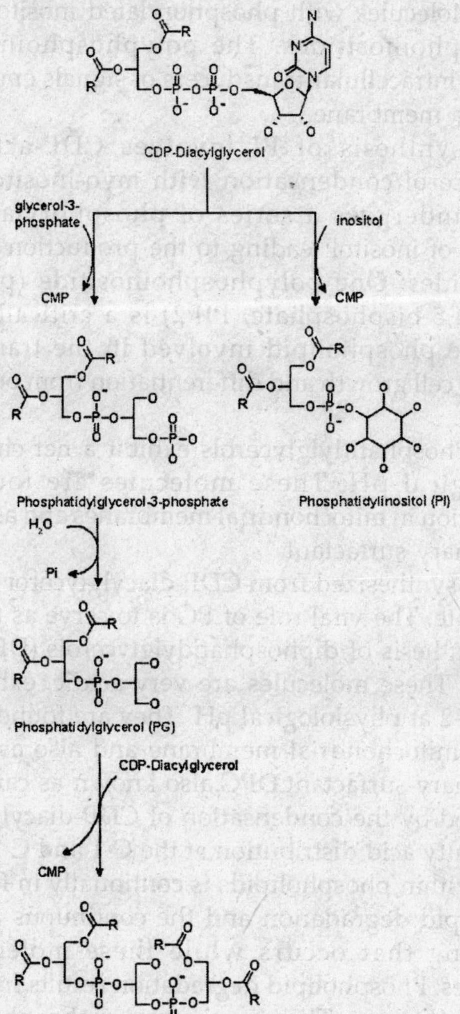


*PI:* These molecules contain almost exclusively stearic acid at carbon 1 and arachidonic acid at carbon 2. Phosphatidylinositols composed exclusively of non-phosphorylated inositol exhibit a net charge of -1 at physiological pH. These molecules exist in membranes with various levels of phosphate esterified to the hydroxyls of the inositol. Molecules with phosphorylated inositol are termed polyphosphoinositides. The polyphosphoinositides are important intracellular transducers of signals emanating from the plasma membrane.

The synthesis of PI involves CDP-activated 1,2-diacylglycerol condensation with myo-inositol. PI subsequently undergoes a series of phosphorylations of the hydroxyls of inositol leading to the production of polyphosphoinositides. One polyphosphoinositide (phosphatidylinositol 4,5-bisphosphate, PIP2) is a critically important membrane phospholipid involved in the transmission of signals for cell growth and differentiation from outside the cell to inside.

*PG:* Phosphatidylglycerols exhibit a net charge of -1 at physiological pH. These molecules are found in high concentration in mitochondrial membranes and as components of pulmonary surfactant.

PG is synthesized from CDP-diacylglycerol and glycerol-3-phosphate. The vital role of PG is to serve as the precursor for the synthesis of diphosphatidylglycerols (DPGs).

*DPG:* These molecules are very acidic, exhibiting a net charge of -2 at physiological pH. They are found primarily in the inner mitochondrial membrane and also as components of pulmonary surfactant.DPG,also known as cardiolipins are synthesized by the condensation of CDP-diacylglycerol with PG. The fatty acid distribution at the C-1 and C-2 positions of glycerol within phospholipids is continually in flux, owing to phospholipid degradation and the continuous phospholipid remodeling that occurs while these molecules are in membranes. Phospholipid degradation results from the action of phospholipases. There are various phospholipases that exhibit substrate specificities for different positions in phospholipids.

In many cases the acyl group which was initially transferred to glycerol, by the action of the acyl transferases, is not the same acyl group present in the phospholipid when it resides within a membrane. The remodeling of acyl groups in phospholipids is the result of the action of phospholipase $A_1$ and phospholipase $A_2$.

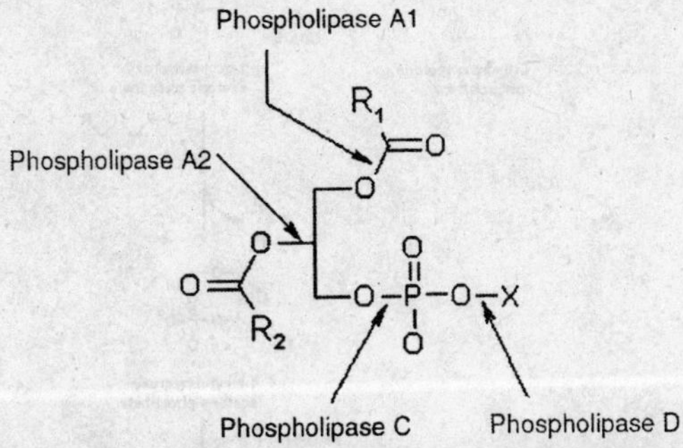


Sites of action of the phospholipases $A_1$, $A_2$, C and D.

The products of these phospholipases are called lysophospholipids and can be substrates for acyl transferases utilizing different acyl-CoA groups. Lysophospholipids can also accept acyl groups from other phospholipids in an exchange reaction catalyzed by lysolecithin:lecithin acyltransferase (LLAT).

Phospholipase A2 is also an important enzyme, whose activity is responsible for the release of arachidonic acid from the C-2 position of membrane phospholipids. The released arachidonate is then a substrate for the synthesis of the prostaglandins and leukotrienes.

## PLASMALOGENS

Plasmalogens are glycerol ether phospholipids. They are of two types, alkyl ether and alkenyl ether. Dihydroxyacetone phosphate serves as the glycerol precursor for the synthesis of glycerol ether phospholipids. Three major classes of plasmalogens have been identified: choline, ethanolamine and serine plasmalogens. Ethanolamine plasmalogen is prevalent in myelin. Choline plasmalogen is abundant in cardiac tissue.

One particular choline plasmalogen (1-alkyl, 2-acetyl phosphatidylcholine) has been identified as an extremely powerful biological mediator, capable of inducing cellular responses at concentrations as low as 10-11 M. This molecule is called platelet activating factor, PAF.

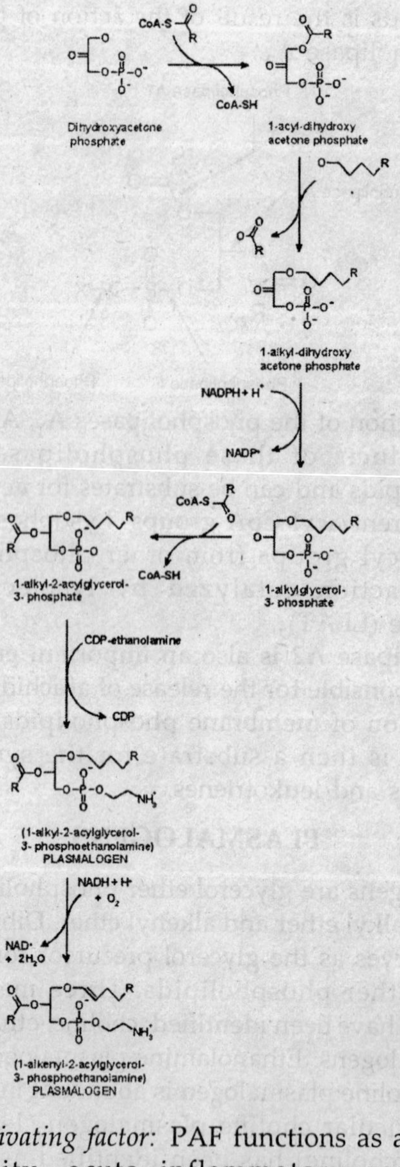


*Platelet activating factor:* PAF functions as a mediator of hypersensitivity, acute inflammatory reactions and anaphylactic shock. PAF is synthesized in response to the formation of antigen-IgE complexes on the surfaces of basophils, neutrophils, eosinophils, macrophages and monocytes. The synthesis and release of PAF from cells leads to platelet aggregation and the release of serotonin from platelets. PAF also produces responses in liver, heart, smooth muscle, and uterine and lung tissues.

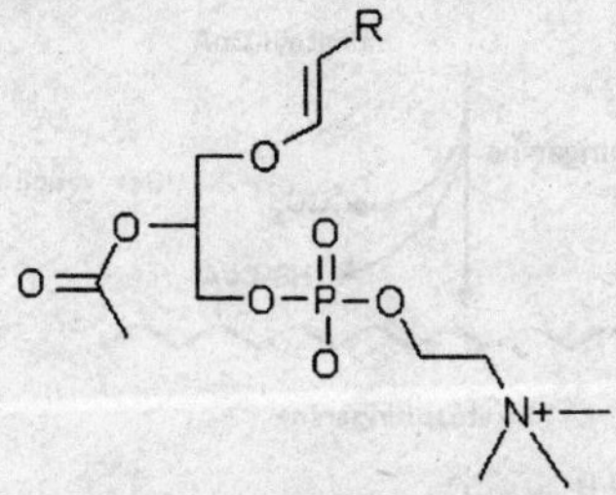

## METABOLISM OF THE SPHINGOLIPIDS

The sphingolipids, like the phospholipids, are composed of a polar head group and two nonpolar tails. The core of sphingolipids is the long-chain amino alcohol, sphingosine.

The sphingolipids include sphingomyelins and glycosphingolipids (cerebrosides, sulfatides, globosides and gangliosides). Sphingomyelins are the only sphingolipid that are also phospholipids. Sphingolipids are a component of all membranes but are particularly abundant in the myelin sheath.Ceramide ,the molecule present both in sphingomyelins and glycosphingolipids derives from the condensation of serine and palmitoyl-CoA the following the pathway:

Sphingomyelins are sphingolipids that are also phospholipids. Sphingomyelins are important structural lipid components of nerve cell membranes. The predominant sphingomyelins contain palmitic or stearic acid N-acylated at carbon 2 of sphingosine.

The sphingomyelins are synthesized by the transfer of phosphorylcholine from phosphatidylcholine to a ceramide in a reaction catalyzed by sphingomyelin synthase. Defects in the enzyme acid sphingomyelinase result in the lysosomal storage disease known as Niemann-Pick disease. There are at least 4 related disorders identified as Niemann-Pick disease Type A and B (both of which result from defects in acid sphingomyelinase ), Type C1 and a related C2 and Type D.

Types C1, C2 and D do not result from defects in acid sphingomyelinase.

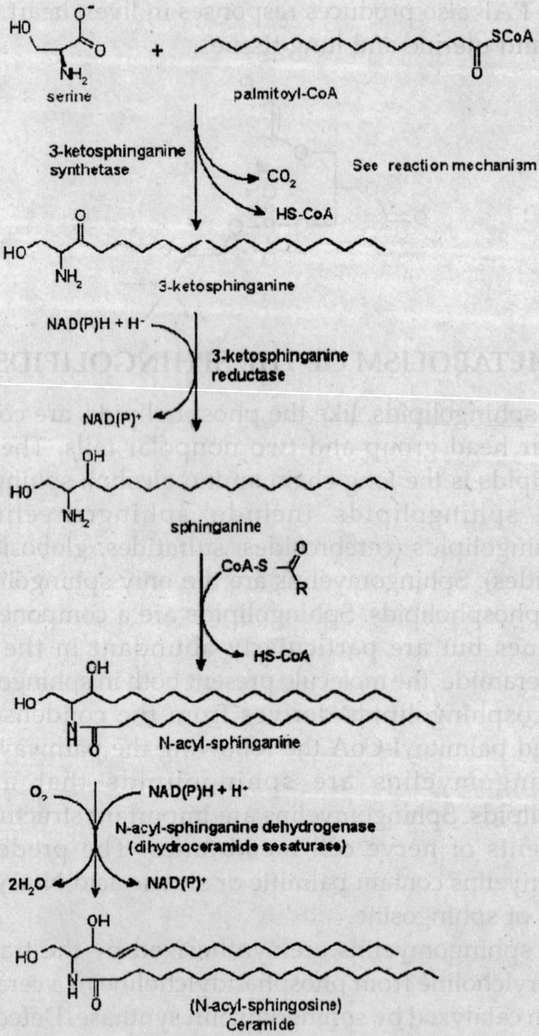


Glycosphingolipids , or glycolipids, are composed of a ceramide backbone with a wide variety of carbohydrate groups (mono- or oligosaccharides) attached to carbon 1 of sphingosine. The four principal classes of glycosphingolipids are the cerebrosides, sulfatides, globosides and gangliosides.

Cerebrosides have a single sugar group linked to ceramide. The most common of these is galactose (galactocerebrosides), with a minor level of glucose (glucocerebrosides).

Galactocerebrosides are found predominantly in neuronal cell membranes. By contrast glucocerebrosides are not normally found in membranes, especially neuronal membranes; instead, they represent intermediates in the synthesis or degradation of more complex glycosphingolipids.

Galactocerebrosides are synthesized from ceramide and UDP-galactose. Excess accumulation of glucocerebrosides is observed in Gaucher's disease.

*Sulfatides:* The sulfuric acid esters of galactocerebrosides are the sulfatides. Sulfatides are synthesized from galactocerebrosides and activated sulfate, 3'-phosphoadenosine 5'-phosphosulfate (PAPS). Excess accumulation of sulfatides is observed in sulfatide lipidosis (metachromatic leukodystrophy).

*Globosides:* Globosides represent cerebrosides that contain additional carbohydrates, predominantly galactose, glucose or GalNAc. Lactosyl ceramide is a globoside found in erythrocyte plasma membranes. Globotriaosylceramide (also called ceramide trihexoside) contains glucose and two moles of galactose and accumulates, primarily in the kidneys, of patients suffering from Fabry's disease.

*Gangliosides:* Gangliosides are very similar to globosides except that they also contain NANA in varying amounts. The specific names for gangliosides are a key to their structure. The letter G refers to "ganglioside" and the subscripts M, D, T and Q indicate that the molecule contains mono-, di-, tri and quatra(tetra)-sialic acid. The numerical subscripts 1, 2 and 3 refer to the carbohydrate sequence that is attached to ceramide;1 stands for

Gal-GalNAc-Gal-Glc-ceramide, 2 for GalNAc-Gal-Glc-ceramide and 3 for Gal-Glc-ceramide. Deficiencies in lysosomal enzymes, which normally are responsible for the degradation of the carbohydrate portions of various gangliosides, underlie the symptoms observed in rare autosomally inherited diseases termed lipid storage diseases , many of which are listed below.

GANGLIOSIDE BIOSYNTHESIS

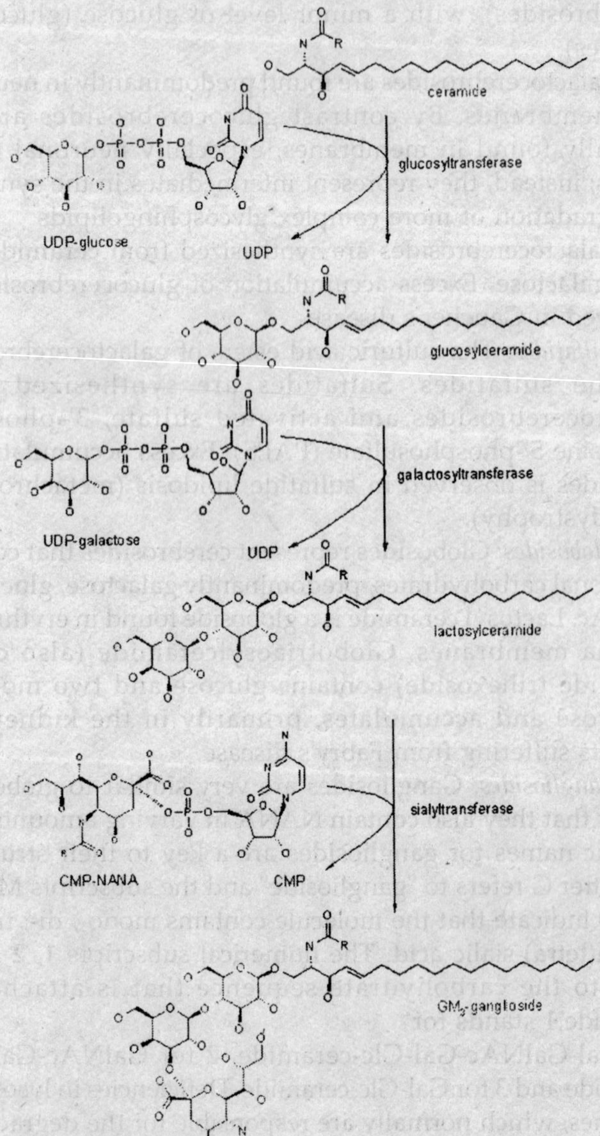

## CLINICAL SIGNIFICANCES OF SPHINGOLIPIDS

One of the most clinically important classes of sphingolipids are those that confer antigenic determinants on the surfaces of cells, particularly the erythrocytes. The ABO blood group antigens are the carbohydrate moieties of glycolipids on the surface of cells as well as the carbohydrate portion of serum glycoproteins. When present on the surface of cells the ABO carbohydrates are linked to sphingolipid and are therefore of the glycosphingolipid class. When the ABO carbohydrates are associated with protein in the form of glycoproteins they are found in the serum and are referred to as the seThe eicosanoids consist of the

creted forms. Some individuals produce the glycoprotein forms of the ABO antigens while others do not. This property distinguishes secretors from non-secretors, a property that has forensic importance such as in cases of rape.

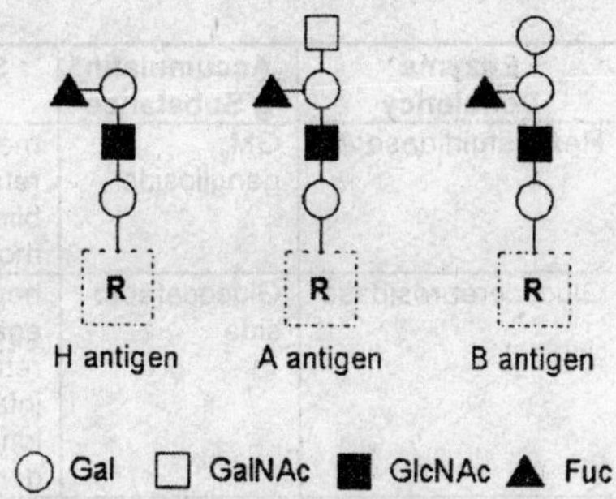


**Fig.** Strucures of the ABO Blood Group Carbohydrates:

A significant cause of death in premature infants and, on occasion, in full term infants is respiratory distress syndrome (RDS) or hyaline membrane disease. This condition is caused by an insufficient amount of pulmonary surfactant. Under normal conditions the surfactant is synthesized by type II endothelial cells and is secreted into the alveolar spaces to prevent atelectasis following expiration during breathing. Surfactant is comprised primarily of dipalmitoyllecithin; additional lipid components include phosphatidylglycerol and phosphatidylinositol along with proteins of 18 and 36 kDa (termed surfactant proteins). During the third trimester the fetal lung synthesizes primarily sphingomyelin, and type II endothelial cells convert the majority of their stored glycogen to fatty acids and then to dipalmitoyllecithin. Fetal lung maturity can be determined by measuring the ratio of lecithin to sphingomyelin (L/S ratio) in the amniotic fluid. An L/S ratio less than 2.0 indicates a potential risk of RDS. The risk is nearly 75-80% when the L/S ratio is 1.5. The carbohydrate portion of the ganglioside, GM 1, present on the surface of intestinal epithelial cells, is the site of attachment of cholera toxin, the protein secreted by Vibrio cholerae. These are just a few examples of how sphingolipids and glycosphingolipids are involved in various recognition functions at the surface of cells. As with the complex glycoproteins, an understanding of all of the functions of the glycolipids is far from complete.

## DISORDERS ASSOCIATED WITH ABNORMAL SPHINGOLIPID METABOLISM

| Disorder | Enzyme Deficiency | Accumulating Substance | Symptoms |
|---|---|---|---|
| Tay-Sachs disease | Hexosaminidase A | $GM_2$ ganglioside | mental retardation, blindness, early mortality |
| Gaucher's disease | Glucocerebrosidase | Glucocerebroside | hepatosplenomegaly, mental retardation in infantile form, long bone degeneration |
| Fabry's disease | α-Galactosidase A | Globotriaosylceramide; also called ceramide trihexoside (CTH) | kidney failure, skin rashes |
| Niemann-Pick disease, more info below<br>Types A and B<br>Type C1<br>Type C2<br>Type D | Sphingomyelinase<br>see info below<br>see info below | Sphingomyelin<br>LDL-derived cholesterol<br>LDL-derived cholesterol | all types lead to mental retardation, hepatosplenomegaly, early fatality potential |
| Krabbe's disease; globoid leukodystrophy | Galactocerebrosidase | Galactocerebroside | mental retardation, myelin deficiency |

| Disorder | Enzyme Deficiency | Accumulating Substance | Symptoms |
|---|---|---|---|
| Sandhoff-Jatzkewitz disease | Hexosaminidase A and B | Globoside, $GM_2$ ganglioside | same symptoms as Tay-Sachs,progresses more rapidly |
| $GM_1$ gangliosidosis | $GM_1$ ganglioside: β - galactosidase | $GM_1$ ganglioside | mental retardation, skeletal abnormalities, hepatomegaly |
| Sulfatide lipodosis; metachromatic leukodystrophy | Arylsulfatase A | Sulfatide | mental retardation, metachromasia of nerves |
| Fucosidosis | α-L-Fucosidase | Pentahexosylfucoglycolipid | cerebral degeneration, thickened skin, muscle spasticity |
| Farber's lipogranulomatosis | Acid ceramidase | Ceramide | hepatosplenomegaly, painful swollen joints |

Based upon genetic linkage analyses as well as enzyme studies and the characterization of accumulating lysosomal substances, Niemann Pick disease should be divided into type I and type II; type I has 2 subtypes, A and B (257200), which show deficiency of acid sphingomyelinase. Niemann Pick disease type II likewise has 2 subtypes, type C1 and C2 (NPC) and type D (NPD). It is obviously confusing to use the abbreviation NPD for Niemann Pick disease in some cases and for subtype D of Niemann Pick disease in other cases.

This gene contains regions of homology to mediators of cholesterol homeostasis suggesting why LDL-cholesterol accumulates in lysosomes of afflicted individuals. The encoded protein product of NPC1 gene is 1278 amino acids long. Within the protein are regions of homology to the transmembrane domain of the morphogen receptor PATCHED (of Drosophila melanogaster ), the putative sterol-sensing regions of SREBP (sterol regulatory element binding protein) cleavage-activating protein (SCAP) and HMG-CoA reductase.

## METABOLISM OF THE EICOSANOIDS

The eicosanoids consist of the prostaglandins (PGs), thromboxanes (TXs) and leukotrienes (LTs). The PGs and TXs are collectively identified as prostanoids. Prostaglandins were originally shown to be synthesized in the prostate gland, thromboxanes from platelets (thrombocytes) and leukotrienes from leukocytes, hence the derivation of their names.

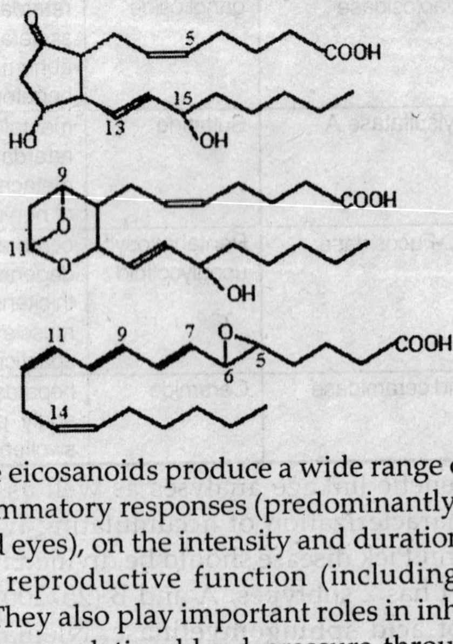

$PGE_2$

$TXA_2$

$LTA_4$

The eicosanoids produce a wide range of biological effects on inflammatory responses (predominantly those of the joints, skin and eyes), on the intensity and duration of pain and fever, and on reproductive function (including the induction of labor). They also play important roles in inhibiting gastric acid secretion, regulating blood pressure through vasodilation or constriction, and inhibiting or activating platelet aggregation and thrombosis.

The principal eicosanoids of biological significance to humans are a group of molecules derived from the C20 fatty acid, arachidonic acid. Minor eicosanoids are derived from dihomo-g-linoleic acid and eicosopentaenoic acid. The major source of arachidonic acid is through its release from cellular stores. Within the cell, it resides predominantly at the C-2 position of membrane phospholipids and is released from there upon the activation of phospholipase A2 The immediate dietary precursor of arachidonate is linoleate. Linoleate is also the precursor for dihomo-g-linoleic acid and eicosopentaenoic acid synthesis. Therefore, the absence of linoleic acid from the diet would seriously threaten the body's ability to synthesize eicosanoids.

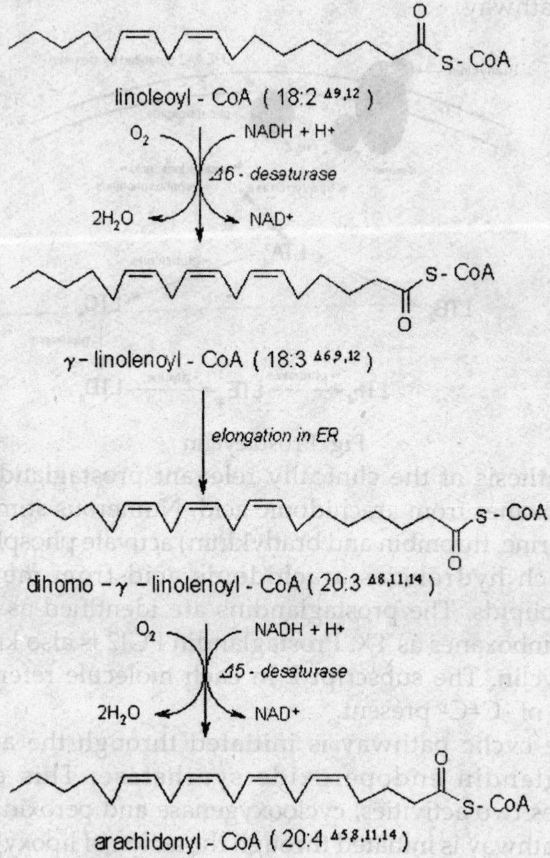


Pathway from linoleic acid to arachidonic acid. Numbers in parentheses refer to the fatty acid length and the number and positions of unsaturations

All mammalian cells except erythrocytes synthesize eicosanoids. These molecules are extremely potent, able to cause profound physiological effects at very dilute concentrations. All eicosanoids function locally at the site of synthesis, through receptor-mediated G-protein linked signaling pathways leading to an increase in cAMP levels.

Two main pathways are involved in the biosynthesis of eicosanoids. The prostaglandins and thromboxanes are synthesized by the cyclic pathway, the leukotrienes by the linear pathway.

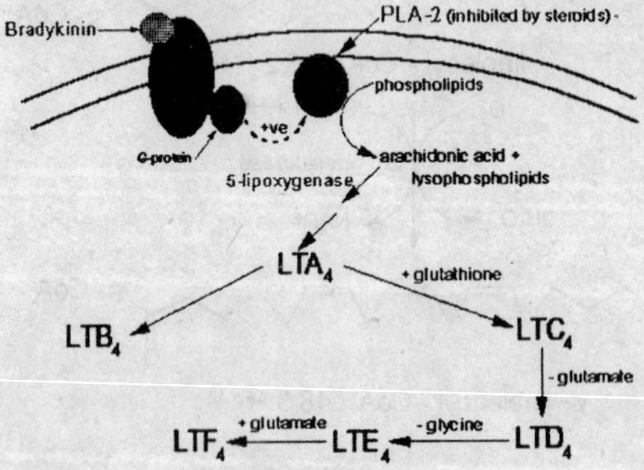


**Fig.** Prostacyclin

Synthesis of the clinically relevant prostaglandins and thromboxanes from arachidonic acid. Numerous stimuli (e.g. epinephrine, thrombin and bradykinin) activate phospholipase A2 which hydrolyzes arachidonic acid from membrane phospholipids. The prostaglandins are identified as PG and the thromboxanes as TX. Prostaglandin PGI2 is also known as prostacyclin. The subscript 2 in each molecule refers to the number of -C=C- present.

The cyclic pathway is initiated through the action of prostaglandin endoperoxide synthetase. This enzyme possesses two activities, cyclooxygenase and peroxidase. The linear pathway is initiated through the action of lipoxygenases. It is the enzyme, 5-lipoxygenase that gives rise to the leukotrienes.

A widely used class of drugs, the non-steroidal anti-inflammatory drugs (NSAIDs such as ibuprofen, indomethacin, naproxen, phenylbutazone and aspirin, all act upon the cyclooxygenase activity. Another class, the corticosteroidal drugs, act to inhibit phospholipase A2, thereby inhibiting the release of arachidonate from membrane phospholipids and the subsequent synthesis of eicosinoids.

## FATTY ACID OXIDATION

Utilization of dietary lipids requires that they first be absorbed through the intestine. As these molecules are oils they would be essentially insoluble in the aqueous intestinal environment. Solubilization (emulsification) of dietary lipid is accomplished via bile salts that are synthesized in the liver and secreted from the gallbladder.

The emulsified fats can then be degraded by pancreatic lipases (lipase and phospholipase A2). These enzymes, secreted into the intestine from the pancreas, generate free fatty acids and a mixtures of mono- and diacylglycerols from dietary triacylglycerols. Pancreatic lipase degrades triacylglycerols at the 1 and 3 positions sequentially to generate 1,2-diacylglycerols and 2-acylglycerols. Phospholipids are degraded at the 2 position by pancreatic phospholipase A2 releasing a free fatty acid and the lysophospholipid.

Following absorption of the products of pancreatic lipase by the intestinal mucosal cells, the resynthesis of triacylglycerols occurs. The triacylglycerols are then solubilized in lipoprotein complexes (complexes of lipid and protein) called chylomicrons. A chylomicron contains lipid droplets surrounded by the more polar lipids and finally a layer of proteins. Triacylglycerols synthesized in the liver are packaged into VLDLs and released into the blood directly. Chylomicrons from the intestine are then released into the blood via the lymph system for delivery to the various tissues for storage or production of energy through oxidation.

The triacylglycerol components of VLDLs and chylomicrons are hydrolyzed to free fatty acids and glycerol in the capillaries of adipose tissue and skeletal muscle by the action of lipoprotein lipase. The free fatty acids are then absorbed by the cells and the glycerol is returned via the blood to the liver (and kidneys). The glycerol is then converted to the glycolytic intermediate DHAP.

The classification of blood lipids is distinguished based upon the density of the different lipoproteins. As lipid is less dense than protein, the lower the density of lipoprotein the less protein there is.

## MOBILIZATION OF FAT STORES

The primary sources of fatty acids for oxidation are dietary and mobilization from cellular stores. Fatty acids from the diet can are delivered from the gut to cells via transport in the blood. Fatty acids are stored in the form of triacylglycerols primarily within adipocytes of adipose tissue. In response to energy demands, the fatty acids of stored triacylglycerols can be mobilized for use by peripheral tissues. The release of metabolic energy, in the form of fatty acids, is controlled by a complex series of interrelated cascades that result in the activation of hormone-sensitive lipase.

The stimulus to activate this cascade, in adipocytes, can be glucagon, epinephrine or b-corticotropin. These hormones bind cell-surface receptors that are coupled to the activation of adenylate cyclase upon ligand binding. The resultant increase in cAMP leads to activation of PKA, which in turn phosphorylates and activates hormone-sensitive lipase. This enzyme hydrolyzes fatty acids from carbon atoms 1 or 3 of triacylglycerols. The resulting diacylglycerols are substrates for either hormone-sensitive lipase or for the non-inducible enzyme diacylglycerol lipase. Finally the monoacylglycerols are substrates for monoacylglycerol lipase. The net result of the action of these enzymes is three moles of free fatty acid and one mole of glycerol. The free fatty acids diffuse from adipose cells, combine with albumin in the blood, and are thereby transported to other tissues, where they passively diffuse into cells.

### Hormone Induced Fatty Acid Mobilization in Adipocytes

Model for the activation of hormone-sensitive lipase by epinephrine. Epinephrine binds its' receptor and leads to the activation of adenylate cyclase. The resultant increase in cAMP activates PKA which then phosphorylates and activates hormone-sensitive lipase. Hormone-sensitive lipase hydrolyzes fatty acids from triacylglycerols and diacylglycerols. The final fatty acid is released from monoacylglycerols through the action of monoacylglycerol lipase, an enzyme active in the absence of hormonal stimulation.

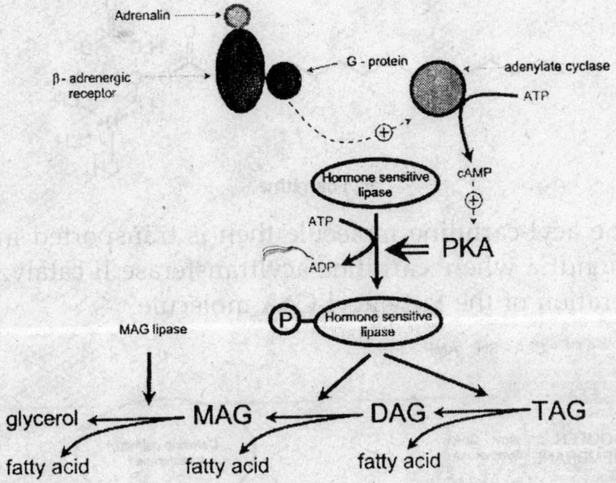


In contrast to the hormonal activation of adenylate cyclase and (subsequently) hormone-sensitive lipase in adipocytes, the mobilization of fat from adipose tissue is inhibited by numerous stimuli. The most significant inhibition is that exerted upon adenylate cyclase by insulin. When an individual is well fed state, insulin released from the pancreas prevents the inappropriate mobilization of stored fat. Instead, any excess fat and carbohydrate are incorporated into the triacylglycerol pool within adipose tissue.

## ACTIVATION AND TRANSFER OF FATTY ACID

Fatty acids must be activated in the cytoplasm before being oxidized in the mitochondria. Activation is catalyzed by fatty acyl-CoA ligase (also called acyl-CoA synthetase or thiokinase). The net result of this activation process is the consumption of 2 molar equivalents of ATP.

Fatty acid + ATP + CoA → Acyl-CoA + PPi + AMP

Oxidation of fatty acids occurs in the mitochondria. The transport of fatty acyl-CoA into the mitochondria is accomplished via an acyl-carnitine intermediate, which itself is generated by the action of carnitine acyltransferase I, an enzyme that resides in the outer mitochondrial membrane, from acyl-CoA and carnitine.

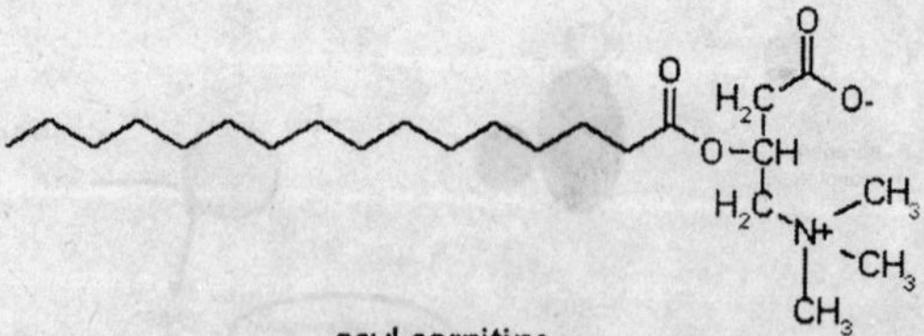

acyl carnitine

The acyl-carnitine molecule then is transported into the mitochondria where carnitine acyltransferase II catalyzes the regeneration of the fatty acyl-CoA molecule.

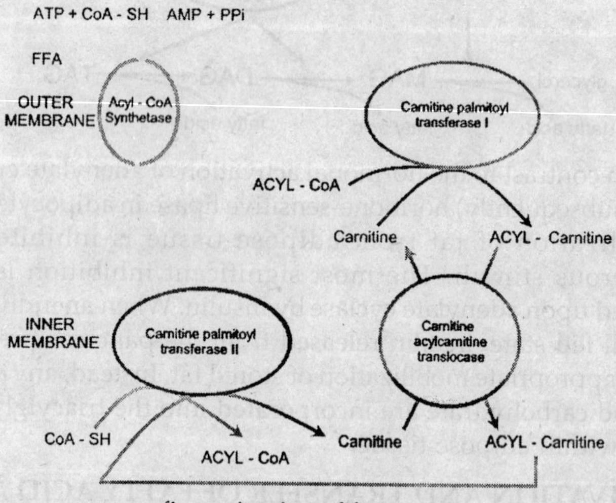


fig. activation of fatty asid

Transport of fatty acids from the cytoplasm to the inner mitochondrial space for oxidation. Following activation to a fatty-CoA, the CoA is exchanged for carnitine by carnitine-palmitoyltransferase I.

The fatty-carnitine is then transported to the inside of the mitochondrion where a reversal exchange takes place through the action of carnitine-palmitoyltransferase II. Once inside the mitochondrion the fatty-CoA is a substrate for the b-oxidation machinery.

The human body produces carnitine through an enzymatically catalyzed biosynthetic pathway out of lysine and methionine.

## REACTION SEQUENCE IN THE -OXIDATION

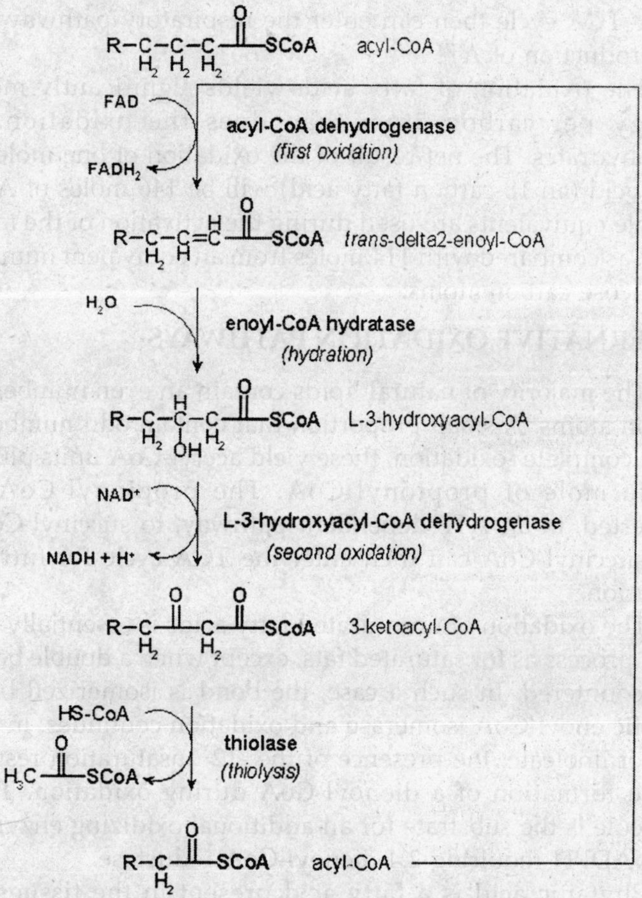


The process of fatty acid oxidation is termed -oxidation since it occurs through the sequential removal of 2-carbon units by oxidation at the b-carbon position of the fatty acyl-CoA molecule.

Each round of -oxidation produces one mole of NADH, one mole of FADH2 and one mole of acetyl-CoA. The acetyl-CoA--- the end product of each round of -oxidation--- then enters the TCA cycle, where it is further oxidized to CO2 with the concomitant generation of three moles of NADH, one mole of FADH2 and one mole of ATP. The NADH and FADH2 generated during the fat oxidation and acetyl-CoA oxidation in the TCA cycle then can enter the respiratory pathway for the production of ATP.

The oxidation of fatty acids yields significantly more energy per carbon atom than does the oxidation of carbohydrates. The net result of the oxidation of one mole of oleic acid (an 18-carbon fatty acid) will be 146 moles of ATP (2 mole equivalents are used during the activation of the fatty acid), as compared with 114 moles from an equivalent number of glucose carbon atoms.

## ALTERNATIVE OXIDATION PATHWAYS

The majority of natural lipids contain an even number of carbon atoms. A small proportion that contain odd numbers; upon complete -oxidation, these yield acetyl-CoA units plus a single mole of propionyl-CoA. The propionyl-CoA is converted, in an ATP-dependent pathway, to succinyl-CoA. The succinyl-CoA can then enter the TCA cycle for further oxidation.

The oxidation of unsaturated fatty acids is essentially the same process as for saturated fats, except when a double bond is encountered. In such a case, the bond is isomerized by a specific enoyl-CoA isomerase and oxidation continues. In the case of linoleate, the presence of the -12 unsaturation results in the formation of a dienoyl-CoA during oxidation. This molecule is the substrate for an additional oxidizing enzyme, the NADPH requiring 2,4-dienoyl-CoA reductase.

Phytanic acid is a fatty acid present in the tissues of ruminants and in dairy products and is, therefore, an important dietary component of fatty acid intake. Because phytanic acid is methylated, it cannot act as a substrate for the first enzyme of the -oxidation pathway (acyl-CoA dehydrogenase).

An additional mitochondrial enzyme, -hydroxylase, adds a hydroxyl group to the -carbon of phytanic acid, which then serves as a substrate for the remainder of the normal oxidative enzymes. This process is termed -oxidation.

## REGULATION OF FATTY ACID METABOLISM

In order to understand how the synthesis and degradation of fats needs to be exquisitely regulated, one must consider the energy requirements of the organism as a whole. The blood is the carrier of triacylglycerols in the form of VLDLs and chylomicrons, fatty acids bound to albumin, amino acids, lactate, ketone bodies and glucose. The pancreas is the primary organ involved in sensing the organism's dietary and energetic states by monitoring glucose concentrations in the blood. Low blood glucose stimulates the secretion of glucagon, whereas, elevated blood glucose calls for the secretion of insulin.

The metabolism of fat is regulated by two distinct mechanisms. One is short-term regulation, which can come about through events such as substrate availability, allosteric effectors and/or enzyme modification. The other mechanism, long-term regulation, is achieved by alteration of the rate of enzyme synthesis and turn-over.

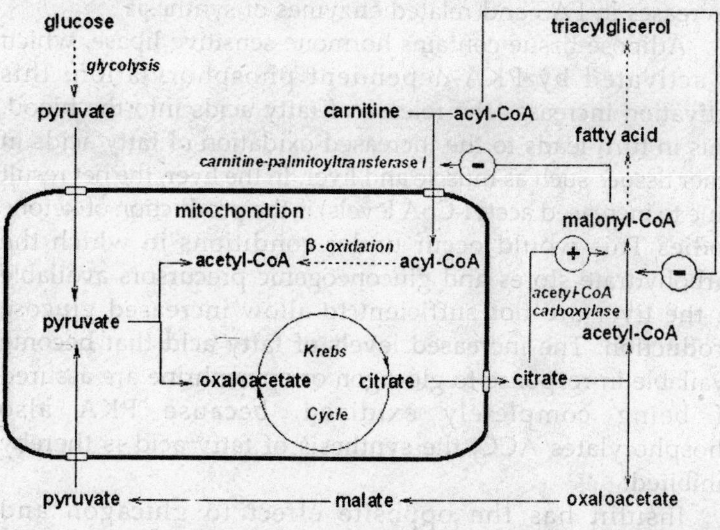


**Fig.** Metabolism of Fatty Asid

ACC (acetyl-CoA carboxylase) is the rate-limiting (committed) step in fatty acid synthesis. This enzyme is activated by citrate and inhibited by palmitoyl-CoA and other long-chain fatty acyl-CoAs. ACC activity can also be affected by phosphorylation. For instance, glucagon-stimulated increases in PKA activity result in the phosphorylation of certain serine residues in ACC leading to decreased activity of the enzyme. By contrast, insulin leads to PKA-independent phosphorylation of ACC at sites distinct from glucagon, which bring about increased ACC activity. Both of these reaction chains are examples of short-term regulation.

Insulin, a product of the well-fed state, stimulates ACC and FAS synthesis, whereas starvation leads to a decrease in the synthesis of these enzymes. Adipose tissue levels of lipoprotein lipase also are increased by insulin and decreased by starvation. However, the effects of insulin and starvation on lipoprotein lipase in the heart are just the inverse of those in adipose tissue. This sensitivity allows the heart to absorb any available fatty acids in the blood in order to oxidize them for energy production. Starvation also leads to increases in the levels of cardiac enzymes of fatty acid oxidation, and to decreases in FAS and related enzymes of synthesis.

Adipose tissue contains hormone-sensitive lipase, which is activated by PKA-dependent phosphorylation; this activation increases the release of fatty acids into the blood. This in turn leads to the increased oxidation of fatty acids in other tissues such as muscle and liver. In the liver, the net result (due to increased acetyl-CoA levels) is the production of ketone bodies This would occur under conditions in which the carbohydrate stores and gluconeogenic precursors available in the liver are not sufficient to allow increased glucose production. The increased levels of fatty acid that become available in response to glucagon or epinephrine are assured of being completely oxidized, because PKA also phosphorylates ACC; the synthesis of fatty acid is thereby inhibited.

Insulin has the opposite effect to glucagon and epinephrine: it increases the synthesis of triacylglycerols (and glycogen). One of the many effects of insulin is to lower cAMP levels, which leads to increased dephosphorylation through the enhanced activity of protein phosphatases such as PP-1.

With respect to fatty acid metabolism, this yields dephosphorylated and inactive hormone-sensitive lipase. Insulin also stimulates certain phosphorylation events. This occurs through activation of several cAMP-independent kinases, one of which phosphorylates and thereby stimulates the activity of ACC.

Fat metabolism can also be regulated by malonyl-CoA-mediated inhibition of carnitine acyltransferase I. Such regulation serves to prevent de novo synthesized fatty acids from entering the mitochondria and being oxidized.

## CLINICAL SIGNIFICANCE OF FATTY ACIDS

The majority of clinical problems related to fatty acid metabolism are associated with processes of oxidation.

These disorders fall into four main groups:

- Deficiencies in Carnitine: Deficiencies in carnitine lead to an inability to transport fatty acids into the mitochondria for oxidation. This can occur in newborns and particularly in pre-term infants. Carnitine deficiencies also are found in patients undergoing hemodialysis or exhibiting organic aciduria. Carnitine deficiencies may manifest systemic symptomology or may be limited to only muscles. Symptoms can range from mild occasional muscle cramping to severe weakness or even death. Treatment is by oral carnitine administration.
- Carnitine Palmitoyltransferase I (CPT I) Deficiency: Deficiencies in this enzyme affect primarily the liver and lead to reduced fatty acid oxidation and ketogenesis. Carnitine Palmitoylransferase II (CPT II) deficiency results in recurrent muscle pain and fatigue and myoglobinuria following strenuous exercise. Carnitine acyltransferases may also be inhibited by sulfonylurea drugs such as tolbutamide and glyburide.
- Deficiencies in Acyl-CoA Dehydrogenases: A group of inherited diseases that impair -oxidation result from deficiencies in acyl-CoA dehydrogenases. The enzymes affected may belong to one of four categories:

- Very long-chain acyl-CoA dehydrogenase (VLCAD)
- Long-chain acyl-CoA dehydrogenase (LCAD)
- Medium-chain acyl-CoA dehydrogenase (MCAD)
- Short-chain acyl-CoA dehydrogenase (SCAD)

MCAD deficiency is the most common form of this disease. In the first years of life this deficiency will become apparent following a prolonged fasting period. Symptoms include vomiting, lethargy and frequently coma. Excessive urinary excretion of medium-chain dicarboxylic acids as well as their glycine and carnitine esters is diagnostic of this condition. In the case of this enzyme deficiency. taking care to avoid prolonged fasting is sufficient to prevent clinical problems.

- Refsum's Disease: Refsum's disease is a rare inherited disorder in which patients lack the mitochondrial - oxidizing enzyme. As a consequence, they accumulate large quantities of phytanic acid in their tissues and serum. This leads to severe symptoms, including cerebellar ataxia retinitis pigmentosa, nerve deafness and peripheral neuropathy. As expected, the restriction of dairy products and ruminant meat from the diet can ameliorate the symptoms of this disease.

## KETOGENESIS

During high rates of fatty acid oxidation, primarily in the liver, large amounts of acetyl-CoA are generated. These exceed the capacity of the TCA cycle, and one result is the synthesis of ketone bodies, or ketogenesis. The ketone bodies are acetoacetate, -hydroxybutyrate, and acetone.

The formation of acetoacetyl-CoA occurs by condensation of two moles of acetyl-CoA through a reversal of the thiolase catalyzed reaction of fat oxidation. Acetoacetyl-CoA and an additional acetyl-CoA are converted to -hydroxy- -methylglutaryl-CoA (HMG-CoA) by HMG-CoA synthase, an enzyme found in large amounts only in the liver. Some of the HMG-CoA leaves the mitochondria, where it is converted to mevalonate (the precursor for cholesterol synthesis) by HMG-CoA reductase. HMG-CoA in the mitochondria is converted to acetoacetate by the action of HMG-CoA lyase.

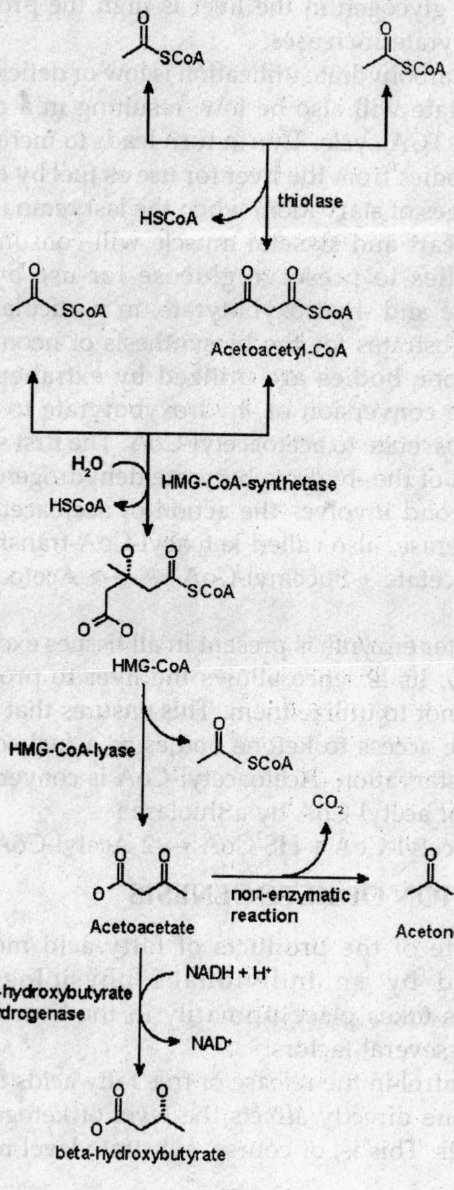

Acetoacetate can undergo spontaneous decarboxylation to acetone, or be enzymatically converted to -hydroxybutyrate through the action of -hydroxybutyrate dehydrogenase. When the level of glycogen in the liver is high the production of -hydroxybutyrate increases.

When carbohydrate utilization is low or deficient, the level of oxaloacetate will also be low, resulting in a reduced flux through the TCA cycle. This in turn leads to increased release of ketone bodies from the liver for use as fuel by other tissues. In early stages of starvation, when the last remnants of fat are oxidized, heart and skeletal muscle will consume primarily ketone bodies to preserve glucose for use by the brain. Acetoacetate and -hydroxybutyrate, in particular, also serve as major substrates for the biosynthesis of neonatal cerebral lipids. Ketone bodies are utilized by extrahepatic tissues through the conversion of -hydroxybutyrate to acetoacetate and of acetoacetate to acetoacetyl-CoA. The first step involves the reversal of the -hydroxybutyrate dehydrogenase reaction, and the second involves the action of acetoacetate:succinyl-CoA transferase, also called ketoacyl-CoA-transferase.

Acetoacetate + Succinyl-CoA <------> Acetoacetyl-CoA + succinate

The latter enzyme is present in all tissues except the liver. Importantly, its absence allows the liver to produce ketone bodies but not to utilize them. This ensures that extrahepatic tissues have access to ketone bodies as a fuel source during prolonged starvation. Acetoacetyl-CoA is converted into two molecules of acetyl-CoA by a thiolase :

Acetoacetyl-CoA + HS-CoA $\leftrightarrow$ 2 Acetyl-CoA

## REGULATION OF KETOGENESIS

The fate of the products of fatty acid metabolism is determined by an individual's physiological status. Ketogenesis takes place primarily in the liver and may by affected by several factors:

- Control in the release of free fatty acids from adipose tissue directly affects the level of ketogenesis in the liver. This is, of course, substrate-level regulation.

- Once fats enter the liver, they have two distinct fates. They may be activated to acyl-CoAs and oxidized, or esterified to glycerol in the production of triacylglycerols. If the liver has sufficient supplies of glycerol-3-phosphate, most of the fats will be turned to the production of triacylglycerols.
- The generation of acetyl-CoA by oxidation of fats can be completely oxidized in the TCA cycle. Therefore, if the demand for ATP is high the fate of acetyl-CoA is likely to be further oxidation to $CO_2$.
- The level of fat oxidation is regulated hormonally through phosphorylation of ACC, which may activate it (in response to glucagon) or inhibit it (in the case of insulin).

## CLINICAL SIGNIFICANCE OF KETOGENESIS

The production of ketone bodies occurs at a relatively low rate during normal feeding and under conditions of normal physiological status. Normal physiological responses to carbohydrate shortages cause the liver to increase the production of ketone bodies from the acetyl-CoA generated from fatty acid oxidation. This allows the heart and skeletal muscles primarily to use ketone bodies for energy, thereby preserving the limited glucose for use by the brain. The most significant disruption in the level of ketosis, leading to profound clinical manifestations, occurs in untreated insulin-dependent diabetes mellitus.

This physiological state, diabetic ketoacidosis (DKA), results from a reduced supply of glucose (due to a significant decline in circulating insulin) and a concomitant increase in fatty acid oxidation (due to a concomitant increase in circulating glucagon).The increased production of acetyl-CoA leads to ketone body production that exceeds the ability of peripheral tissues to oxidize them. Ketone bodies are relatively strong acids (pKa around 3.5) and their increase lowers the pH of the blood. This acidification of the blood is dangerous chiefly because it impairs the ability of hemoglobin to bind oxygen.

# CHOLESTEROL AND BILE METABOLISM

## CHOLESTEROL METABOLISM

Cholesterol is an extremely important biological molecule that has roles in membrane structure as well as being a precursor for the synthesis of the steroid hormones and bile acids. Both dietary cholesterol and that synthesized de novo are transported through the circulation in lipoprotein particles. The same is true of cholesteryl esters, the form in which cholesterol is stored in cells.

The synthesis and utilization of cholesterol must be tightly regulated in order to prevent over-accumulation and abnormal deposition within the body. Of particular importance clinically is the abnormal deposition of cholesterol and cholesterol-rich lipoproteins in the coronary arteries. Such deposition, eventually leading to atherosclerosis, is the leading contributory factor in diseases of the coronary arteries.

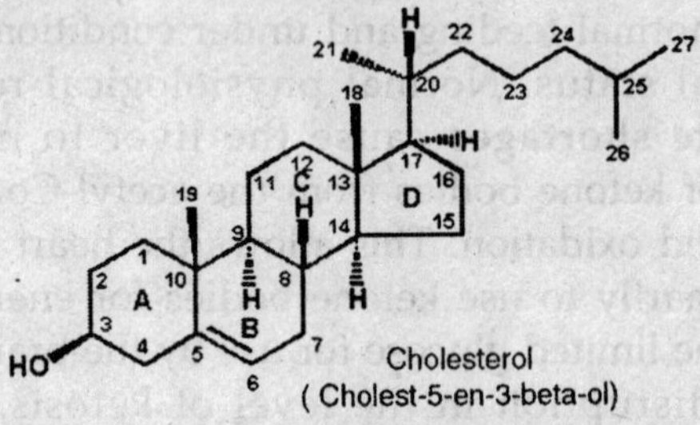

Cholesterol
( Cholest-5-en-3-beta-ol)

## BIOSYNTHESIS OF CHOLESTEROL

Slightly less than half of the cholesterol in the body derives from biosynthesis de novo. Biosynthesis in the liver accounts for approximately 10% and in the intestines approximately 15%, of the amount produced each day. Cholesterol synthesis occurs in the cytoplasm and microsomes from the two-carbon acetate group of acetyl-CoA.

The process has nine major steps:

- Acetyl-CoAs are converted to ω3-hydroxy- 3-methyl glutaryl-CoA (HMG-CoA).Unlike the HMG-CoA formed during ketone body synthesis in the mitochondria, this form is synthesized in the cytoplasm. However, the pathway and the necessary enzymes are the same as those in the mitochondria. Two moles of acetyl-CoA are condensed in a reversal of the thiolase reaction, forming acetoacetyl-CoA. Acetoacetyl-CoA and a third mole of acetyl-CoA are converted to HMG-CoA by the action of HMG-CoA synthase.

- HMG-CoA is converted to mevalonate by HMG-CoA reductase (this enzyme is bound to the endoplasmic reticulum). HMG-CoA reductase absolutely requires NADPH as a cofactor and two moles of NADPH are consumed during the conversion of HMG-CoA to mevalonate.

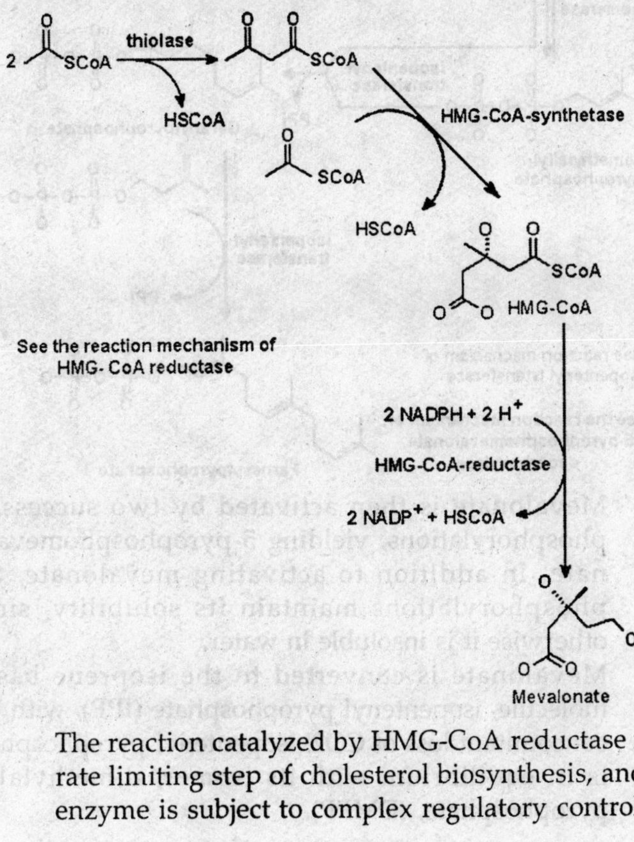


The reaction catalyzed by HMG-CoA reductase is the rate limiting step of cholesterol biosynthesis, and this enzyme is subject to complex regulatory controls.

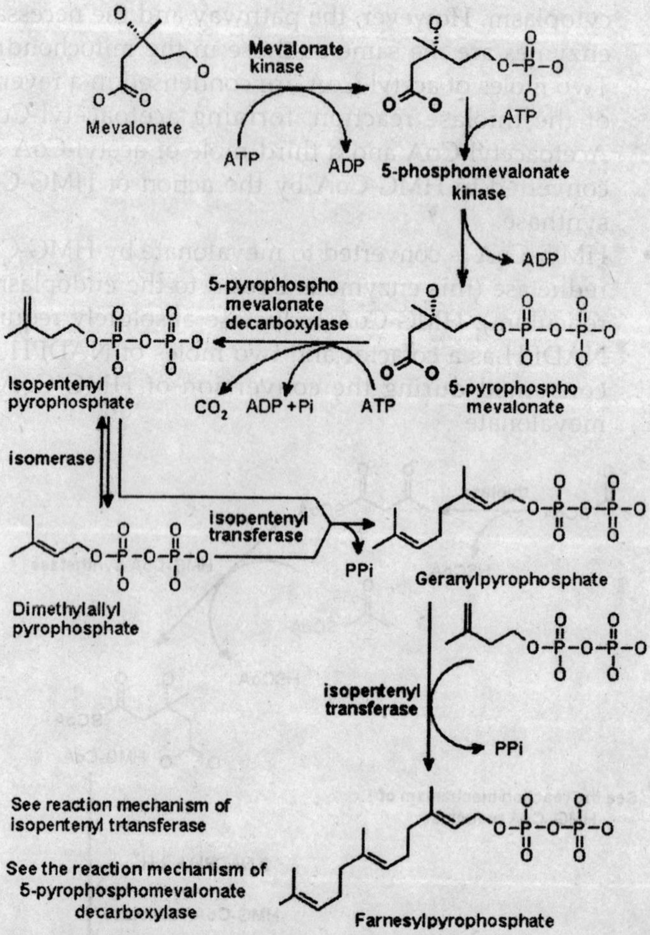


- Mevalonate is then activated by two successive phosphorylations, yielding 5-pyrophosphomevalonate. In addition to activating mevalonate, the phosphorylations maintain its solubility, since otherwise it is insoluble in water.
- Mevalonate is converted to the isoprene based molecule, isopentenyl pyrophosphate (IPP), with the concomitant loss of CO2. Isopentenyl pyrophosphate is in equilibrium with its isomer, dimethylallyl pyrophosphate, (DMPP).

- One molecule of IPP condenses with one molecule of DMPP to generate geranyl pyrophosphate, (GPP).
- GPP further condenses with another IPP molecule to yield farnesyl pyrophosphate, (FPP).

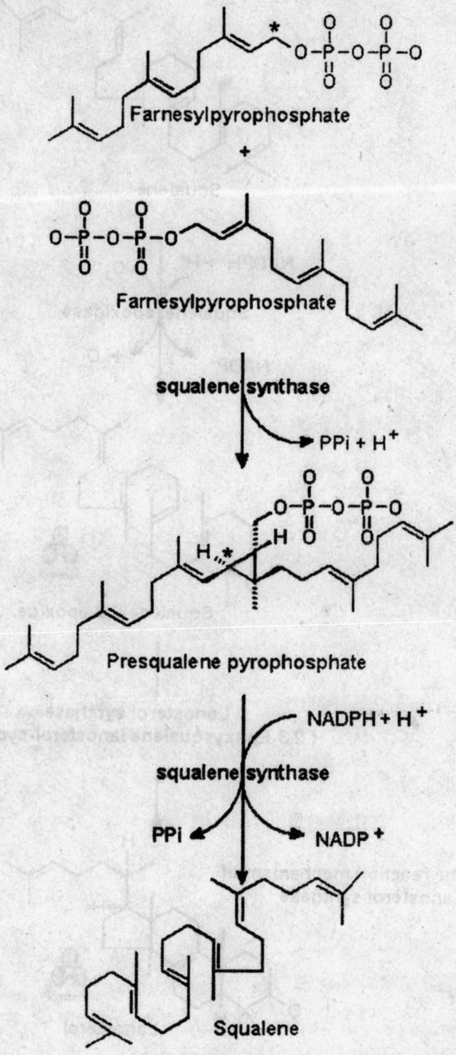


- Finally, the NADPH-requiring enzyme, squalene

synthase catalyzes the head-to-tail condensation of two molecules of FPP, yielding squalene (squalene synthase also is tightly associated with the endoplasmic reticulum).

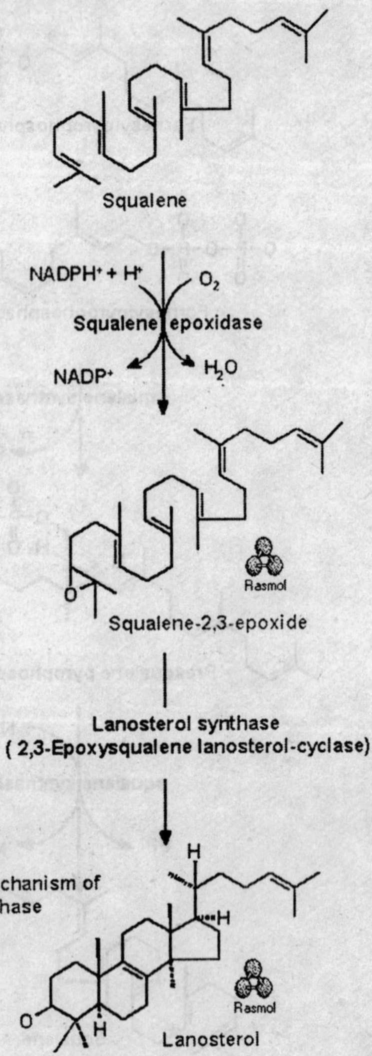


See the reaction mechanism of Lanosterol synthase

- Squalene undergoes a two step cyclization to yield

lanosterol. The first reaction is catalyzed by squalene monooxygenase. This enzyme uses NADPH as a cofactor to introduce molecular oxygen as an epoxide at the 2,3 position of squalene.

- Through a series of 19 additional reactions, lanosterol is converted to cholesterol.

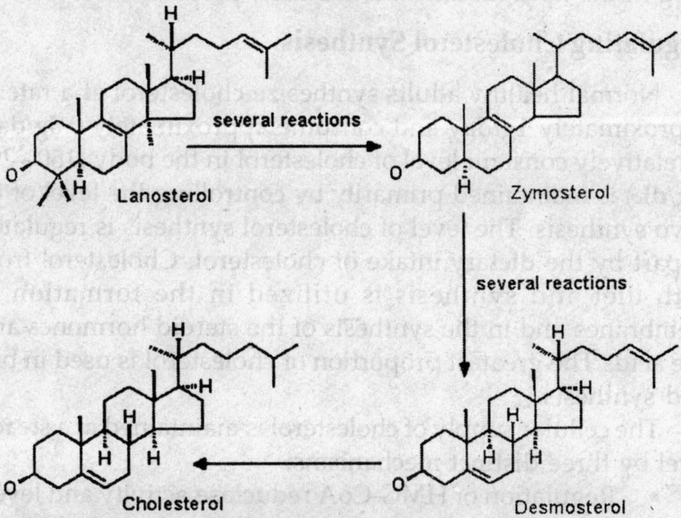


## PATHWAY OF CHOLESTEROL BIOSYNTHESIS

Synthesis begins with the transport of acetyl-CoA from the mitochondrion to the cytosol. The rate limiting step occurs at the 3-hydroxy-3-methylglutaryl-CoA (HMG-CoA) reducatse catalyzed step. The phosphorylation reactions are required to solubilize the isoprenoid intermediates in the pathway. Intermediates in the pathway are used for the synthesis of prenylated proteins, dolichol, coenzyme Q and the side chain of heme a.

The acetyl-CoA utilized for cholesterol biosynthesis is derived from an oxidation reaction (eg, fatty acids or pyruvate) in the mitochondria and is transported to the cytoplasm by the same process as that described for fatty acid synthesis. Acetyl-CoA can also be derived from cytoplasmic oxidation of ethanol by acetyl-CoA synthetase. All the reduction reactions of cholesterol biosynthesis use NADPH as a cofactor. The isoprenoid intermediates of cholesterol biosynthesis can be diverted to other synthesis reactions, such as those for dolichol (used in the synthesis of N-linked glycoproteins, coenzyme Q (of the oxidative phosphorylation) pathway or the side chain of heme a. Additionally, these intermediates are used in the lipid modification of some proteins.

## Regulating Cholesterol Synthesis

Normal healthy adults synthesize cholesterol at a rate of approximately 1g/day and consume approximately 0.3g/day. A relatively constant level of cholesterol in the body (150 - 200 mg/dL) is maintained primarily by controlling the level of de novo synthesis. The level of cholesterol synthesis is regulated in part by the dietary intake of cholesterol. Cholesterol from both diet and synthesis is utilized in the formation of membranes and in the synthesis of the steroid hormones and bile acids The greatest proportion of cholesterol is used in bile acid synthesis.

The cellular supply of cholesterol is maintained at a steady level by three distinct mechanisms:

- Regulation of HMG-CoA reductase activity and levels
- Regulation of excess intracellular free cholesterol through the activity of acyl-CoA:cholesterol acyltransferase, ACAT
- Regulation of plasma cholesterol levels via LDL receptor-mediated uptake and HDL-mediated reverse transport.

Regulation of HMG-CoA reductase activity is the primary means for controlling the level of cholesterol biosynthesis.

Regulation of HMG-CoA reductase by covalent modification. HMG-CoA reductase is most active in the dephosphorylated state. Phosphorylation is catalyzed by reductase kinase (RK), an enzymes whose activity is also regulated by phosphorylation. Phosphorylation of RK is catalyzed by reductase kinase kinase (RKK). Hormones such as glucagon and epinephrine negatively affect cholesterol biosynthesis by increasing the activity of the inhibitor of phosphoprotein phosphatase-1, PPI-1. Conversely, insulin stimulates the removal of phosphates and, thereby, activates HMG-CoA reductase activity. Additional regulation of HMG-CoA reductase occurs through an inhibition of synthesis of the enzyme by elevation in intracellular cholesterol levels.

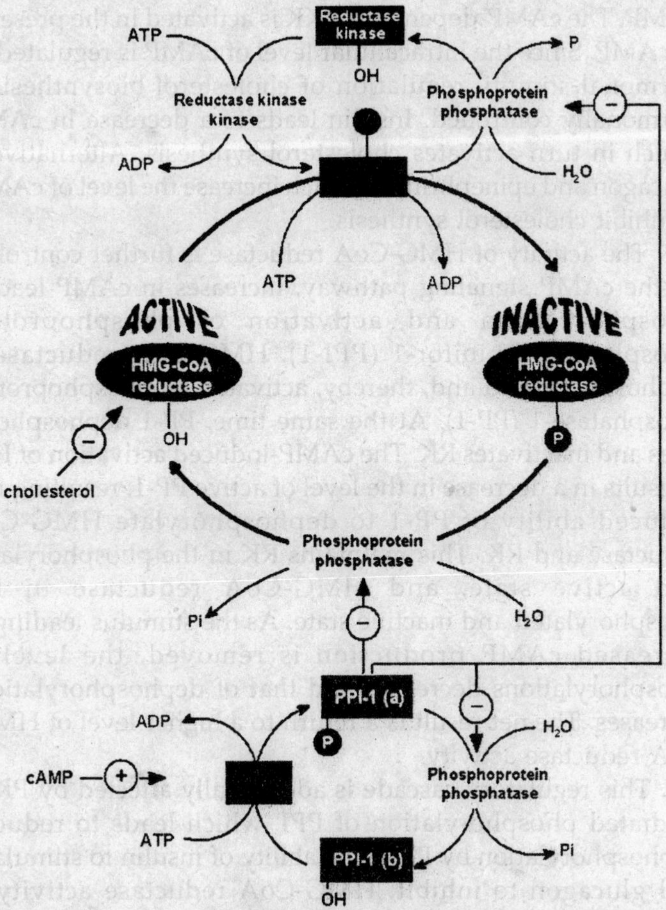


**Fig.** Cholesterol Synthesis

Regulation of HMG-CoA reductase takes place both through covalent modification and through control of the absolute level of the enzyme within cells. HMG-CoA reductase is a single polypeptide embedded in microsomal membranes;

it is most active in its unmodified form. Phosphorylation of the enzyme decreases its activity.HMG-CoA reductase is phosphorylated by reductase kinase, RK. RK itself is activated via phosphorylation. The phosphorylation of RK is catalyzed by reductase kinase kinase, RKK. There are two isoforms of RKK, one independent of cAMP and one dependent upon cAMP. The cAMP-dependent RKK is activated in the presence of cAMP. Since the intracellular level of cAMP is regulated by hormonal stimuli, regulation of cholesterol biosynthesis is hormonally controlled. Insulin leads to a decrease in cAMP, which in turn activates cholesterol synthesis. Alternatively, glucagon and epinephrine--- which increase the level of cAMP--- inhibit cholesterol synthesis.

The activity of HMG-CoA reductase is further controlled by the cAMP signaling pathway. Increases in cAMP lead to phosphorylation and activation of phosphoprotein phosphatase inhibitor-1 (PPI-1). HMG-CoA reductase is dephosphorylated and, thereby, activated by phosphoprotein phosphatase-1 (PP-1). At the same time, PP-1 dephosphorylates and inactivates RK. The cAMP-induced activation of PPI-1 results in a decrease in the level of active PP-1, resulting in a reduced ability of PP-1 to dephosphorylate HMG-CoA reductase and RK. This maintains RK in the phosphorylated and active state, and HMG-CoA reductase in the phosphorylated and inactive state. As the stimulus leading to increased cAMP production is removed, the level of phosphorylations decreases and that of dephosphorylations increases. The net result is a return to a higher level of HMG-CoA reductase activity.

This regulatory cascade is additionally affected by PKA-mediated phosphorylation of PPI, which leads to reduced dephosphorylation by PP-1.The ability of insulin to stimulate, and glucagon to inhibit, HMG-CoA reductase activity is consistent with the effects of these hormones on other metabolic pathways. The basic function of these two hormones is to control the availability and delivery of energy to all cells of the body.Long-term control of HMG-CoA reductase activity is exerted primarily through control over the synthesis and degradation of the enzyme. When levels of cholesterol are high, the level of expression of the HMG-CoA gene is reduced. Conversely, reduced levels of cholesterol activate expression of the gene. Insulin also brings about long-term regulation of cholesterol metabolism by increasing the level of HMG-CoA reductase synthesis. The rate of HMG-CoA turn-over is also regulated by the supply of cholesterol. When cholesterol is abundant, the rate of HMG-CoA reductase degradation increases.

## PROTEOLYTIC REGULATION OF HMG-COA REDUCTASE

The stability of HMGR is regulated as the rate of flux through the mevalonate synthesis pathway changes. When the flux is high the rate of HMGR degradation is also high. When the flux is low, degradation of HMGR decreases. This phenomenon can easily be observed in the presence of the statin drugs.

HMGR is localized to the ER and like SREBP contains a sterol-sensing domain,SSD. When sterol levels increase in cells there is an concommitant increase in the rate of HMGR degradation. The degradation of HMGR occurs within the proteosome, a multiprotein complex dedicated to protein degradation. The primary signal directing proteins to the proteosome is ubiquitination. Ubiquitin is a 7.6kDa protein that is covalently attached to proteins targeted for degradation by ubiqitin ligases. These enzymes attach multiple copies of ubiquitin allowing for recognition by the proteosome. HMGR has been shown to be ubiquitinated prior to its degradation. The primary sterol regulating HMGR degradation is cholesterol itself. As the levels of free cholesterol increase in cells, the rate of HMGR degradation increases.

### The Utilization of Cholesterol

Cholesterol is transported in the plasma predominantly as cholesteryl esters associated with lipoproteins. Dietary cholesterol is transported from the small intestine to the liver within chylomicrons. Cholesterol synthesized by the liver, as well as any dietary cholesterol in the liver that exceeds hepatic needs, is transported in the serum within LDLs. The liver synthesizes VLDLs and these are converted to LDLs through the action of endothelial cell-associated lipoprotein lipase. Cholesterol found in plasma membranes can be extracted by HDLs and esterified by the HDL-associated enzyme LCAT. The cholesterol acquired from peripheral tissues by HDLs can then be transferred to VLDLs and LDLs via the action of cholesteryl ester transfer protein (apo-D) which is associated with HDLs. Reverse cholesterol transport allows peripheral cholesterol to be returned to the liver in LDLs. Ultimately, cholesterol is excreted in the bile as free cholesterol or as bile salts following conversion to bile acids in the liver.

## Regulation of Cellular Sterol Content

The continual alteration of the intracellular sterol content occurs through the regulation of key sterol synthetic enzymes as well as by altering the levels of cell-surface LDL receptors. As cells need more sterol they will induce their synthesis and uptake, conversely when the need declines synthesis and uptake are decreased. Regulation of these events is brought about primarily by sterol-regulated transcription of key rate limiting enzymes and by the regulated degradation of HMGR. Activation of transcriptional control occurs through the regulated cleavage of the membrane-bound transcription factor sterol regulated element binding protein, SREBP. Degradation of HMGR is controlled by the ubiquitin-mediated pathway for proteolysis.

Sterol control of transcription affects more than 30 genes involved in the biosynthesis of cholesterol, triacylglycerols, phospholipids and fatty acids. Transcriptional control requires the presence of an octamer sequence in the gene termed the sterol regulatory element, SRE-1. It has been shown that SREBP is the transcription factor that binds to SRE-1 elements. It turns out that there are 2 distinct SREBP gene, SREBP-1 and SREBP-2. In addition, the SREBP-1 gene encodes 2 proteins, SREBP-1a and SREBP-1c as a consequence of alternative exon usage. All 3 proteins are proteolytically regulated by sterols. Full-length SREBPs have several domains and are embedded in the membrane of the endoplasmic reticulum (ER). The N-terminal domain contains a transcription factor motif of the basic helix-loop-helix (bHLH) type that is exposed to the cytoplasmic side of the ER.

There are 2 transmembrane spanning domains followed by a large C-terminal domain also exposed to the cytosolic side. When sterols are scarce, cleavage of the full-length SREBP takes place with the result being that the N-terminla bHLH motif is released into the cytosol.

The bHLH domian then migrates to the nucleus to direct transcription. Conversely, when sterols are abundant, cleavage of SREBP is inhibited. To control the level of SREBP-mediated transcription, the soluble bHLH domain is itself subject to rapid proteolysis.

The cleavage of SREBP is carried out by 2 distinct enzymes, one of which is regulated by sterols. The regulated cleavage occurs in the lumenal loop between the 2 transmembrane domains. This cleavage is catalyzed by site-1 protease, S1P. High sterol content blocks the activity of S1P. The second cleavage, catalyzed by site-2 protease, S2P, occurs in the first transmembrane span, leading to release of active SREBP. In order for S2P to act on SREBP site-1 must already have been cleaved.

Additional studies on sterol-regulated gene expression demonstrated that cleavage of SREBP by S1P is controlled by the level and action of an additional protein termed, SREBP cleavage-activating protein, SCAP. SCAP is a large protein also found in the ER membrane and contains at least 8 transmembrane spans. The C-terminal portion, which extends into the cytosol, has been shown to interact with the C-terminal domain of SREBP. This C-terminal region of SCAP contains 4 motifs called WD40 repeats.

The WD40 repeats are required for interaction of SCAP with SREBP. Interestingly, the N-terminus of SCAP, including membrane spans 2-6, resembles HMGR which itself is subject to sterol-stimulated degradation This shared motif is called the sterol sensing domain, SSD. Several proteins whose functions involve streols also contain the SSD. These include patched, an important development regulating receptor whose ligand, hedgehog, is modified by attachment of cholesterol and the Niemann Pick C1 (NPC1) protein which is involved in cholesterol transport in the secretory pathway.

The function of SCAP is to positively stimulate S1P-mediated cleavage of SREBP. The function of sterols is to inhibit this positive action of SCAP. The activity of SCAP involves movement from the ER to the Golgi and back. Because the C-terminus of SCAP interacts with SREBP, movement of SCAP takes SREBP along for the ride. When sterols are low, SCAP and SREBP move to the Golgi. This transit is required for SREBP cleavage as S1P is Golgi-localized. When sterols are high, movement of SCAP is halted. Thus, the overall effect of sterols is to regulate the ability of SCAP to present SREBP to S1P.

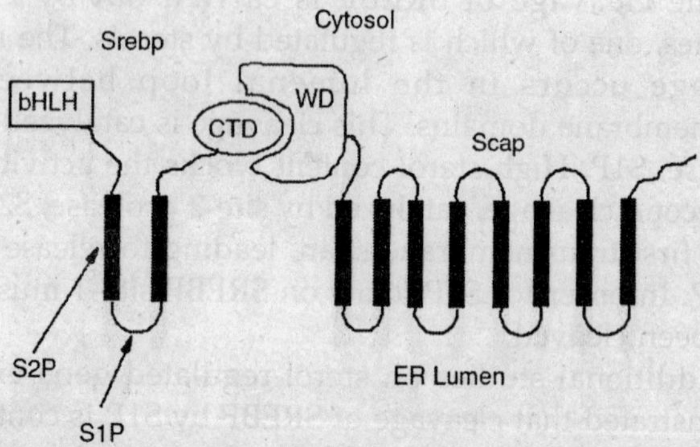


**Fig.** Interactions between SREBP and SCAP in the Membrane of the ER.

CTD = C-terminal Domain.

## BILE ACIDS SYNTHESIS AND UTILIZATION

The end products of cholesterol utilization are the bile acids, synthesized in the liver. Synthesis of bile acids is the predominant mechanisms for the excretion of excess cholesterol. However, the excretion of cholesterol in the form of bile acids is insufficient to compensate for an excess dietary intake of cholesterol.

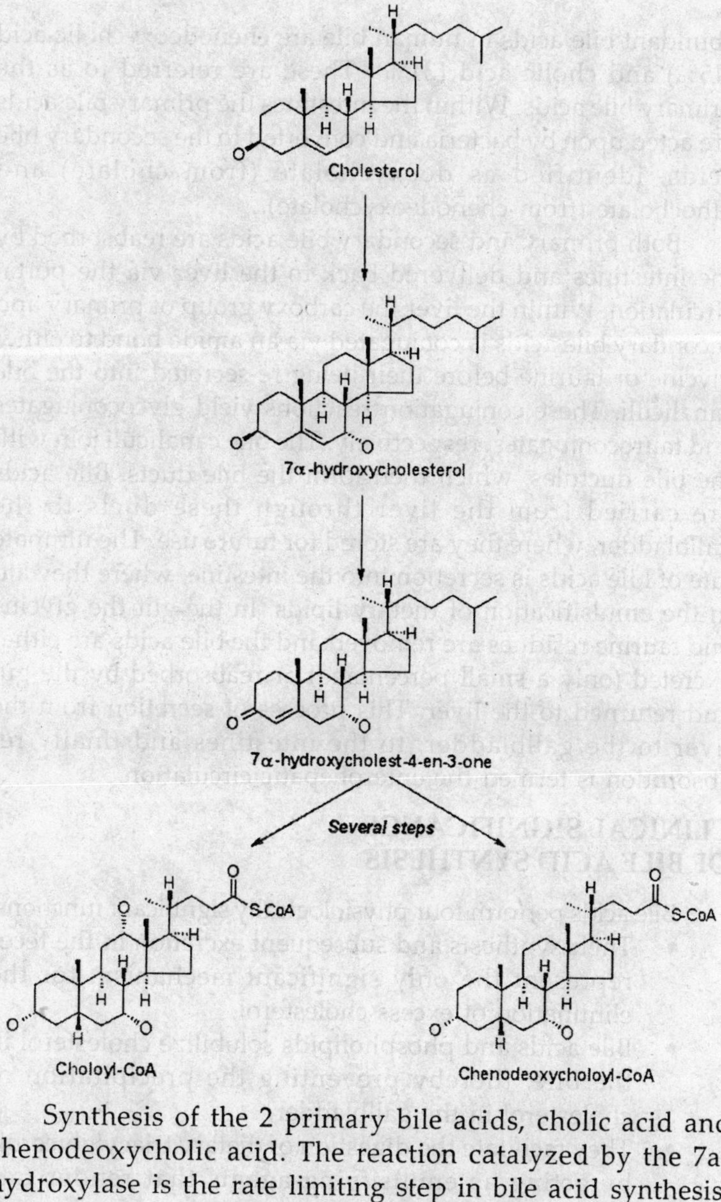


Synthesis of the 2 primary bile acids, cholic acid and chenodeoxycholic acid. The reaction catalyzed by the 7a-hydroxylase is the rate limiting step in bile acid synthesis. Conversion of 7 -hydroxycholesterol to the bile acids requires several steps not shown in detail in this image. The most abundant bile acids in human bile are chenodeoxycholic acid (45%) and cholic acid (31%). These are referred to as the primary bile acids. Within the intestines the primary bile acids are acted upon by bacteria and converted to the secondary bile acids, identified as deoxycholate (from cholate) and lithocholate (from chenodeoxycholate).

Both primary and secondary bile acids are reabsorbed by the intestines and delivered back to the liver via the portal circulation. Within the liver the carboxy group of primary and secondary bile acids is conjugated via an amide bond to either glycine or taurine before their being re-secreted into the bile canaliculi. These conjugation reactions yield glycoconjugates and tauroconjugates, respectively. The bile canaliculi join with the bile ductules, which then form the bile ducts. Bile acids are carried from the liver through these ducts to the gallbladder, where they are stored for future use. The ultimate fate of bile acids is secretion into the intestine, where they aid in the emulsification of dietary lipids. In the gut the glycine and taurine residues are removed and the bile acids are either excreted (only a small percentage) or reabsorbed by the gut and returned to the liver. This process of secretion from the liver to the gallbladder, to the intestines and finally re-absorbtion is termed the enterohepatic circulation.

## CLINICAL SIGNIFICANCE OF BILE ACID SYNTHESIS

Bile acids perform four physiologically significant functions:

- Their synthesis and subsequent excretion in the feces represent the only significant mechanism for the elimination of excess cholesterol.
- Bile acids and phospholipids solubilize cholesterol in the bile, thereby preventing the precipitation of cholesterol in the gallbladder.
- They facilitate the digestion of dietary triacylglycerols by acting as emulsifying agents that render fats accessible to pancreatic lipases.
- They facilitate the intestinal absorption of fat-soluble vitamins.

## MASTER METABOLIC REGULATOR

AMP-activated protein kinase (AMPK) was first discovered as an activity that inhibited preparations of acetyl-CoA carboxylase (ACC) and 3-hydroxy-3-methylglutaryl-CoA reductase (HMG-CoA reductase, HMGR) and was induced by AMP. AMPK induces a cascade of events within cells in response to the ever changing energy charge of the cell. The role of AMPK in regulating cellular energy charge places this enzyme at a central control point in maintaining energy homeostasis. More recent evidence has shown that AMPK activity can also be regulated by physiological stimuli, independent of the energy charge of the cell, including hormones and nutrients.

Once activated, AMPK-mediated phosphorylation events switch cells from active ATP consumption (e.g. fatty acid and cholesterol biosynthesis) to active ATP production (e.g. fatty acid and glucose oxidation). These events are rapidly initiated and are referred to as short-term regulatory processes. The activation of AMPK also exerts long-term effects at the level of both gene expression and protein synthesis. Other important activities attributable to AMPK are regulation of insulin synthesis and secretion in pancreatic islet b-cells and modulation of hypothalamic functions involved in the regulation of satiety.

## STRUCTURE OF AMPK

The mammalian AMPK is a trimeric enzyme composed of a catalytic subunit and the non-catalytic and subunits. There are two genes encoding isoforms of both the and subunits (1, 2, 1 and 2) and three genes encoding isoforms of the subunit (1- 3). The 2 isoform is the subunit of AMPK found predominantly within skeletal and cardiac muscle, whereas, approximately equal distribution of both the 1 and 2 isoforms are present in hepatic AMPK. Within pancreatic islet -cells the 1 isoform predominates.

The N-terminal half of the a subunits contains a typical serine/threonine kinase catalytic domain. Interaction with the and subunits occurs via the C-terminal half of the subunits.

The yeast AMPK subunits are lipid modified with myristic acid. Myristoylation may account for the membrane association mammalian AMPK. The core of the subunits have a glycogen-binding domain (GBD). This domain is closely related to the isoamylase N domain subfamily and weakly related to domains in the glycogen-targeting phosphatase subunits and several starch-binding proteins.

The close proximity of AMPK to cellular glycogen stores allows it to rapidly affect changes in glycogen metabolism in response to changes in metabolic demands. The subunits of AMPK have been shown to contain nucleotide binding sites with similarity to cystathionine -synthase (CBS) domains. Indeed, direct AMP-binding studies have shown that AMP is bound to the subunits by a pair of CBS domains. Of clinical significance is the observation that mutations in the CBS domains of the 2 subunit are associated with Wolff-Parkinson-White syndrome and familial hypertrophic cardiomyopathy.

## REGULATION OF AMPK

AMPK is activated by phosphorylation by one or more upstream AMPK kinases (AMPKKs). In the absence of phosphorylation there is no detectable activity of AMPK towards any substrates. Phosphorylation of AMPK occurs in the a subunit at threonine 172 (Thr-172) which lies in the activation loop. One kinase activator of AMPK is Ca2+-calmodulin-dependent kinase kinase (CAMKK) which phosphorylates and activates AMPK in response to increased calcium. Recent evidence has demonstrated that the threonine kinase, LKB1, encoded by the Peutz-Jegher syndrome tumor suppressor gene, is required for activation of AMPK in response to stress. Loss of LKB1 activity in adult mouse liver leads to near complete loss of AMPK activity and is associated with hyperglycemia.

The hyperglycemia is, in part, due to an increase in the transcription of gluconeogenic genes. Of particular significance is the increased expression of the peroxisome proliferator-activated receptor- (PPAR-) coactivator 1 (PGC-1 ) which drives gluconeogenesis. Reduction in PGC-1 activity results in normalized blood glucose levels in LKB1-deficient mice. As the name implies, AMPK is also regulated by AMP. The effects of AMP are two-fold: a direct allosteric activation and making AMPK a poorer substrate for dephosphorylation. Because AMP affects both the rate of AMPK phoshorylation in the positive direction and dephosphorylation in the negative direction, the cascade is ultrasensitive. This means that a very small rise in AMP levels can induce a dramatic increase in the activity of AMPK. The activity of adenylate kinase, catalyzing the reaction shown below, ensures that AMPK is highly sensitive to small changes in the intracellular [ATP]/[ADP] ratio.

2 ADP $\rightarrow$ ATP + AMP

Negative allosteric regulation of AMPK also occurs and this effect is exerted by phosphocreatine. The subunits of AMPK have a glycogen-binding domain, GBD. In muscle, a high glycogen content represses AMPK activity and this is likely the result of interaction between the GBD and glycogen, although this has not been shown directly. The GBD of AMPK allows association of the enzyme with the regulation of glycogen metabolism by placing AMPK in close proximity to one of its substrates glycogen synthase. AMPK has also been shown to be activated by receptors that are coupled to phospholipase C- (PLC-) and by cytokines secreted by adipose tissue (adipocytokines) such as leptin and adiponectin.

## TARGETS OF AMPK

The signaling cascades initiated by the activation of AMPK exert effects on glucose and lipid metabolism, gene expression and protein synthesis. These effects are most important for regulating metabolic events in the liver, skeletal muscle, heart, adipose tissue, and pancreas.

Demonstration of the central role of AMPK in the regulation of metabolism in response to events such as nutrient- or exercise-induced stress. Several of the known physiologic targets for AMPK are included as well as several pathways whose flux is affected by AMPK activation. Arrows indicate positive effects of AMPK, whereas, T-lines indicate inhibitory effects.

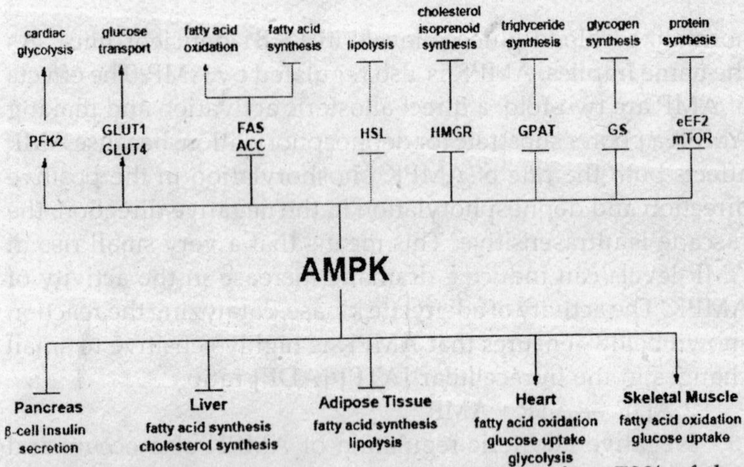


The uptake, by skeletal muscle, accounts for >70% of the glucose removal from the serum in humans. Therefore, it should be obvious that this event is extremely important for overall glucose homeostasis, keeping in mind, of course, that glucose uptake by cardiac muscle and adipocytes cannot be excluded from consideration. An important fact related to skeletal muscle glucose uptake is that this process is markedly impaired in individuals with type 2 diabetes. The uptake of glucose increases dramatically in response to stress (such as ischemia) and exercise and is stimulated by insulin-induced recruitment of glucose transporters to the plasma membrane, primarily GLUT4. Insulin-independent recruitment of glucose transporters also occurs in skeletal muscle in response to contraction

The activation of AMPK plays an important, albeit not an exclusive, role in the induction of GLUT4 recruitment to the plasma membrane. In addition, there is some demonstration that AMPK may regulate glucose transport through GLUT1. Increased glucose uptake will result in an increase in glycolysis and ATP production. Under ischemic conditions in the heart the activation of AMPK leads to the phosphorylation and activation of phosphofructokinase-2, PFK-2 (6-phosphofructo-2-kinase). PFK-2 is one of the most potent regulators of the rate of flux through glycolysis and gluconeogenesis.

It is important to note that like many enzymes, there are multiple isoforms of PFK-2 and neither the liver nor the skeletal muscle isoforms contain the AMPK phosphorylation sites of the inducible form of PFK-2. Of particular significance is the fact that the inducible form of PFK-2 is commonly found in tumor cells and this may allow AMPK to play an important role in protecting tumor cells from hypoxic stress. Indeed, techniques for depleting AMPK in tumor cells have shown that these cells become sensitized to nutritional stress upon loss of AMPK activity. Whereas, stress and exercise are powerful inducers of AMPK activity in skeletal muscle, additional regulators of its activity have been identified. Insulin-sensitizing drugs of the thiazolidinedione family as well as the hypoglycemia drug metformin exert a portion of their effects through regulation of the activity of AMPK.

The activity of the AMPK activating kinase LKB1, is critical for regulation of gluconeogenic flux and consequent glucose homeostasis. The action of metformin in reducing blood glucose levels requires the activity of LKB1 in the liver for this function. Also, several adipocytokines either stimulate or inhibit AMPK activation: leptin and adiponectin have been shown to stimulate AMPK activation, whereas, resistin inhibits AMPK activation. Within skeletal muscle and heart activation of AMPK leads to the phosphorylation and inhibition of acetyl-CoA carboxylase (ACC). This inhibition results in a drop in the level of malonyl-CoA which itself is an inhibitor of carnitine palmitoyltransferase I (CPT I). With a drop in the inhibition of CPT I a concomitant increase in -oxidation of fatty acid will occur within the mitochondria. An increase in fatty acid oxidation, like increases in glycolysis, will lead to increases in ATP production.

In addiiton to ACC, AMPK has been shown to phosphorylate and thus regulate the activities of HMG-CoA reductase, (HMGR); hormone-sensitive lipase, (HSL); glycerophosphate acyltransferase, (GPAT); malonyl-CoA decarboxylase, (MCD); glycogen synthase, (GS) and creatine kinase, (CK). Therefore, the effects of AMPK activation are exerted on not only glucose homeostasis and fatty acid metabolism but overall energy homeostasis including glycogen metabolism, cholesterol metabolism and phosphocreatine metabolism.

Not only does activation of AMPK exerts effects on enzyme activity through phosphorylation, there are marked effects on the expression of a number of glycolytic and lipogenic enzymes in the liver and adipose tissue. Included in this list are the genes for the liver isoform of pyruvate kinase (L-PK), fatty acid synthase (FAS) and ACC. Activation of AMPK leads to a reduction in the level of SREBP a transcription factor that is a key regulator of the expression of numerous lipogenic enzymes. Another transcription factor reduced in response to AMPK activation is hepatocyte nuclear factor 4 , HNF4 which is a member of the steroid/thyroid hormone superfamily. HNF4 in known to regulate the expression of several liver and pancreatic b-cell genes such as GLUT2, L-PK and preproinsulin.

Of clinical significance is that mutations in HNF4a are responsible for maturity-onset diabetes of the young, MODY-1. Recent evidence indicates that the newly discovered carbohydrate-response-element-binding protein (ChREBP) is a target for AMPK in the liver. ChREBP is implicated in the transcriptional regulation of L-PK. The target of the thiazolidinedione class of drugs used to treat type 2 diabetes is the peroxisome proliferator-activated receptor , PPAR which itself may be a target for the action of AMPK. The transcription co-activator, p300, is phosphorylated by AMPK which inhibits interaction of p300 with not only PPAR but also the retinoic acid receptor, retinoid X receptor, and thyroid receptor. Another transcription factor target of AMPK is the forkhead protein, FKHR FoxO1a is involved in the activation of glucose-6-phosphatase expression and, therefore, loss of FoxO1a activity in response to AMPK activation will lead to reduced hepatic output of glucose.

AMPK activation in response to hypoxia exerts effects on rates of protein synthesis. Hepatic translation elongation factor, eEF2 is a target for phosphorylation in repsonse to AMPK activation. AMPK phosphorylates and activates the kinase that phosphorylates eEF2 (eEF2K) leading to inhibition of protein synthesis. Another indirect substrate for AMPK that plays a role in protein synthesis is the mammalian target of rapamycin, mTOR. A detailed description of the role of mTOR in regulating protein synthesis

## RELEVANCE OF AMPK TO TYPE 2 DIABETES

An impairment in fuel metabolism occurs in obesity and this impairment is a leading pathogenic factor in type 2 diabetes. The insulin resistance associated with type 2 diabetes is most profound at the level of skeletal muslce as this is the primary site of glucose and fatty acid utilization. Therefore, an understanding of how to activate AMPK in skeletal muscle would offer significant pharmacologic benefits in the treatment of type 2 diabetes. In the non-pharmacologic context, activation of AMPK occurs in response to exercise, an activity known to have significant benefit for type 2 diabetics. Future targets for type 2 diabetes treatments will likely be those that can effect beneficial changes in the activity of AMPK.

# LIPOPROTEINS

## NTESTINAL UPTAKE OF LIPIDS

In order for the body to make use of dietary lipids, they must first be absorbed from the small intestine. Since these molecules are oils, they are essentially insoluble in the aqueous environment of the intestine. The solubilization (or emulsification) of dietary lipids is therefore accomplished by means of bile salts, which are synthesized from cholesterol in the liver and then stored in the gallbladder; they are secreted following the ingestion of fat.

Mixed micelle formed by bile salts, triacylglycerols and pancreatic lipase.

The emulsification of dietary fats renders them accessible to pancreatic lipases (primarily lipase and phospholipase A2). These enzymes, secreted into the intestine from the pancreas, generate free fatty acids and a mixtures of mono- and diacylglycerols from dietary triacylglycerols. Pancreatic lipase degrades triacylglycerols at the 1 and 3 positions sequentially to generate 1,2-diacylglycerols and 2-acylglycerols. Phospholipids are degraded at the 2 position by pancreatic phospholipase A2 releasing a free fatty acid and the lysophospholipid. The products of pancreatic lipases then diffuse into the intestinal epithelial cells, where the re-synthesis of triacyglycerols occurs.

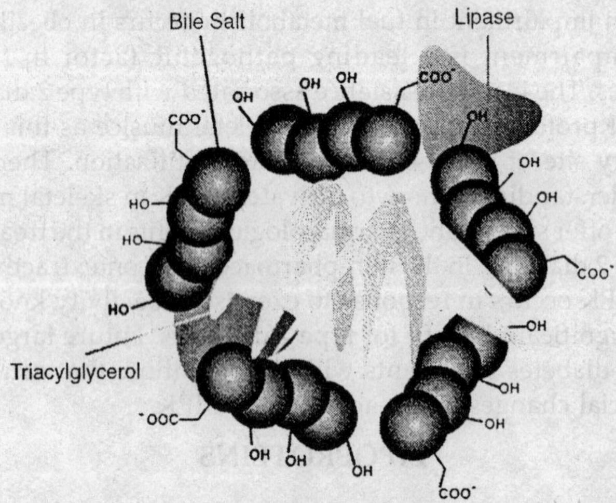


Fig. Lipoprotiens

Dietary triacylglycerols and cholesterol, as well as triacylglycerols and cholesterol synthesized by the liver, are solubilized in lipid-protein complexes. These complexes contain triacylglycerol lipid droplets and cholesteryl esters surrounded by the polar phospholipids and proteins identified as apolipoproteins. These lipid-protein complexes vary in their content of lipid and protein.

## INTERMEDIATE DENSITY LIPOPROTEINS, IDLS

IDLs are formed as triacylglycerols are removed from VLDLs. The fate of IDLs is either conversion to LDLs or direct uptake by the liver. Conversion of IDLs to LDLs occurs as more triacylglycerols are removed. The liver takes up IDLs after they have interacted with the LDL receptor to form a complex, which is endocytosed by the cell. For LDL receptors in the liver to recognize IDLs requires the presence of both apo-B-100 and apo-E (the LDL receptor is also called the apo-B-100/apo-E receptor). The importance of apo-E in cholesterol uptake by LDL receptors has been demonstrated in transgenic mice lacking functional apo-E genes. These mice develop severe atherosclerotic lesions at 10 weeks of age.

## LOW DENSITY LIPOPROTEINS, LDLS

The cellular requirement for cholesterol as a membrane component is satisfied in one of two way: either it is synthesized de novo within the cell, or it is supplied from extra-cellular sources, namely, chylomicrons and LDLs. The dietary cholesterol that goes into chylomicrons is supplied to the liver by the interaction of chylomicron remnants with the remnant receptor. In addition, cholesterol synthesized by the liver can be transported to extra-hepatic tissues if packaged in VLDLs. In the circulation VLDLs are converted to LDLs through the action of lipoprotein lipase. LDLs are the primary plasma carriers of cholesterol for delivery to all tissues.

The exclusive apolipoprotein of LDLs is apo-B-100. LDLs are taken up by cells via LDL receptor-mediated endocytosis, for IDL uptake. The uptake of LDLs occurs predominantly in liver (75/%), adrenals and adipose tissue. As with IDLs, the interaction of LDLs with LDL receptors requires the presence of apo-B-100. The endocytosed membrane vesicles (endosomes) fuse with lysosomes, in which the apoproteins are degraded and the cholesterol esters are hydrolyzed to yield free cholesterol. The cholesterol is then incorporated into the plasma membranes as necessary. Excess intracellular cholesterol is re-esterified by acyl-CoA-cholesterol acyltransferase (ACAT), for intracellular storage. The activity of ACAT is enhanced by the presence of intracellular cholesterol.

Insulin and tri-iodothyronine (T3) increase the binding of LDLs to liver cells, whereas glucocorticoids (e.g., dexamethasone) have the opposite effect. The precise mechanism for these effects is unclear but may be mediated through the regulation of apo-B degradation. The effects of insulin and T3 on hepatic LDL binding may explain the hypercholesterolemia and increased risk of athersclerosis that have been shown to be associated with uncontrolled diabetes or hypothyroidism.

An abnormal form of LDL, identified as lipoprotein-X (Lp-X), predominates in the circulation of patients suffering from lecithin-cholesterol acyl transferas deficiency or cholestatic liver disease. In both cases there is an elevation in the level of circulating free cholesterol and phospholipids.

## HIGH DENSITY LIPOPROTEINS, HDLS

HDLs are synthesized de novo in the liver and small intestine, as primarily protein-rich disc-shaped particles. These newly formed HDLs are nearly devoid of any cholesterol and cholesteryl esters. The primary apoproteins of HDLs are apo-A-I, apo-C-I, apo-C-II and apo-E. In fact, a major function of HDLs is to act as circulating stores of apo-C-I, apo-C-II and apo-E.

HDLs are converted into spherical lipoprotein particles through the accumulation of cholesteryl esters. This accumulation converts nascent HDLs to HDL2 and HDL3. Any free cholesterol present in chylomicron remnants and VLDL remnants (IDLs) can be esterified through the action of the HDL-associated enzyme, lecithin-cholesterol acyl transferase, LCAT. LCAT is synthesized in the liver and so named because it transfers a fatty acid from the C-2 position of lecithin to the C-3-OH of cholesterol, generating a cholesteryl ester and lysolecithin. The activity of LCAT requires interaction with apo-A-I, which is found on the surface of HDLs.

Cholesterol-rich HDLs return to the liver, where they are endocytosed. Hepatic uptake of HDLs, or reverse cholesterol transport, may be mediated through an HDL-specific apo-A-I receptor or through lipid-lipid interactions. Macrophages also take up HDLs through apo-A-I receptor interaction. HDLs can then acquire cholesterol and apo-E from the macrophages; cholesterol-enriched HDLs are then secreted from the macrophages. The added apo-E in these HDLs leads to an increase in their uptake and catabolism by the liver. HDLs also acquire cholesterol by extracting it from cell surface membranes. This process has the effect of lowering the level of intracellular cholesterol, since the cholesterol stored within cells as cholesteryl esters will be mobilized to replace the cholesterol removed from the plasma membrane.

The cholesterol esters of HDLs can also be transferred to VLDLs and LDLs through the action of the HDL-associated enzyme, cholesterol ester transfer protein (CETP, also identified as apo-D). This has the added effect of allowing the excess cellular cholesterol to be returned to the liver through the LDL-receptor pathway as well as the HDL-receptor pathway.

## LDL RECEPTORS

LDLs are the principal plasma carriers of cholesterol delivering cholesterol from the liver (via hepatic synthesis of VLDLs) to peripheral tissues, primarily the adrenals and adipose tissue. LDLs also return cholesterol to the liver. The cellular uptake of cholesterol from LDLs occurs following the interaction of LDLs with the LDL receptor . The sole apoprotein present in LDLs is apo-B-100, which is required for interaction with the LDL receptor.

The LDL receptor is a polypeptide of 839 amino acids that spans the plasma membrane. An extracellular domain is responsible for apo-B-100/apo-E binding. The intracellular domain is responsible for the clustering of LDL receptors into regions of the plasma membrane termed coated pits. Once LDL binds the receptor, the complexes are rapidly internalized (endocytosed). ATP-dependent proton pumps lower the pH in the endosomes, which results in dissociation of the LDL from the receptor. The portion of the endosomal membranes harboring the receptor are then recycled to the plasma membrane and the LDL-containing endosomes fuse with lysosomes. Acid hydrolases of the lysosomes degrade the apoproteins and release free fatty acids and cholesterol. The free cholesterol is either incorporated into plasma membranes or esterified (by ACAT) and stored within the cell.

The level of intracellular cholesterol is regulated through cholesterol-induced suppression of LDL receptor synthesis and cholesterol-induced inhibition of cholesterol synthesis. The increased level of intracellular cholesterol that results from LDL uptake has the additional effect of activating ACAT, thereby allowing the storage of excess cholesterol within cells. However, the effect of cholesterol-induced suppression of LDL receptor synthesis is a decrease in the rate at which LDLs and IDLs are removed from the serum. This can lead to excess circulating levels of cholesterol and cholesteryl esters when the dietary intake of fat and cholesterol exceeds the needs of the body. The excess cholesterol tends to be deposited in the skin, tendons and (more gravely) within the arteries, leading to atherosclerosis.

## CLINICAL SIGNIFICANCES OF LIPOPROTEIN METABOLISM

Fortunately, few individuals carry the inherited defects in lipoprotein metabolism that lead to hyper- or hypolipoproteinemias Persons suffering from diabetes mellitus, hypothyroidism and kidney disease often exhibit abnormal lipoprotein metabolism as a result of secondary effects of their disorders. For example, because lipoprotein lipase (LPL) synthesis is regulated by insulin, LPL deficiencies leading to Type I hyperlipoproteinemia may occur as a secondary outcome of diabetes mellitus. Additionally, insulin and thyroid hormones positively affect hepatic LDL-receptor interactions; therefore, the hypercholesterolemia and increased risk of athersclerosis associated with uncontrolled diabetes or hypothyroidism is likely due to decreased hepatic LDL uptake and metabolism.

Of the many disorders of lipoprotein metabolism, familial hypercholesterolemia (FH) may be the most prevalent in the general population. Heterozygosity at the FH locus occurs in 1:500 individuals, whereas, homozygosity is observed in 1:1,000,000 individuals. FH is an inherited disorder comprising four different classes of mutation in the LDL receptor gene. The class 1 defect (the most common) results in a complete loss of receptor synthesis. The class 2 defect results in the synthesis of a receptor protein that is not properly processed in the Golgi apparatus and therefore is not transported to the plasma membrane. The class 3 defect results in an LDL receptor that is incapable of binding LDLs. The class 4 defect results in receptors that bind LDLs but do not cluster in coated pits and are, therefore, not internalized.

FH sufferers may be either heterozygous or homologous for a particular mutation in the receptor gene. Homozygotes exhibit grossly elevated serum cholesterol (primarily in LDLs). The elevated levels of LDLs result in their phagocytosis by macrophages. These lipid-laden phagocytic cells tend to deposit within the skin and tendons, leading to xanthomas. A greater complication results from cholesterol deposition within the arteries, leading to atherosclerosis, the major contributing factor of nearly all cardiovascular diseases.

## PHARMACOLOGIC INTERVENTION

Drug treatment to lower plasma lipoproteins and/or cholesterol is primarily aimed at reducing the risk of athersclerosis and subsequent coronary artery disease that exists in patients with elevated circulating lipids. Drug therapy usually is considered as an option only if non-pharmacologic interventions (altered diet and exercise) have failed to lower plasma lipids.

- Mevinolin, Mevastatin, Lovastatin: These drugs are fungal HMG-CoA reductase inhibitors. The net result of treatment is an increased cellular uptake of LDLs, since the intracellular synthesis of cholesterol is inhibited and cells are therefore dependent on extracellular sources of cholesterol. However, since mevalonate (the product of the HMG-CoA reductase reaction) is required for the synthesis of other important isoprenoid compounds besides cholesterol, long-term treatments carry some risk of toxicity.
- Nicotinic acid: Nicotinic acid reduces the plasma levels of both VLDLs and LDLs by inhibiting hepatic VLDL secretion, as well as suppressing the flux of FFA

release from adipose tissue by inhibiting lipolysis. Because of its ability to cause large reductions in circulating levels of cholesterol, nicotinic acid is used to treat Type II, III, IV and V hyperlipoproteinemias.

- Clofibrate, Gemfibrozil, Fenofibrate: These compounds are derivatives of fibric acid and promote rapid VLDL turnover by activating lipoprotein lipase. They also induce the diversion of hepatic free fatty acids from esterification reactions to those of oxidation, thereby decreasing the liver's secretion of triacylglycerol- and cholesterol-rich VLDLs.
- Probucol: Probucol increases the rate of LDL metabolism and may block the intestinal transport of cholesterol. The net result is a significant reduction in plasma cholesterol levels.
- Cholestyramine or colestipol (resins): These compounds are nonabsorbable resins that bind bile acids which are then not reabsorbed by the liver but excreted. The drop in hepatic reabsorption of bile acids releases a feedback inhibitory mechanism that had been inhibiting bile acid synthesis. As a result, a greater amount of cholesterol is converted to bile acids to maintain a steady level in circulation. Additionally, the synthesis of LDL receptors increases to allow increased cholesterol uptake for bile acid synthesis, and the overall effect is a reduction in plasma cholesterol. (This treatment is ineffective in homozygous FH patients, since they are completely deficient in LDL receptor

## MEMBRANE LIPIDS AND BILAYER MEMBRANES

The simpler lipids are used primarily for energy storage by living organisms, but more complex lipids are used to form biological membranes. The most important and abundant class of biological membrane lipids is the class of phospholipids. Most phospholipids are derived from glycerol and so they are called phosphoglycerides. The amino alcohol sphingosine gives rise to sphingomyelin, the only significant membrane phospholipid which is not a phosphoglyceride. In phosphoglycerides, the hydroxy groups at C-1 and C-2 of glycerol are esterified to fatty acids while the hydroxy group at C-3 of glycerol is esterified to phosphoric acid; this forms diacylglycerol-3-phosphates.

The two fatty acids found in a diacylglycerol-3-phosphate molecule need not be the same. Diacylglycerol-3-phosphates are only a minor constituent of membranes. The major constituents of membranes are derivatives of diacylglycerol-3-phosphates in which the esterified phosphate also forms an ester linkage with an alcohol. The alcohols commonly found in phosphoglycerides include glycerol, ethanolamine, and choline. The glycolipids are lipids which contain sugars. Glycolipids are derived from sphingosine, as is the phospholipid sphingomyelin. Glycolipids contain no esterified phosphate.

Both phospholipids and glycolipids are amphipathic molecules, that is, molecules in which one end is hydrophobic and one end is hydrophilic. The fatty acid chains form the hydrophobic end of the molecule while the polar glycerol-phosphate-alcohol or sphingosine-sugar portion forms the hydrophilic end of the molecule.

## BILAYER MEMBRANES

When amphipathic molecules are placed in aqueous solution, their hydrophobic tails attempt to orient themselves toward each other. The two ways in which they do so are to form small spherical micelles or to form a planar lipid bilayer. A micelle is a small structure, generally less than two micrometer in diameter, while a lipid bilayer typically has a thickness of about 0.5 micrometer and can have an area of several square millimeters. Sonication or other treatment of lipids can produce spherical liposomes, or lipid vesicles, which contain aqueous solution within an enclosing lipid bilayer. Liposomes are considerably larger than micelles.

Both liposomes and planar bilayer membranes can be prepared from simple solutions of phospholipids such as the diacylglycerol-3-phosphatidyl cholines. Studies of such artificially prepared membranes have given us much insight into the functions and operation of membranes in living cells. Simple bilayer membranes are highly permeable to water molecules, while ions such as sodium ion and potassium ion can traverse them only more slowly (by nine orders of magnitude!). For small molecules, the rates of permeation through simple bilayer membranes increase with their solubility in nonpolar solvents relative to their solubility in water. These observations strongly suggest that transfer of substances through a membrane requires desolvation, transfer through the anhydrous membrane interior, and then resolvation on the other side of the membrane.

The differences in transport behaviour between biological membranes and simple bilayer membranes are now believed to be due to the incorporation of protein molecules or other specifically transporting molecules directly into or onto the surface of a bilayer cell membrane. Considerable research in biology and in chemistry is being directed toward the transport properties of membranes.

## LIPIDS AND FATTY ACIDS

Lipids are the generic names assigned to a group of fat soluble compounds found in the tissues of plants and animals,: and are broadly classified as:

- Fats,
- Phospholipids,
- Sphingomyelins,
- Waxes,
- Sterols.

Fats are the fatty acid esters of glycerol and are the primary energy depots of animals. These are used for long-term energy requirements during periods of extensive exercise or during periods of inadequate food and energy intake. Fish have the unique capability of metabolizing these compounds readily and, as a result, can exist for long periods of time under conditions of food deprivation. A typical example is the many weeks of migration by salmon in their return upstream to spawn; stored lipid deposits are burned for fuel to enable body processes to continue during the strenuous journey.

Phospholipids are the esters of fatty acids and phosphatidic acid. These are the main constituent lipids of cellular membranes allowing the membrane surfaces to be hydrophobic or hydrophylic depending on the orientation of the lipid compounds into the intra or extracellular spaces. Sphingomyelins are the fatty acid esters of sphingosine and are present in brain and nerve tissue compounds. Waxes are fatty acid esters of long-chain alcohols. These compounds can be metabolized for energy and to impart physical and chemical characteristics through the stored lipids of some plant and several animal compounds. Sterols are polycyclic, long-chain alcohols and function as components of several hormone systems, especially in sexual maturation and sex-related physiological functions.

Fatty acids can exist as straight chain or branch chain components; many of the fish fats contain numerous unsaturated double bonds in the fatty acid structures. A short bond designation for. fatty acids will be used throughout where the number identifies the position of the first double bond counting from the methyl end. Linolenic acid would be written 18:3ω3.

The first number identifies the number of carbons; the second number, the number of double bonds; and the last number, the position of the double bonds. Most work on lipid requirements of fish has been with salmonids. Rainbow trout have an essential fatty acid (EFA) requirement for the linolenic of ω3[1] series rather than for linolenic or ω6 as required by most mammals. The main emphasis on lipid requirements has been on EFA and on the energy value of lipids.

## FATTY ACID COMPOSITION OF FISH

## ENVIRONMENTAL INFLUENCES

### Salinity

The difference between fatty acid compositions of marine and freshwater fish has been noted by several authors. Some examples of fatty acid patterns Although these fish lipids are higher in ω 3 fatty acids, it is clear that freshwater fish have higher levels of ω 6 fatty acids than marine species. The average ω6/ω3 ratios are 0.37 and 0.16 for freshwater and marine fish, respectively. Fish in general contain more ω 3 than ω 6 polyunsaturated fatty acids (PUFA) and should have a higher dietary requirement for ω 3 PUFA; thus the dietary EFA requirement of marine fish for ω 3 PUFA may be higher than that of freshwater fish. The same type of difference in the ω6/ω3 ratio between freshwater and seawater is seen when some species of fish migrate from oceans to streams or vice versa. The PUFA ratio of sweet smelt (Plecoglosus altivelis) changes drastically in only one month as they migrate from the sea to a freshwater river.

A similar but reverse change occurs in the masu salmon (Oncorhynchus masu) as they migrate from freshwater to seawater. Even within the same species of fish, the salinity of the water seems to cause a dramatic change in the fatty acid pattern. The difference between marine and freshwater fish may be due simply to differences in the fatty acid content in the diet or it may be related to a specific requirement in fish related to physiological adaptations to the environments.

The phospholipids are generally considered to be structural or functional lipid, being incorporated to a large extent in the membrane structure of cell and subcellular particles. The triglycerides are more often storage lipids and reflect the fatty acid composition of the diet to a greater extent than do the phospholipids. The fatty acid compositions of the triglyceride and phospholipid fractions of fish lipids are presented. It can be seen that the effect of changing environment on the fatty acid composition of the phospholipid is as great in the case of salmon, and considerably greater in the case of the sweet smelt, than it is on the triglyceride composition. Rainbow trout on diets containing either corn oil, which is high in ω6 but low in ω6 PUFA, showed a higher mortality and growth reduction in seawater than in freshwater over the twelve-week feeding period.

## Temperature

There are several other factors besides the salinity of the

water which affect the fatty acid composition and especially the PUFA of fish. The salmonids are mostly cold-water fish. The fatty acids from a number of marine animals from temperate and arctic waters show some significant differences in the general pattern; unfortunately analysis included fatty acids longer than 20:1.

There are a number of other experiments demonstrating the effect of environmental temperature on fatty acid composition of aquatic animals. The general trend toward higher content of long chain PUFA at lower temperatures is quite clear. The ω6/ω3 ratio decreases with a decrease in temperature If the trends in fatty acid composition can be taken as clues to the EFA requirements of fish, the ω3 requirement would be greater for fish raised at lower temperatures. Fish raised in warmer waters, such as common carp, channel catfish, and tilapia may do better with a mixture of ω6 and ω3 fatty acids.

## Effects of Diet

Some of the fatty acid compositions may be seriously affected by the dietary lipids. The mosquito fish and guppies were fed trout pellets which had an ω6/ω3 ratio of 2.75. The catfish were fed diets supplemented with either beef tallow or menhaden oil, with ω6/ω3 ratios of 18.13 and 0.15, respectively. These fish were able to alter the dietary ω6/ω3 ratio in favour of ω3 fatty acid incorporation into the flesh lipids even at the highest temperature.

Commercially available trout pellets are often low in ω3 PUFA and high in ω6 fatty acids. It is important not to ignore the effect of dietary lipid composition on fatty acid composition of fish fed artificial diets. When the dietary ratio is very high in ω6 fatty acids supplied by animal lard or vegetable oils, there is a tendency for fish to alter the ratio of PUFA incorporated in favour of ω3 fatty acids. When the dietary oil is a fish oil high in to 3 fatty acids, there is little change in the ω6/ω3 ratio of lipids incorporated into the fish. This is further suggestive evidence of an EFA requirement of fish for ω3 PUFA.

## Seasonal Variation

Seasonal variations in the fatty acid composition of fish species have often been reported. Seasonal changes have been observed in total lipid and iodine values of herring oils. The iodine value or degree of unsaturation of the oil was minimal in April and maximal in June. The great increase in unsaturation corresponded to the onset of feeding in spring. The absence of a gas liquid chromatograph (GLC) at the time precluded identification of changes in individual fatty acids. Flesh and viscera lipid content of the sardine Sardinops melanosticta vary from 3.9 to 10.77 percent and from 10.9 to 38.3 percent, respectively.

The fatty acids of principal interest with respect to EFA metabolism are 20:4ω6, 20:5ω3, and 22:6ω3. There was considerable variation in all of these fatty acids in both neutral and polar lipid from both tissues. In the flesh, the 20:4ω6 was consistently higher in the neutral lipid than in the polar lipid. The total 20:5 3 plus 22:6 3 was consistently higher in polar lipid than in the neutral lipid. Thus, in spite of the major fluctuations in fatty acids caused by changes in diet and temperature throughout the seasons, there was a consistent preferential incorporation of PUFA of the ω3 series into the polar or phospholipid fraction of the lipids.

One of the best clues to the EFA requirements of a species can be gained from the fatty acid composition of the lipids incorporated into the offspring or egg. The act of reproduction or spawning also has a significant effect on the seasonal fluctuation of lipids in fish. Fatty acid composition of fish egg lipids is probably distinctive for each species and contains increased levels of 16:0, 20:4ω6, 20:5ω3 and 22:6ω3 compared to the liver lipids of the same female fish Elevated levels of 16:0, 20:5 3, and 22:6ω3 and reduced 18:1 in the ovary occurred compared to mesenteric fat of Pacific sardine fed a natural copepod diet.

The blood fatty acids of the sardine fed the natural diet were similar to those of the ovary. When the sardines were fed trout food, both the blood and mesenteric fat responded to the diet with elevated 18:2ω6 and reduced 20:5ω3 arid

22:6ω3. The effect of the diet on ovary fatty acid content was considerably less, as relatively high levels of 20:5ω3 and 22:6ω3 were retained. The ovary lipids of the sweet smelt show an increase in 16:0, and a reduction in the PUFA, especially in the phospholipids, compared to the lipids from the flesh of fish caught at the same time of year. The ω6/ω3 ratio of the ovary was lower than that of the flesh lipids, 0.21 and 0.17 for the ovary compared to 0.31 and 0.20 for the triglycerids and phospholipids of the flesh, respectively.

The hatchability of eggs from common carp fed several different formulated feeds is greatly reduced when the 22:6ω3 of the egg lipids is less than 10 percent. Further, the muscle, plasma, and erythrocyte fatty acid compositions are more affected by dietary lipid than those of the eggs. The EFA requirements of a number of species of fish have been investigated in nutritional studies.

The fish themselves have given ample evidence for EFA preference by the types of fatty acids they incorporate into their lipids. Fish, in general, tend to utilize ω3 over ω6. This is especially observed when the dietary lipids are high in ω6, as the fish tend to alter the ω6/ 3 ratio toward the ω3 fatty acids in the tissue lipids. The lipids of the egg must satisfy the EFA requirement of the embryo until it is able to feed. The fatty acid composition data suggest that the ω3 requirement is greater in seawater than in freshwater and higher in cold water than in warm water.

## BODY LIPID COMPOSITION AND DIETARY LIPID REQUIREMENTS

Detailed information is still lacking on the dietary lipid requirements of many species of fish, but there is an abundance of information on the fatty acid composition of fish oils. Information on the lipid composition of fish can be used to make some guesses about dietary lipid requirements. Linolenic acid (18:3ω3) resulted in some sparing action and growth promotion in rats, and fatty acids of the ω6 EFA prevented all of the EFA-deficiency symptoms. Research with homeothermic land-dwelling animals showed that the ω6 series of fatty acids are the "essential fatty acids", while the ω3 series are considered to be non-essential or only have a partial sparing action on EFA-deficiency. The ω6 series of fatty acids have been shown to be essential to enough species that it began to become accepted that these are the essential fatty acids for all animals.

It was assumed by many that fish also required ω6 fatty acids. Many researchers began by supplementing fish diets with vegetable oils, such as corn, peanut, or sunflower oil, which were rich in linoleic acid. The main sympton observed during the development of EFA deficiency in chinook salmon fed fat-free diets was a marked depigmentation that can be prevented by addition of 1 percent trilinolein, but not by 0.1 percent linolenic acid. Although the ω6 fatty acids are considered to be essential, one of the general characteristics of fish oils is the low levels of ω6 series fatty acids and the higher levels of ω3 type fatty acids. There is evidence that polyunsaturated fatty acids (PUPA) of the ω3 series, which are present in relatively large concentrations in fish oil, play the role of essential fatty acid for fish.

## POLYUNSATURATED FATTY ACID REQUIREMENTS OF FISH

When a test diet containing 13 percent corn oil and 2 percent cod-liver oil was fed to rainbow trout, subsequent deletion of the cod-liver oil from the diet depressed growth and produced some kidney degeneration which might be attributed to a lack of sufficient ω3 PUFA present in significant quantities in cod-liver oil Dietary fish oil is superior to corn oil in promoting growth of rainbow trout (Salmo gairdneri) and the yellow-tail (Seriola guingueradiata). Dietary linolenic acid or ethyl linolenate (18:3ω3) gives a positive growth response for rainbow trout which may be attributed to a dietary requirement for ω3 fatty acids.

## ESSENTIAL FATTY ACID REQUIREMENTS OF FISH

One of the most widely accepted theories explaining the presence of such high levels of 20:5ω3 and 22:6ω3 fatty acids in fish oils is related to the effect of unsaturation on the melting point of a lipid. The greater degree of unsaturation of fatty acids in the fish phospholipids allows for flexibility of cell membrane at lower temperatures. The ω3 structure allows a greater degree of unsaturation than the ω6 or ω9. This theory is consistent with the fact that cold water fish have a greater nutritional requirement for ω3 fatty acids, while the EFA requirement of some warm water fish can be satisfied by a mixture of ω6 plus ω3.

## Rainbow Trout

Rainbow trout, a cold water fish, requires ω3 fatty acids as EFA in the diet. The EFA requirement can be met by 1 percent 18:3ω3 in the diet. Inclusion of 18:2ω6 in the diet may result in some improvement in growth and feed conversion compared to EFA deficient diets; however, the ω6 fatty acids will not prevent some EFA deficiency symptoms such as the "shock syndrome". Although it is clear that rainbow trout require ω3 fatty acids, it remains to be shown conclusively whether some dietary level of ω6 fatty acid is essential.

EFA value for rainbow trout. Either 20:5ω3 or 22:6ω3 is superior to 18:3ω3 in an EFA value for rainbow trout, and the former two fatty acids in combination are superior to either alone. This is consistent with data for mammals, where 20:4ω6 has higher EFA value than 18:2ω6. The superior nutritional value of C-20 and C-22 carbon ω3-PUFA is further supported by the excellent growth promoting effects of dietary fish oils such as pollock liver oil and salmon oil for rainbow trout.

## Channel Catfish

One of the most important warm water fish in North America is the channel catfish (Ictalurus punctatus). The quantitative EFA requirement of the catfish has not yet been determined. However, the evidence is convincing that the ω3 requirement is not as great as that of rainbow trout. Analysis of fatty acids of lipids from catfish purchased at five processing plants showed very low levels of 20:4ω6, 20:5ω3, and 22:6ω3; 0.8 - 5.5, 0.2 - 1.3, and 0.6 - 6.1 percent of the total fatty acids, respectively. It was shown that corn oil added to a semipurified casein based diet initially resulted in a positive growth response and protein sparing, but later growth inhibition was observed. The apparent repressive effects of corn oil may be due to its 18:2ω6 content since 20:5ω3 and 22:6ω3 present in menhaden oil had no apparent detrimental effects. The growth suppressing effects of 18:2ω6 were also noted when 3 percent corn oil was added to 3 percent beef tallow and 3 percent menhaden oil. The growth suppression caused by unsaturated fatty acids does not appear to be limited to ω6 fatty acids. Linseed oil (high in 18:3ω3) in the diet of catfish resulted in growth suppression similar to that caused by corn oil compared with dietary beef tallow, olive oil and menhaden oil.

## The Common Carp

The picture for another warm water fish, the common carp (Cyprinus carpio) is much clearer than that for the channel catfish. This fish has an EFA requirement for both ω3 and ω6 fatty acids. The best weight gains and feed conversions are obtained in fish receiving a diet containing both 1 percent 18:2ω6 and 1 percent 18:3ω3. With the carp, 20:5ω3 and 22:6ω3 at 0.5 percent of the diet are superior to 1 percent of 18:3u3. Carp fed a fat-free, or EFA deficient, diet incorporated high levels of 20:3 9 in their lipids, especially in the phospholipids.

## The Eel

The eel (Anguilla japonica), another warm water fish, has a requirement for both ω3 and ω6 fatty acids. Corn oil (high in ω6) and cod liver oil (high in ω3)in a 2:1 mixture are most favourable for the growth of eels. The eel requires ω6 and ω3 in the same proportion as the carp, but at a lower level in the diet; namely, 0.5 percent of each, rather than 1.0 percent of each PUFA.

## The Plaice

The plaice becomes depleted of both ω3 and ω6 PUFA when fed a fat-free diet. The addition of 12:0 and 14:0 to the diet result in the synthesis of saturated and monoenoic fatty acids of chain lengths up to C18; however, increased levels of 20:3ω9 noted in trout and mammals have not been reported in plaice. Plaice fed dietary 18:2ω6 and 18:3ω3 will not produce significant amounts of 20:4 ω6, 20:5ω3, or 22:6ω3.

## The Turbot

The growth of turbot (Scophthalmus matimus) is much better with ω3 PUFA than with ω6 or saturated fat (hydrogenated coconut oil) in the diet. The turbot also appears to be unable to convert dietary 18:2ω6 to 20:4ω6 when fed corn oil, or to convert endogenous 18:1ω9 to 20:3ω9 when fed the EFA deficient diet.

Although it appears to have an EFA requirement for ω3 fatty acids such as are present in cod liver oil, this requirement is not satisfied by 18:3ω3. The chain elongation and desaturation of 18:lω9, 18:2ω6, or 18:3ω3 has been found to be very limited (3-15 percent) in turbot compared to the rainbow trout where 70 percent of the 18:3ω3 was converted to 22:6ω3. The required level of long-chain ω3 fatty acids for turbot is at least 0.8 percent of the diet.

## The Red Sea Bream

The red sea bream (Chrysophyrys major) grows better when the dietary lipid is of marine origin (pollock residual oil) rather than a vegetable oil (such as corn oil). The EFA requirement of the red sea bream is not satisfied by either linoleic acid of corn oil or supplemented linolenate. A mixture of 20:5ω3 and 22:6ω3 supplemented to the corn oil diet has been shown to be effective in improving growth and condition of these fish. Thus, even in warm water, marine fish seem to require not just ω3 fatty acids but 0)3 fatty acids of 20 to 22 carbon-chain length. A direct correlation between feed efficiency and the 18:1 level in the lipids of the red sea bream has been postulated.

## Other Species

Among warm water marine fish, mullet and fundulus possess the ability to chain, elongate, and desaturate 18:2ω6 or 18:3ω3 PUFAs. The process is, however, inhibited in fundulus by high levels (about 5 percent) of these PUFAs of 18:2ω6 or 18:3ω3 in the diet.

## FATTY ACID METABOLISM IN FISH

It appears that high levels of 18-carbon ω6 or ω3 fatty acids inhibit the synthesis and metabolism of 18:1ω9. It is interesting to note that the channel catfish, which also exhibits negative growth response to dietary 18:2ω6 or 18:3ω3, incorporates very high levels of 18:1 into its body lipids. The inclusion of either 18:2ω6 or 18:3ω3 in the diet reduces the levels of 18:1 fatty acids in body lipids. A similar reduction has also been observed in red sea bream liver phospholipid when either of the PUFAs is added to the diet.

The competitive inhibition of chain elongation and desaturation of members from one series of fatty acids for members of another series is well established, with ω3 > 6 > 9 being the usual order of potency for inhibition.

The pathways of fatty acid metabolism have been reviewed by Mead and Kayama (1967). Fish are able to synthesize, de novo from acetate, the even-chain, saturated fatty acids. Radio tracer studies have shown that fish can convert 16:0 to the ω7 monoene and 18:0 to the ω9 monoene. The ω5, 11 and ω3 monoenes are proposed based on the identification of these isomers in the monoenes of herring oil.

Fish are unable to synthesize any fatty acids of the ω6 and u3 series unless a precursor with this structure is present in the diet. Fish are able to desaturate and elongate fatty acids of the ω9, 6, or ω3 series There is competitive inhibition of the elongation desaturation of fatty acids of one series by members of the other series.

The ω3 fatty acids are the most potent inhibitors, the ω9 are the least. The ability to elongate and desaturate fatty acids is not the same in all species of fish, as was noted earlier. The turbot was able to desaturate and elongate only 3-15 percent of 18:1ω9, 18:2ω6, or 18:3ω3, when given the C14 labelled fatty acid; in the rainbow trout, 70 percent of the label from 18:3ω3 (C14) was found in 22:6ω3.

The essential fatty acids are not unique in their ability to supply energy. The -oxidation of fatty acids in fish is basically the same as in mammals. The EFA and saturated and monoenoic fatty acids are all equally utilized by fish for energy production.

Increased swelling rates of liver mitochondria occur in rainbow trout fed diets deficient in ω3 fatty acids. It is possible that EFA plays an important role in the permeability as well as the plasticity of membranes. The role of ω3 fatty acids in membrane permeability may be one of the factors accounting for differences in content of this family of fatty acids between freshwater and marine fish.

Fish mitochondria with high levels of the ω3 PUFA and very low levels of ω6 fatty acids are very similar to mammalian mitochondria with respect to cytochrome content, -oxidation of fatty acids, operation of the tricarboxylic acid cycle, electron transport, and oxidative phosphorylation. The ω3 PUFA may play the same role in fish that the ω6 fatty acids play in rats. The EFA play another role in the mitochondria. In addition to their importance in membrane structure, the EFA are important in some enzyme systems. Unsaturated fatty acids play an important role in the transportation of other lipids. It has been repeatedly shown that feeding PUFA will lower the cholesterol levels in animals with above-normal blood lipid and cholesterol levels.

Fish oils are more effective in lowering cholesterol levels than are most dietary lipids. The major portion of the fatty acids absorbed across the intestinal mucosa are transported as protein-lipid complexes stabilized by phospholipids. The low body temperature in fish probably results in a greater importance for unsaturation in transport of lipids than in homeothermic animals.

## NEGATIVE ASPECTS OF LIPIDS IN FISH NUTRITION

The requirement by fish for PUFA of the ω3 series creates problems with respect to feed storage. These types of fatty acids are very labile on oxidation. The products of lipid oxidation may react with other nutrients such as proteins, vitamins, etc. and reducing the available dietary levels or the oxidation products may be toxic. The effect of oxidized lipids on dietary proteins, enzymes and amino acids have been demonstrated.

The use of oxidized menhaden oil in the diets of swine and rats caused decreased appetite, reduced growth, yellowish-brown pigmentation of depot fat, and decreased haemoglobin and haematocrit levels. The negative effects of the oxidized fish oils were reversed by the addition of alpha-tocopherol acetate or ethoxyoquin to the diet.

Much of the use of vegetable oils in fish diets in the 1950s and 1960s might, in part, have been based on their greater stability in prepared diets.

It has been demonstrated that rancid herring and hake meals in fish feeds caused dark colouration, anaemia, lethargy, brown-yellow pigmented liver, abnormal kidneys, and small gill clubbing in chinook salmon. The symptons can be alleviated by addition of alpha-tocopherol to the diets containing rancid fish meals. The addition of vitamin E would prevent the toxic or negative effects of adding 5 percent highly oxidized salmon oil to the diet of rainbow trout. This same sparing effect of alpha-tocopherol can also apply to rancid carp feed.

The positive nutritional value of ω3 fatty acids in fish lipids for fish feeds can become a negative factor if adequate care is not taken in the preparation and storage of feeds. Only fresh oils with low peroxide values should be used in feeds. Fish feed ingredients such as fish meals should be protected against oxidation. The level of vitamin E added to the diet should be increased as the PUFA level is increased. The finished feed, if possible, should be stored in air tight containers at reduced temperatures with minimum exposure to UV radiation and other factors accelerating the rate of lipid oxidation. The problems of rancidity or antioxidation of lipids in fish feeds should not be ignored.

## GLYCOSPHINGOLIPIDS

The glycosphingolipids, describes the structure and

classification of the different core series of these molecules, and outlines general principles regarding their biological roles in membrane structure, modulation of membrane protein function, cell-cell recognition, and host-pathogen interactions. The major emphasis is on higher animal glycosphingolipids, about which the most information is currently available. The glycosphingolipids were originally discovered in the 1940s as lipid-rich substances from normal tissues that accumulated in the tissues of patients with so-called "storage disorders" such as Tay-Sachs disease.

Other investigators also extracted glycosphingolipids from the stroma of the erythrocytes of various animals. Their complex and unusual structure as well as their as yet unknown functions caused them to be called *sphingolipids*, after the enigmatic Egyptian Sphinx. Further structural work showed that these amphipathic molecules consisted of two components: an outer glycan portion attached to an inner lipid tail consisting of a sphingolipid called ceramide. The hydrophobic nature of the sphingolipid component causes these molecules to be embedded in the outer leaflet of cell membranes, along with phospholipids, cholesterol, and glycerolipids.

## Isolation, Purification, and Analysis

Most glycosphingolipids can be obtained in good yield from cells and tissues by sequential organic extractions of increasing polarity. Subsequently, they can be fractionated away from other lipids in the extract and from each other using techniques such as DEAE ion-exchange chromatography and silica gel on either thin-layer plates or columns.

High-performance liquid chromatography adaptations of these traditional methods have been particularly powerful in obtaining complete separations of these complex mixtures. It is noteworthy that the base treatment, which is traditionally used to eliminate phospholipids from crude extracts, can damage labile O-acetyl and O-acyl groups in glycosphingolipids.

The availability of specific endoglyco-ceramidases that

cleave the glycan portion away from the ceramide has introduced an alternative approach to the their analysis.

## Structures, Major Classes, and Nomenclature

The minimal motif that defines the glycosphingolipids is a monosaccharide attached directly to a ceramide unit. In higher-animal cells, this monosaccharide is usually glucose or galactose, giving glucosylceramide or galactosylceramide. Each of these basic units can then be further extended by the stepwise addition of further monosaccharides.

The number of variations in extension of galactosylceramide is somewhat limited. Glucosylceramide typically has an additional galactose added, giving lactosylceramide, which is then extended in many ways, giving rise to several defined core structures. Some of these classes of molecules tend to be enriched in certain tissues types, e.g., the ganglio-series in the brain and the neolacto-series in leukocytes.

Although all existing classifications of glycosphingolipids are based on these variations in core glycan structure, it is important to note that the ceramide moiety can also show substantial structural heterogeneity.

For example, the typical sphingosine base can vary in the number of double bonds and in the length of the acyl chain. Likewise, the fatty acid attached to the sphingosine via an amide linkage can vary in length from C14 to C24, and in its degree of unsaturation and/or hydroxylation. These variations in ceramide structure are also expressed in a cell-type-specific and developmentally regulated manner, implying that they have some specific purpose.

In this regard, one interesting fact is that certain types of glycans are found to be enriched on certain subclasses of ceramides; e.g., in brain gangliosides, the C20 homolog is much more likely to be found in glycosphingolipids with more than one sialic acid residue. Another instance is the ceramide galactosyltransferase that synthesizes GalCer, which displays a marked preference for hydroxy fatty-acid-containing ceramides.

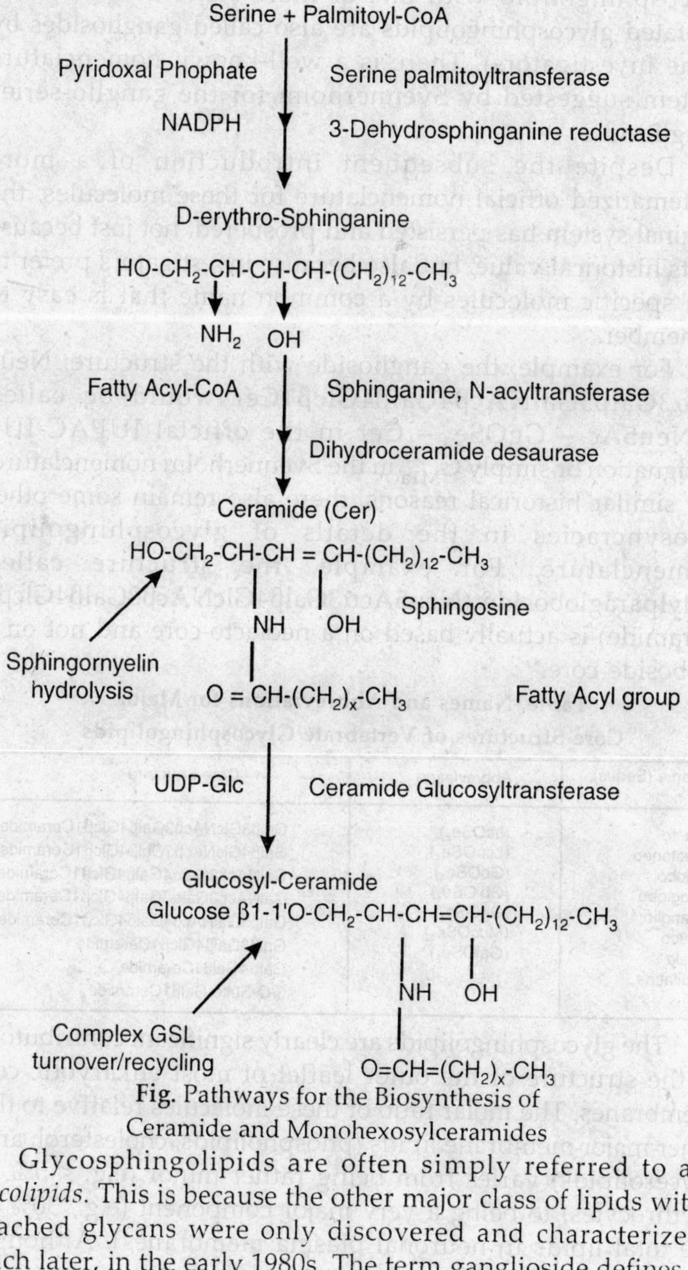


**Fig.** Pathways for the Biosynthesis of Ceramide and Monohexosylceramides

Glycosphingolipids are often simply referred to as *glycolipids*. This is because the other major class of lipids with attached glycans were only discovered and characterized much later, in the early 1980s. The term ganglioside defines a glycosphingolipid with one or more sialic acid residues (sulfated glycosphingolipids are also called gangliosides by some investigators). There is a well-known nomenclature system suggested by Svennerholm for the ganglio-series gangliosides.

Despite the subsequent introduction of a more systematized official nomenclature for these molecules, the original system has persisted and prospered, not just because of its historical value, but also because investigators prefer to call specific molecules by a common name that is easy to remember.

For example, the ganglioside with the structure: Neu5 Acα3Galβ3GalNAcβ4Galβ4Glcβ1Cer would be called $II^2$Neu5Ac – $GgOSe_4$ – Cer in the official IUPAC-IUB designation or simply $G_{M1a}$ in the Svennerholm nomenclature. For similar historical reasons, there also remain some other idiosyncracies in the details of glycosphingolipid nomenclature. For example, the structure called sialylparagloboside (Neu5Acα3Galβ4GlcNAcβ3Galβ4Glcβ1 Ceramide) is actually based on a neolacto-core and not on a globoside core.

**Table. Names and Abbreviations for Major Core Structures of Vertebrate Glycosphingolipids**

| Name (Series) | Abbreviation | Core Structure |
|---|---|---|
| Lacto | ($LcOSe_4$) | Galβ3GlcNAcβ3Galβ4Glcβ1Ceramide |
| Lactoneo | ($LcnOSe_4$) | Galβ4GlcNAcβ3Galβ4Glcβ1Ceramide |
| Globo | ($GbOSe_4$) | GalNAcβ3Galα4Galβ4Glcβ1Ceramide |
| Isoglobo | ($GbiOSe_4$) | GalNAcβ3Galα3Galβ4Glcβ1Ceramide) |
| Ganglio | ($GgOSe_4$) | Galβ3GalNAcβ4Galβ4Glcβ1Ceramide |
| Muco | ($MucOSe_4$) | Galβ3Galβ4Glcβ1Ceramide |
| Gala | ($GalOSe_2$) | Galα4Galβ1Ceramide |
| Sulfatide | | 3-O-Sulfo-Galβ1Ceramide |

The glycosphingolipids are clearly significant contributors to the structure of the outer leaflet of most eukaryotic cell membranes. The molar ratio of these molecules relative to the other major membrane lipids (phospholipids, cholesterol, and glycerolipids) varies from being rather minor (e.g. $< 5\%$ in erythrocytes) to being a very major component (e.g., 30% of the total lipids in neuronal plasma membranes). Although glycosphingolipids are found throughout all eukaryotes, the nature of the core structures can vary in different taxa.

## TERPENE

Terpenes are a large and varied class of hydrocarbons, produced primarily by a wide variety of plants, particularly conifers though also by some insects such as termites or swallowtail butterflies, which emit terpenes from their osmeterium.

They are the major components of resin, and of turpentine produced from resin. The name "terpene" is derived from the word "turpentine".

In addition to their roles as end-products in many organisms, terpenes are major biosynthetic building blocks within nearly every living creature.

Steroids, for example, are derivatives of the triterpene squalene. When terpenes are modified chemically, such as by oxidation or rearrangement of the carbon skeleton, the resulting compounds are generally referred to as terpenoids. Some authors will use the term terpene to include all terpenoids.

Terpenoids are also known as Isoprenoids. Terpenes and terpenoids are the primary constituents of the essential oils of many types of plants and flowers. Essential oils are used widely as natural flavor additives for food, as fragrances in perfumery, and in traditional and alternative medicines such as aromatherapy. Synthetic variations and derivatives of natural terpenes and terpenoids also greatly expand the variety of aromas used in perfumery and flavors used in food additives.

### Structure and Biosynthesis

Terpenes are derived biosynthetically from units of isoprene, which has the molecular formula $C_5H_8$. The basic molecular formulae of terpenes are multiples of that, $(C_5H_8)_n$ where n is the number of linked isoprene units. This is called the *isoprene rule* or the *C5 rule*. The isoprene units may be linked together "head to tail" to form linear chains or they may be arranged to form rings. Isoprene itself does not undergo the building process, but rather activated forms, isopentenyl pyrophosphate (IPP or also isopentenyl diphosphate) and dimethylallyl pyrophosphate (DMAPP or also dimethylallyl diphosphate), are the components in the biosynthetic pathway.

IPP is formed from acetyl-CoA via the intermediacy of mevalonic acid in the HMG-CoA reductase pathway.

An alternative, totally unrelated biosynthesis pathway of IPP is known in some bacterial groups and the plastids of plants, the so-called MEP(2-Methyl-D-erythritol-4-phosphate)-pathway, which is initiated from C5-sugars. In both pathways, IPP is isomerized to DMAPP by the enzyme isopentenyl pyrophosphate isomerase.

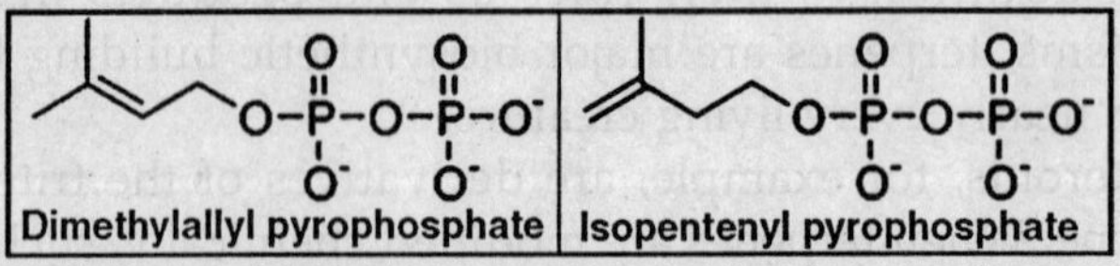


As chains of isoprene units are built up, the resulting terpenes are classified sequentially by size as hemiterpenes, monoterpenes, sesquiterpenes, diterpenes, sesterterpenes, triterpenes, and tetraterpenes.

## STEROL

Sterols are an important class of organic molecules. They occur naturally in both plants and animals, with the most familiar type of animal sterol being cholesterol, which has been shown to contribute to high blood pressure and heart disease.

Within the past decade, interest in plant sterols as a dietary supplement has increased, due to studies showing that they can contribute to lower cholesterol levels.

### Types of Sterols

Sterols of plants are called *phytosterols* and sterols of animals are called *zoosterols*. Important zoosterols are cholesterol and some steroid hormones; notable phytosterols include campesterol, sitosterol, and stigmasterol.

Ergosterol is a sterol present in the cell membrane of fungi, where it serves a role similar to cholesterol in animal cells.

### Role in Biochemistry

Sterols and related compounds play essential roles in the physiology of eukaryotic organisms. For example, cholesterol forms part of the cellular membrane in animals, where it affects the cell membrane's fluidity and serves as second messenger in developmental signaling.

In humans and other animals, corticosteroids, such as cortisol act as signalling compounds in cellular communication and general metabolism.

### As a Nutritional Supplement

Phytosterols, more commonly known as plant sterols, have been shown in clinical trials to block cholesterol absorption sites in the human intestine, thus helping to reduce cholesterol in humans. They are currently approved by the U.S.

Food and Drug Administration for use as a food additive, however, there is some concern that they may block absorption, not only of cholesterol, but of other important nutrients as well. At present the American Heart Association has recommended that supplemental plant sterols be taken only by the those diagnosed with elevated cholesterol, and has particularly recommended that they not be take by pregnant women or nursing mothers.

### Chemical Classification and Structure

Sterols are also known as steroid alcohols. They are a subgroup of steroids with a hydroxyl group at the 3-position of the A-ring. They are amphipathic lipids synthesized from acetyl-coenzyme A via the HMG-CoA reductase pathway. The overall molecule is quite flat. The hydroxyl group on the A ring is polar. The rest of the aliphatic chain is non-polar.

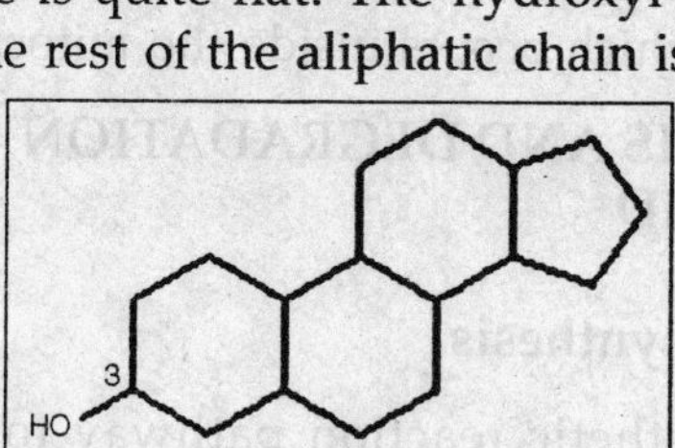

## STEROID

A steroid is a terpenoid lipid characterized by a carbon skeleton with four fused rings, generally arranged in a 6-6-6-5 fashion. Steroids vary by the functional groups attached to these rings and the oxidation state of the rings. Hundreds of distinct steroids are found in plants, animals, and fungi. All steroids are made in cells either from the sterol lanosterol (animals and fungi) or the sterol cycloartenol (plants). Both sterols are derived from the cyclization of the triterpene squalene. IUPAC recommended ring (left) and atom numbering (right) of the steroid skeleton.

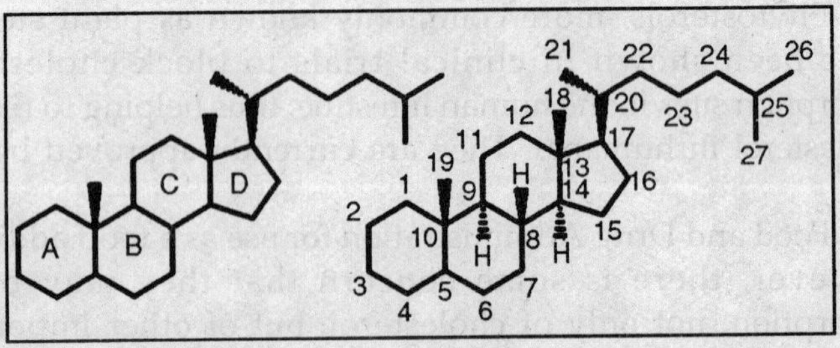


### Origin of Steroid

Steroids include estrogen, cortisol, progesterone and testosterone. Estrogen and progesterone are made primarily in the ovary and in the placenta during pregnancy and testosterone in the testes. Testosterone is also converted into estrogen to regulate the supply of each, in the bodies of both females and males. Certain neurons and glia in the central nervous system (CNS) express the enzymes that are required for the local synthesis of pregnane neurosteroids, either *de novo* or from peripherally derived sources. The rate limiting step of steroid synthesis is the conversion of cholesterol to pregnenolone which occurs inside the mitochondrion.

## BIOSYNTHESIS AND DEGRADATION OF FATTY ACID

### Fatty Acid Biosynthesis

The biosynthetic reaction pathway to a compound is

usually not a simple opposite of its breakdown. In fatty acid synthesis, acetyl-CoA is the direct precursor only of the methyl end of the growing fatty acid chain. All the other carbons come from the acetyl group of acetyl-CoA but only after it is modified to provide the actual substrate for fatty acid synthase, malonyl-CoA.

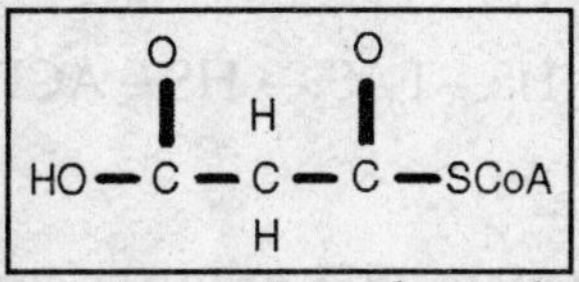

Malonyl-CoA contains a 3-carbon dicarboxylic acid, malonate, bound to Coenzyme A. Malonate is formed from acetyl-CoA by the addition of $CO_2$ using the biotin cofactor of the enzyme acetyl-CoA carboxylase.

$$HCO_3^- \; Acetyl - CoA + HCO_3^- + ATP \rightleftharpoons$$
$$\text{Malonyl-CoA} + ADP + Pi$$

Formation of malonyl-CoA is the commitment step for fatty acid synthesis, because malonyl-CoA has no metabolic role other than serving as a precursor to fatty acids.

Fatty acid synthase (FAS) carries out the chain elongation steps of fatty acid biosynthesis. FAS is a large multienzyme complex. In mammals, FAS contains two subunits, each containing multiple enzyme activities. In bacteria and plants, individual proteins, which associate into a large complex, catalyze the individual steps of the synthesis scheme.

## Initiation

Fatty acid synthesis starts with acetyl-CoA, and the chain grows from the "tail end" so that carbon 1 and the alpha-carbon of the complete fatty acid are added last. The first reaction is the transfer of the acetyl group to a pantothenate group of acyl carrier protein (ACP), a region of the large mammalian FAS protein. (The acyl carrier protein is a small, independent peptide in bacterial FAS, hence the name.) The pantothenate group of ACP is the same as is found on Coenzyme A, so the transfer requires no energy input:

$$Acetyl^{\sim} S - CoA + HS - ACP \longrightarrow HS - CoA + Acetyl^{\sim} S - ACP$$

In the preceding reaction, the *S* and *SH* refer to the thio group on the end of Coenzyme *A* or the pantothenate groups. The ~ is a reminder that the bond between the carbonyl carbon of the acetyl group and the thio group is a "high energy" bond (that is, the activated acetyl group is easily donated to an acceptor). The second reaction is another transfer, this time, from the pantothenate of the ACP to cysteine sulfhydral (–SH) group on FAS.

$$\text{Acetyl}^{\sim}\text{ACP} + \text{HS} - \text{FAS} \rightarrow \text{HS} - \text{ACP} + \text{Acetyl}^{\sim}\text{S} - \text{FAS}$$

## Elongation

The pantothenate –SH group is now ready to accept a malonyl group from malonyl-CoA:

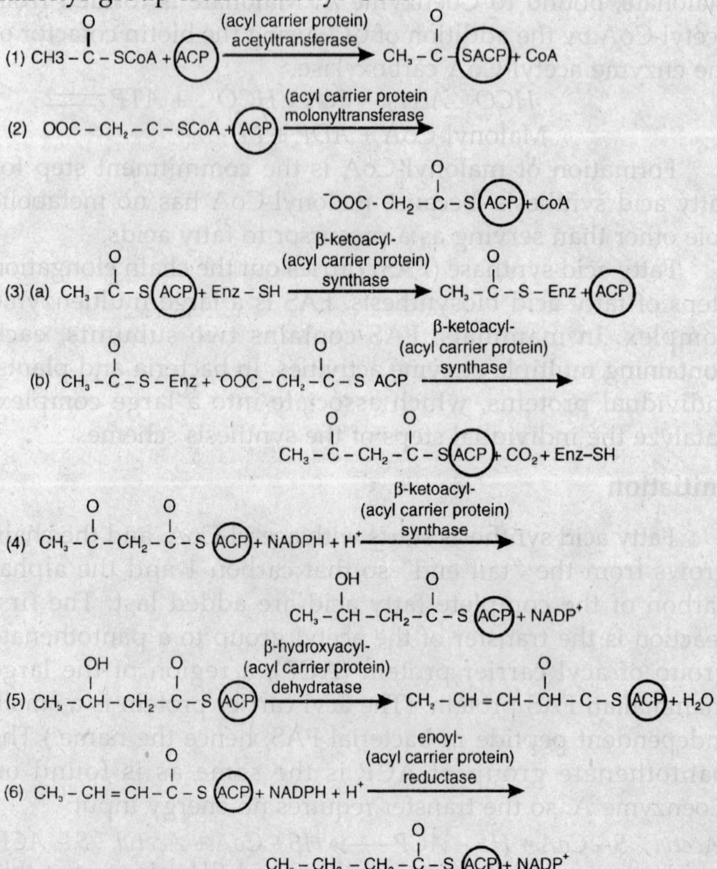

The FAS has two activated substrates, the acetyl group bound on the cysteine –SH and the malonyl group bound on the pantothenate –SH. Transfer of the 2-carbon acetyl unit from Acetyl < S-cysteine to malonyl-CoA has two features:

- Release of the $CO_2$ group of malonyic acid that was originally put on by acetyl-CoA carboxylase
- Generation of a 4-carbon β-keto acid derivative, bound to the pantothenate of the ACP protein

The ketoacid is now reduced to the methylene ($CH_2$) state in a three-step reaction sequence.

1. Reduction by NADPH to form the â- hydroxy acid derivative:

2. Dehydration, that is, the removal of water to make a trans-double bond:

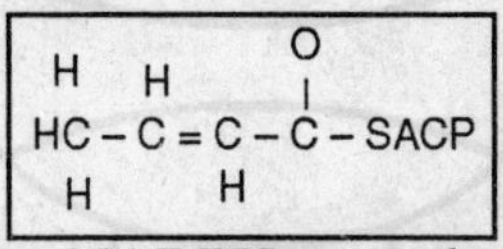

3. Reduction by NADPH to make the saturated fatty acid:

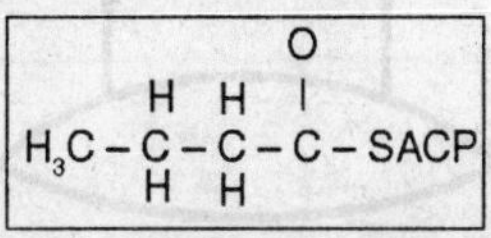

The elongated 4-carbon chain is now ready to accept a new 2-carbon unit from malonyl-CoA. The 2-carbon unit, which is added to the growing fatty acid chain, becomes carbons 1 and 2 of hexanoic acid (6-carbons).

## RELEASE

The cycle of transfer, elongation, reduction, dehydration, and reduction continues until palmitoyl-ACP is made. Then the thioesterase activity of the FAS complex releases the 16-carbon fatty acid palmitate from the FAS. Fatty acid synthesis provides an extreme example of the phenomenon of metabolic channeling: neither free fatty acids with more than four carbons nor their CoA derivatives can directly participate in the synthesis of palmitate.

Instead they must be broken down to acetyl-CoA and reincorporated into the fatty acid. Fatty acids are generated cytoplasmically while acetyl-CoA is made in the mitochondrion by pyruvate dehydrogenase.This implies that a shuttle system must exist to get the acetyl-CoA or its equivalent out of the mitochondrion. The shuttle system operates in the following way: Acetyl-CoA is first converted to citrate by citrate synthase in the TCA-cycle reaction.

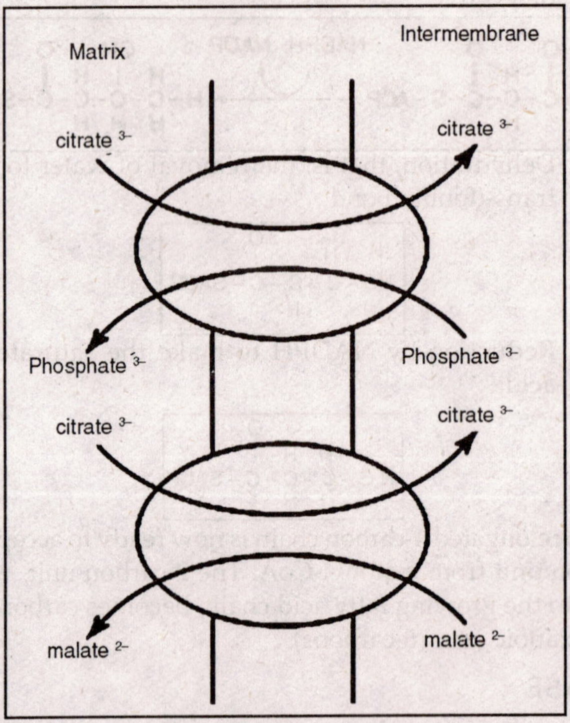


**Fig.** TCA-cycle Reaction

After it is in the cytosol, citrate is cleaved to its 2- and 4-carbon components by citrate lyase to make acetyl-CoA and oxaloacetate. Citrate lyase requires ATP. Fatty acid biosynthesis (and most biosynthetic reactions) requires

NADPH to supply the reducing equivalents. Oxaloacetate is used to generate NADPH for biosynthesis in a two-step sequence. The first step is the malate dehydrogenase reaction found in the TCA cycle. This reaction results in the formation of NAD from NADH (the NADH primarily comes from glycolysis).

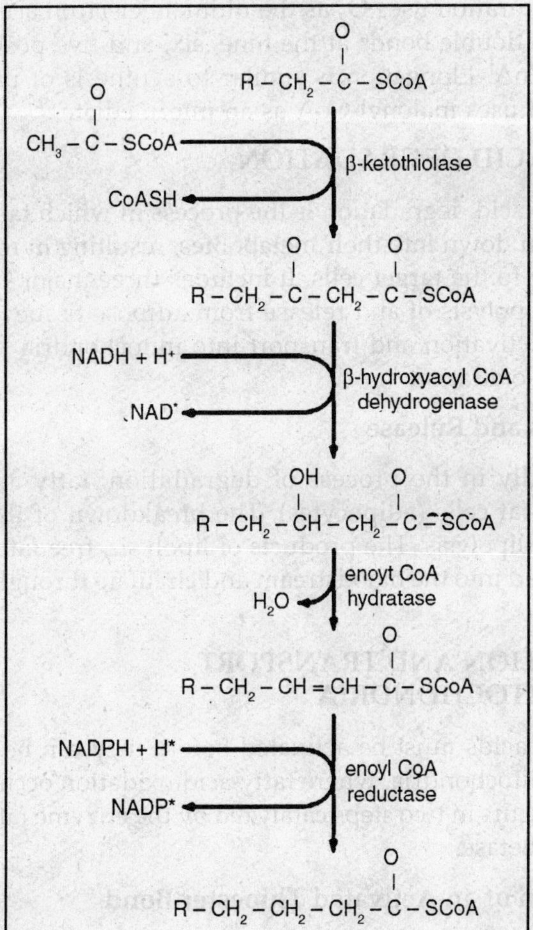


The malate formed is a substrate for the malic enzyme reaction, which makes pyruvate, $CO_2$, and NADPH. Pyruvate is transported into the mitochondria where pyruvate carboxylase uses ATP energy to regenerate oxaloacetate. Palmitate is the starting point for other fatty acids that use a set of related reactions to generate the modified chains and head groups of the lipid classes.

Microsomal enzymes primarily catalyze these chain modifications.

Desaturation uses $O_2$ as the ultimate electron acceptor to introduce double bonds at the nine, six, and five positions of an acyl-CoA. Elongation is similar to synthesis of palmitate because it uses malonyl-CoA as an intermediate.

## FATTY ACID DEGRADATION

Fatty acid degradation is the process in which fatty acids are broken down into their metabolites, resulting in release of its energy to the target cells. It includes three major steps:

- Lipolysis of and release from adipose tissue
- Activation and transport into mitochondria
- β-oxidation

### Lipolysis and Release

Initially in the process of degradation, fatty acids are stored in fat cells (adipocytes). The breakdown of this fat is known as lipolysis. The products of lipolysis, free fatty acids, are released into the bloodstream and circulate throughout the body.

## ACTIVATION AND TRANSPORT INTO MITOCHONDRIA

Fatty acids must be activated before they can be carried into the mitochondria, where fatty acid oxidation occurs. This process occurs in two steps catalyzed by the enzyme fatty acyl-CoA synthetase.

### Formation of an Activated Thioester Bond

The enzyme first catalyzes nucleophilic attack on the α-phosphate of ATP to form pyrophosphate and an acyl chain linked to AMP. The next step is formation of an activated thioester bond between the fatty acyl chain and Coenzyme A.

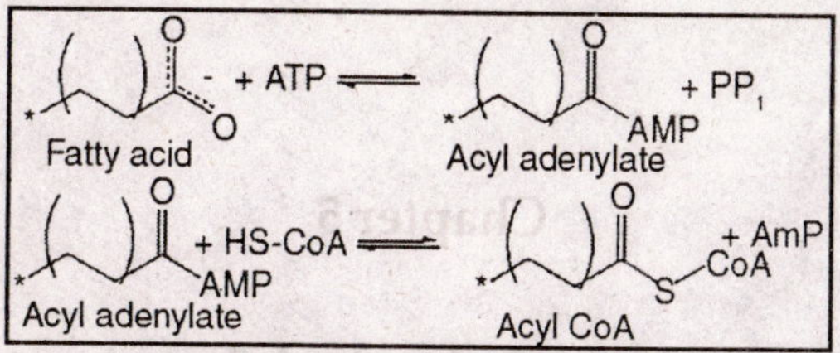


The formula for the above is:

$$RCOO^- + CoA + ATP + H_2O \rightarrow RCO\text{-}CoA + AMP + PP_i + H^+$$

This two-step reaction is freely reversible and its equilibrium lies near 1. To drive the reaction forward, the reaction is coupled to a strongly exergonic hydrolysis reaction: the enzyme inorganic pyrophosphatase cleaves the pyrophosphate liberated from ATP to two phosphate ions. Thus the net reaction becomes:

$$RCOO^- + CoA + ATP + H_2O \rightarrow RCO\text{-}CoA + AMP + 2P_i + 2H^+$$

**Transport into the Mitochondrial Matrix**

The inner mitochondrial membrane is impermeable to fatty acids and a specialized carnitine carrier system operates to transport activated fatty acids from cytosol to mitochondria.

Once activated, the acyl CoA is transported into the mitochondrial matrix. This occurs via a series of similar steps:

1. Acyl CoA is conjugated to carnitine by *carnitine acyltransferase (palmitoyltransferase) I* located on the outer mitochondrial membrane
2. Acyl carnitine is shuttled inside by a translocase
3. Acyl carnitine is converted to acyl CoA by *carnitine acyltransferase (palmitoyltransferase) II* located on the inner mitochondrial membrane. The liberated carnitine returns to the cytosol.

It is important to note that *carnitine acyltransferase I* undergoes allosteric inhibition as a result of malonyl-CoA, an intermediate in fatty acid biosynthesis, in order to prevent futile cycling between beta-oxidation and fatty acid synthesis.

## Chapter 5

# Nucleic Acids

The first isolation of what we now refer to as DNA was accomplished by Johann Friedrich Miescher circa 1870. He reported finding a weakly acidic substance of unknown function in the nuclei of human white blood cells, and named this material "Nuclein".

A few years later, Miescher separated nuclein into protein and nucleic acid components. In the 1920's nucleic acids were found to be major components of chromosomes, small gene-carrying bodies in the nuclei of complex cells. Elemental analysis of nucleic acids showed the presence of phosphorus, in addition to the usual C, H, N & O. Unlike proteins, nucleic acids contained no sulfur.

Complete hydrolysis of chromosomal nucleic acids gave inorganic phosphate, 2-deoxyribose and four different heterocyclic bases. To reflect the unusual sugar component, chromosomal nucleic acids are called deoxyribonucleic acids, abbreviated DNA. Analogous nucleic acids in which the sugar component is ribose are termed ribonucleic acids, abbreviated RNA. The acidic character of the nucleic acids was attributed to the phosphoric acid moiety.

The two monocyclic bases shown here are classified as pyrimidines, and the two bicyclic bases are purines. Each has at least one N-H site at which an organic substituent may be attached. They are all polyfunctional bases, and may exist in tautomeric forms.

Base-catalyzed hydrolysis of DNA gave four nucleoside products, which proved to be N-glycosides of 2'-deoxyribose combined with the heterocyclic amines.The base components are coloured green, and the sugar is black. As noted in the 2'-deoxycytidine structure on the left, the numbering of the sugar carbons makes use of primed numbers to distinguish them from the heterocyclic base sites.

The corresponding N-glycosides of the common sugar ribose are the building blocks of RNA, and are named adenosine, cytidine, guanosine and uridine (a thymidine analog missing the methyl group). From this evidence, nucleic acids may be formulated as alternating copolymers of phosphoric acid (P) and nucleosides (N),

$$\sim P - N - P - N' - P - N'' - P - N''' - P - N \sim$$

At first the four nucleosides, distinguished by prime marks in this crude formula, were assumed to be present in equal amounts, resulting in a uniform structure, such as that of starch. However, a compound of this kind, presumably common to all organisms, was considered too simple to hold the hereditary information known to reside in the chromosomes. This view was challenged in 1944, when Oswald Avery and colleagues demonstrated that bacterial DNA was likely the genetic agent that carried information from one organism to another in a process called "Transformation".

He concluded that "nucleic acids must be regarded as possessing biological specificity, the chemical basis of which is as yet undetermined." Despite this finding, many scientists continued to believe that chromosomal proteins, which differ across species, between individuals, and even within a given organism, were the locus of an organism's genetic information.

It should be noted that single celled organisms like bacteria do not have a well-defined nucleus. Instead, their single chromosome is associated with specific proteins in a region called a "nucleoid". Nevertheless, the DNA from bacteria has the same composition and general structure as that from multicellular organisms, including human beings.

Views about the role of DNA in inheritance changed in the late 1940's and early 1950's. By conducting a careful analysis of DNA from many sources, Erwin Chargaff found its composition to be species specific. In addition, he found that the amount of adenine (A) always equaled the amount of thymine (T) and the amount of guanine (G) always equaled the amount of cytosine (C), regardless of the DNA source.

## NUCLEOSIDE

Nucleosides are glycosylamines consisting of a nucleobase (often referred to simply *base*) bound to a ribose or deoxyribose sugar. Examples of these include cytidine, uridine, adenosine, guanosine, thymidine and inosine. Nucleosides can be phosphorylated by specific kinases in the cell on the sugar's primary alcohol group ($-CH_2 - OH$), producing nucleotides, which are the molecular building blocks of DNA and RNA. Nucleosides can be produced by de novo synthesis pathways, particularly in the liver, but they are more abundantly supplied via ingestion and digestion of nucleic acids in the diet, whereby nucleotidases break down *nucleotides* (such as the thymine nucleotide) into *nucleosides* (such as thymidine) and phosphate. The nucleosides, in turn, are subsequently broken down:

- In the lumen of the digestive system by nucleosidases into nucleobases and ribose or deoxyribose.
- Inside the cell by nucleoside phosphorylases into nitrogenous bases, and ribose-1-phosphate or deoxyribose-1-phosphate.

In medicine several nucleoside analogues are used as antiviral or anticancer agents. The viral polymerase incorporates these compounds with non-canonical bases. These compounds are activated in the cells by being converted into nucleotides, they are administered as nuclosides since charged nucleotides cannot easily cross cell membranes.

In molecular biology several analogues of the sugar back bone exist. Due to the low stability of RNA, which is prone to hydrolysis, several more stable alternative nucleoside/ nucleotide analogues are used which correctly bind to RNA. This is achieved by using a different backbone sugar. These analogues include LNA, morpholino, PNA. In sequencing dideoxynucleotides are used. These nucleotides possess a non-canon sugar, dideoxyribose which lacks 3' hydroxyl group (which accepts the phosphate) and therefore cannot bond with the next base, terminating the chain as DNA polymerases mistake it for a regular deoxyribonucleotide.

## NUCLEOTIDE

Nucleotides are molecules which comprise the structural units of RNA and DNA. Additionally, nucleotides play central roles in metabolism. In that capacity, they serve as sources of chemical energy (adenosine triphosphate and guanosine triphosphate), participate in cellular signaling (cyclic guanosine monophosphate and cyclic adenosine monophosphate), and are incorporated into important cofactors of enzymatic reactions (coenzyme A, flavin adenine dinucleotide, flavin mononucleotide, and nicotinamide adenine dinucleotide phosphate).

### Nucleotide Structure

A nucleotide is composed of a nucleobase (nitrogenous base) and a five-carbon sugar (either ribose or 2'-deoxyribose), and one to three phosphate groups. Together, the nucleobase and sugar comprise a nucleoside. The phosphate groups form bonds with either the 2, 3, or 5-carbon of the sugar, with the 5-carbon site most common. Cyclic nucleotides form when the phosphate group is bound to two of the sugar's hydroxyl groups. Ribonucleotides are nucleotides where the sugar is ribose, and deoxyribonucleotides contain the sugar deoxyribose. Nucleotides can contain either a purine or pyrimidine base.

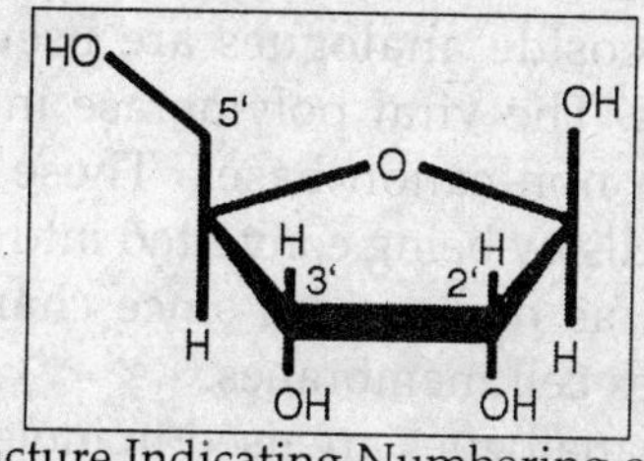


**Fig.** Ribose Structure Indicating Numbering of Carbon Atoms

Nucleic acids are polymeric macromolecules made from nucleotide monomers. In DNA, the purine bases are adenine and guanine, while the pyrimidines are thymine and cytosine. RNA uses uracil in place of thymine.

### Synthesis

Nucleotides can be synthesized by a variety of means both in vitro and in vivo. In vivo, nucleotides can be synthesised de novo or recycled through salvage pathways. Nucleotides undergo breakdown such that useful parts can be reused in synthesis reactions to create new nucleotides. In vitro, protecting groups may be used during laboratory production of nucleotides. A purified nucleoside is protected to create a phosphoramidite, which can then be used to obtain analogues not found in nature and/or to synthesize an oligonucleotide.

## NUCLEIC ACID ANALOGUES

Nucleic acid analogues are compounds structurally similar (analog) to naturally occurring RNA and DNA, used in medicine and in molecular biology research. Nucleic acids are chains of nucleotides, which are composed of three parts: a phosphate backbone, a pucker-shaped pentose sugar, either ribose or deoxyribose, and one of four nucleobases. An analogue may have any of these altered, typically the analogue nucleobases confer, among other things, different base pairing and base stacking proprieties such as universal bases, which can pair with all four canon bases, while the phosphate-sugar backbone analogues affect the properties of the chain, such as PNA, the secondary structure of which differs significantly from DNA, and may form a triplex (a triple stranded helix).

### Medicine

Several nucleoside analogues are used as antiviral or anticancer agents. The viral polymerase incorporates these compounds with non-canon bases. These compounds are activated in the cells by being converted into nucleotides, they are administered as nucleosides since charged nucleotides cannot easily cross cell membranes.

Nucleic acid analogues are used in molecular biology for several purposes:

- As a tool to detect particular sequences
- As a tool with resistance to RNA hydrolysis
- As a tool for another purpose, such as sequencing

- Naturally occurring, such as in tRNA
- Investigation of the mechanisms used by enzyme, such as an Enzyme inhibitor
- Investigation of possible scenarios of the origin of life
- Investigation of the structural features of nucleic acids
- Investigation of the possible alternatives to the natural system in Synthetic biology

## HYDROLYSIS RESISTANT RNA-ANALOGUES

To overcome the fact that ribose's 2' hydroxy group that reacts with the phosphate linked 3' hydroxy group (RNA is too unstable to be used or synthesised reliably), a ribose anologue is used. The most common RNA analogues are locked nucleic acid (LNA), morpholino, peptide nucleic acid (PNA). These oligonucleotides differ as they have a different backbone sugar but still bind according to Watson and Crick pairing with RNA or DNA, but are immune to nuclease activity (They generally cannot be enzymatically synthesised and can only be produced synthetically.

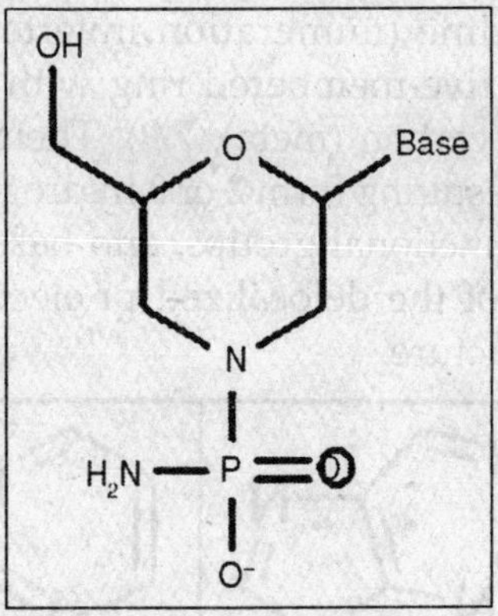


**Fig.** Chemical Structure of Morpholino

## OTHER NOTABLE ANALOGUES USED AS TOOLS

In sequencing dideoxynucleotides are used. These nucleotide triphosphates possess a non-canon sugar, dideoxyribose which lacks 3' hydroxyl group (which accepts the phosphate) and therefore cannot bond with the next base, terminating the chain as the DNA polymerases mistake it for a regular deoxyribonucleotide.

The nucleoside analogue with a ribose lacking both 2' and 3' is called cordycepin, an anticancer drug. Another analogue in sequencing is a nucleobase analogue, 7-deaza-GTP and is used to sequence CG rich regions, instead 7-deaza-ATP is called tubercidin, an antibiotic.

## PRECURSORS TO THE RNA WORLD

RNA may be too complex to be the first nucleic acid, so before the RNA world there may have been one of these several candidate original nucleic acids which differ in the backbone, such as TNA and GNA and PNA.

## BASE ANALOGUES

### Nucleobase Structure and Nomenclature

Natural bases are divided into two classes depending on their structure: pyrimidine (an heterocyclic aromatic six-membered ring with nitrogen atoms in position 1 and 3) and purine (a pyrimidine (numeration inverted) fused with an imidazole ring, a five-membered ring with 2 nitrogen atoms separated by one carbon (meta), 7,9). Their main proprieties are base pairing, resulting form 2 or 3 hydrogen bonds between keto and amino functional groups, and base stacking, caused by the attraction of the delocalized pi electron clouds of the aromatic ring structure.

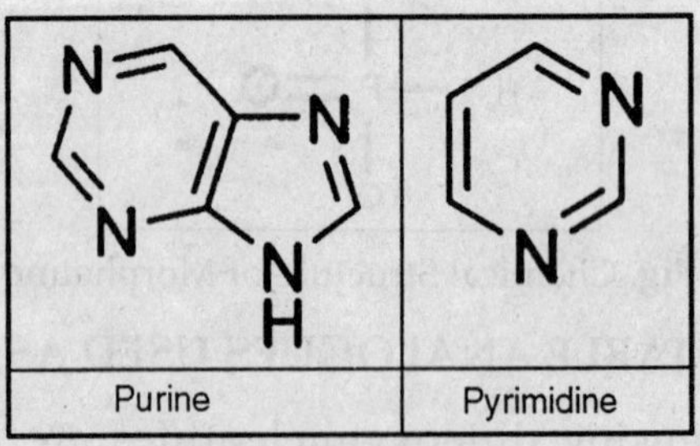


### Fluorophores

Commonly fluorophores (such as rhodamine or

fluorescein) are linked to the ring linked to the sugar (in para) via a flexible arm, presumably extruding form the major groove of the helix. Due to taq polymerases low processivity of the nucleotides linked to bulky adducts such as florophores, the sequence is typically copied using a nucleotide with an arm and later coupled with a reactive fluorophore (indirect labelling):

- *Amine reactive:* Aminoallyl nucleotide contain a primary amine group on a linker that reacts with the amino-reactive dye such as a cyanine or Alexa Fluor dyes, which contain a reactive leaving group, such as a succinimidyl ester (NHS). (base pairing amino groups are not affected).
- *Thiol reactive:* thiol containing nucleotides reacts with the fluorophore linked to a reactive leaving group, such as a maleimide.

biotin linked nucleotides rely on the same indirect labelling principle (+ fluorescent streptavidin) and are used in Affymetrix DNAchips.

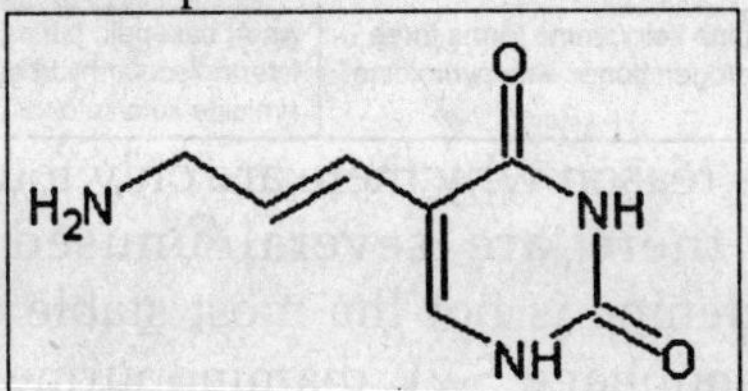


**Fig. Structure of Aminoallyl-uridine**

## Natural Non-canon Bases

In a cell, there are several noncanon bases present: CpG islands in DNA (are often methylated), all eukaryotic mRNA (capped with a methyl-7-guanosine), and several bases of rRNAs (are methylated). Often, tRNAs are heavily modified postranscriptionally in order to improve their conformation or base pairing in particular in/near the anticodon: inosine can base pair with C, U, and even with A, whereas thiouridine (with A) is more specific than uracil (with a purine). Other common tRNA base modifications are pseudorindine (which gives its name to the T$\psi$C loop), dihydrouridine (which does not stack as it is not aromatic), queosine, wyosine and so forth. Nevertheless these are all modifications to normal bases and are not placed by a polymerase.

## Base-pairing

Canonical bases may have either a keto or amino group on the carbons surrounding the nitrogen atom furthest away from the glycosidic bond, which allows them to base pair (Watson-Crick base pairing) via hydrogen bonds (amine with keto, purine with pyrimidine). A and T are amine only and keto only, while C and G are mixed (inverted in respect to each other).

Natural Basepairs

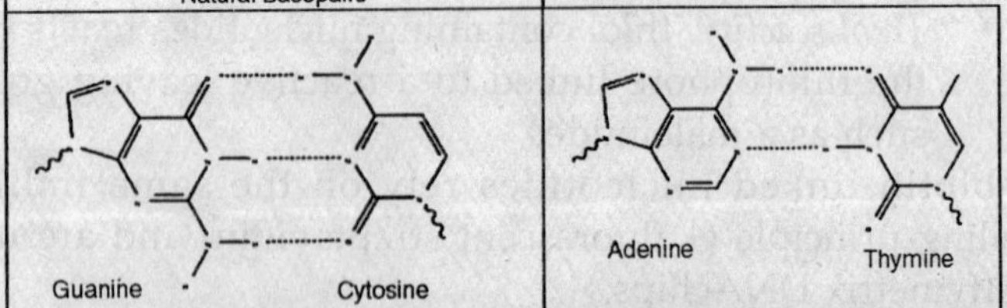


A GC basepair: purine keto/amine forms three intermolecular hydrogen bonds with pyrimidine amine/keto

An At basepair: purine amine/- forms two intermolecular hydrogen bonds with pyrimidine keto/keto.

The precise reason why there are only four nucleotides is debated, but there are several unused possibilities. Furthermore adenine is not the most stable choice for base pairing: in Cyanophage S-2L diaminopurine (DAP) is used instead of adenine (host evasion).

Diaminopurine basepairs perfectly with thymine as it is identical to adenine but has an amine group at position 2 forming 3 intramolecular hydrogen bonds, eliminating the major difference between the two types of basepairs (Weak:A-T and Strong:C-G). This improved stability affects protein binding ineractions which rely on those differences. Other combination include,

- Isoguanosine and isocytosine, which are have their amino and keto inverted, (not used probably as tautomers are problematic for base pairing, but isoG and isoG can be amplified correctly with PCR even in the presence of the 4 canon bases)

- Diaminopyrimidine and a xanthine (not used as xanthine is a deamination product)

However correct DNA structure can form even when the bases are not paired via hydrogen bonding, as studies have shown using DNA isosteres (analogues with same number of atoms), such as the thymine analogue 2,4-difluorotoluene (F) or the adenine analogue 4-methylbenzimidazole (Z).

Other noteworthy basepairs:

- Several fluorescent bases have also been made, such as the 2-amino-6-(2-thienyl)purine and pyrrole-2-carbaldehyde base pair.
- Metal coordinated bases, such as two 2,6-bis(ethyl thiomethyl) pyridine (SPy) with a silver ion or pyridine-2,6-dicarboxamide (Dipam) and a mondentate pyridine (Py) with a copper ion.
- Universal bases may pair indiscriminately with any other base, but generally lower the melting temperature of the sequence considerably, examples include 2'-deoxyinosine derivatives, nitroazole analogues and hydrophobic aromatic non-hydrogen bonding bases (strong stacking effects). These are used as proof of concept and are not generally utilised in degenerate primers (which are a mixture of primers).

The numbers of possible base pairs is doubled when xDNA is considered. xDNA contains expanded bases, in which a benzine ring has been added, which may pair with canon bases, resulting in four possible base-pairs (8 bases:xA-T,xT-A,xC-G,xG-C, 16 bases if the unused arrangements are used). Another form of benzine added bases is yDNA, in which the base is widened by the benzine.

## NUCLEOSIDE BASE DISTRIBUTION IN DNA

In a second critical study, Alfred Hershey and Martha Chase showed that when a bacterium is infected and genetically transformed by a virus, at least 80percent of the viral DNA enters the bacterial cell and at least 80percent of the viral protein remains outside.

Together with the Chargaff findings this work established

DNA as the repository of the unique genetic characteristics of an organism.

| *Organism* | *Base Composition (mole %)* | | | | *Base Ratios* | | *Ratio (A+T)/ (G+C)* |
|---|---|---|---|---|---|---|---|
| | A | G | T | C | A/T | G/C | |
| Human | 30.9 | 19.9 | 29.4 | 19.8 | 1.05 | 1.00 | 1.52 |
| Chicken | 28.8 | 20.5 | 29.2 | 21.5 | 1.02 | 0.95 | 1.38 |
| Yeast | 31.3 | 18.7 | 32.9 | 17.1 | 0.95 | 1.09 | 1.79 |
| Clostridium perfringens | 36.9 | 14.0 | 36.3 | 12.8 | 1.01 | 1.09 | 2.70 |
| Sarcina lutea | 13.4 | 37.1 | 12.4 | 37.1 | 1.08 | 1.00 | 0.35 |

## THE CHEMICAL NATURE OF DNA

The polymeric structure of DNA may be in terms of monomeric units of increasing complexity. In the top shaded box of the following illustration, the three relatively simple components mentioned earlier are shown.

Below that on the left, formulas for phosphoric acid and a nucleoside are drawn.

Condensation polymerization of these leads to the DNA formulation outlined above.

Finally, a 5'- monophosphate ester, called a nucleotide may be drawn as a single monomer unit, shown in the shaded box to the right. Since a monophosphate ester of this kind is a strong acid (pKa of 1.0), it will be fully ionized at the usual physiological pH.

Isomeric 3'-monophospate nucleotides are also known, and both isomers are found in cells. They may be obtained by selective hydrolysis of DNA through the action of nuclease enzymes.

Anhydride-like di- and tri-phosphate nucleotides have been identified as important energy carriers in biochemical reactions, the most common being ATP (adenosine 5'-triphosphate).

**Table: Names of DNA Base Derivatives**

| *Base* | *Nucleoside* | *5′-Nucleotide* |
|---|---|---|
| Adenine | 2'-Deoxyadenosine | 2'-Deoxyadenosine- |
| | 5'-monophos-phate | |
| Cytosine | 2'-Deoxycytidine | 2'-Deoxycytidine- |
| | | 5'-monophos-phate |
| Guanine | 2'-Deoxyguanosine | 2'-Deoxyguanosine- |
| | 5'-monophos-phate | |
| Thymine | 2'-Deoxythymidine | 2'-Deoxythymidine- |
| | 5'-monophos-phate | |

A complete structural representation of a segment of the DNA polymer formed from 5'-nucleotides. Several important characteristics of this formula should be noted.

- First, the remaining P-OH function is quite acidic and is completely ionized in biological systems.
- Second, the polymer chain is structurally directed.
- Third, although this appears to be a relatively simple polymer, the possible permutations of the four nucleosides in the chain become very large as the chain lengthens.
- Fourth, the DNA polymer is much larger than originally believed. Molecular weights for the DNA from multicellular organisms are commonly 109 or greater.

Information is stored or encoded in the DNA polymer by the pattern in which the four nucleotides are arranged. To access this information the pattern must be "read" in a linear fashion, just as a bar code is read at a supermarket checkout. Because living organisms are extremely complex, a correspondingly large amount of information related to this complexity must be stored in the DNA. Consequently, the DNA itself must be very large,. Even the single DNA molecule from an E. coli bacterium is found to have roughly a million nucleotide units in a polymer strand, and would reach a millimeter in length if stretched out.

The nuclei of multicellular organisms incorporate

chromosomes, which are composed of DNA combined with nuclear proteins called histones. The fruit fly has 8 chromosomes, humans have 46 and dogs 78 (note that the amount of DNA in a cell's nucleus does not correlate with the number of chromosomes). The DNA from the smallest human chromosome is over ten times larger than E. coli DNA, and it has been estimated that the total DNA in a human cell would extend to 2 meters in length if unraveled. Since the nucleus is only about 5?m in diameter, the chromosomal DNA must be packed tightly to fit in that small volume.

In addition to its role as a stable informational library, chromosomal DNA must be structured or organized in such a way that the chemical machinery of the cell will have easy access to that information, in order to make important molecules such as polypeptides. Furthermore, accurate copies of the DNA code must be created as cells divide, with the replicated DNA molecules passed on to subsequent cell generations, as well as to progeny of the organism.

## RNA, A DIFFERENT NUCLEIC ACID

The high molecular weight nucleic acid, DNA, is found chiefly in the nuclei of complex cells, known as eucaryotic cells, or in the nucleoid regions of procaryotic cells, such as bacteria. It is often associated with proteins that help to pack it in a usable fashion.

In contrast, a lower molecular weight, but much more abundant nucleic acid, RNA, is distributed throughout the cell, most commonly in small numerous organelles called ribosomes. Three kinds of RNA are identified, the largest subgroup being ribosomal RNA, rRNA, the major component of ribosomes, together with proteins. The size of rRNA molecules varies, but is generally less than a thousandth the size of DNA. The other forms of RNA are messenger RNA , mRNA, and transfer RNA , tRNA. Both have a more transient existence and are smaller than rRNA.

All these RNA's have similar constitutions, and differ from DNA in two important respects.The sugar component of RNA is ribose, and the pyrimidine base uracil replaces the thymine base of DNA. The RNA's play a vital role in the transfer of information (transcription) from the DNA library to the protein factories called ribosomes, and in the interpretation of that information (translation) for the synthesis of specific polypeptides.

## THE SECONDARY STRUCTURE OF DNA

In the early 1950's the primary structure of DNA was well established, but a firm understanding of its secondary structure was lacking. Indeed, the situation was similar to that occupied by the proteins a decade earlier, before the alpha helix and pleated sheet structures were proposed by Linus Pauling. Many researchers grappled with this problem, and it was generally conceded that the molar equivalences of base pairs discovered by Chargaff would be an important factor. Rosalind Franklin, working at King's College, London, obtained X-ray diffraction evidence that suggested a long helical structure of uniform thickness. Francis Crick and James Watson, at Cambridge University, considered hydrogen bonded base pairing interactions, and arrived at a double stranded helical model that satisfied most of the known facts, and has been confirmed by subsequent findings.

## BASE PAIRING

Careful examination of the purine and pyrimidine base components of the nucleotides reveals that three of them could exist as hydroxy pyrimidine or purine tautomers, having an aromatic heterocyclic ring. Despite the added stabilization of an aromatic ring, these compounds prefer to adopt amide-like structures. These options are shown in the following diagram, with the more stable tautomer drawn in blue.

A simple model for this tautomerism is provided by 2-hydroxypyridine. A compound having this structure might be expected to have phenol-like characteristics, such as an acidic hydroxyl group. However, the boiling point of the actual substance is 100º C greater than phenol and its acidity is 100 times less than expected (pKa = 11.7). These differences agree with the 2-pyridone tautomer, the stable form of the zwitterionic internal salt. Note that this tautomerism reverses the hydrogen bonding behaviour of the nitrogen and oxygen functions,the N-H group of the pyridone becomes a hydrogen bond donor and the carbonyl oxygen an acceptor.

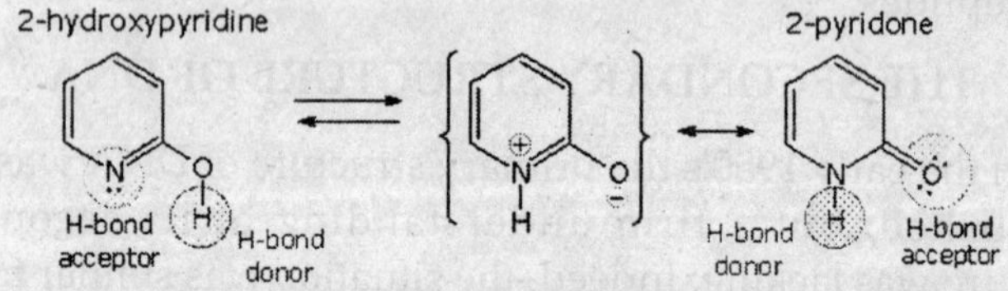


The additional evidence for the pyridone tautomer,consists of infrared and carbon nmr absorptions associated with and characteristic of the amide group.

Once they had identified the favored base tautomers in the nucleosides, Watson and Crick were able to propose a complementary pairing, via hydrogen bonding, of guanosine (G) with cytidine (C) and adenosine (A) with thymidine (T). This pairing, which is shown in the following diagram, explained Chargaff's findings beautifully, and led them to suggest a double helix structure for DNA.

Before viewing this double helix structure itself, it is instructive to examine the base pairing interactions in greater detail. The G#C association involves three hydrogen bonds (coloured pink) and is therefore stronger than the two-hydrogen bond association of A#T. These base pairings might appear to be arbitrary, but other possibilities suffer destabilizing steric or electronic interactions. The C#T pairing on the left suffers from carbonyl dipole repulsion, as well as steric crowding of the oxygens. The G#A pairing on the right is also destabilized by steric crowding (circled hydrogens).

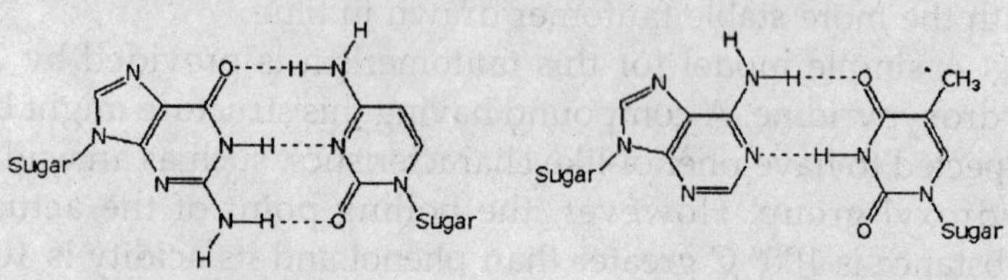


G:::C A:::T

A simple mnemonic device for remembering which bases

are paired comes from the line construction of the capital letters used to identify the bases. A and T are made up of intersecting straight lines. In contrast, C and G are largely composed of curved lines. The RNA base uracil corresponds to thymine, since U follows T in the alphabet.

## The Double Helix

After many trials and modifications, Watson and Crick conceived an ingenious double helix model for the secondary structure of DNA. Two strands of DNA were aligned anti-parallel to each other.Complementary primary nucleotide structures for each strand allowed intra-strand hydrogen bonding between each pair of bases.Coiling these coupled strands then leads to a double helix structure.

### THE DOUBLE HELIX STRUCTURE FOR DNA

The helix shown here has ten base pairs per turn, and rises 3.4 Å in each turn. This right-handed helix is the favored conformation in aqueous systems, and has been termed the B-helix. Two alternating grooves are evident, a wide and deep major groove, and a shallow and narrow minor groove. Other molecules, including polypeptides, may insert into these grooves, and in so doing perturb the chemistry of DNA. Other helical structures of DNA have also been observed, and are designated by letters.

## DNA REPLICATION

A double helix structure for DNA, Watson and Crick stated, "It has not escaped our notice that the specific pairing we have postulated immediately suggests a possible copying mechanism for the genetic material.". The essence of this suggestion is that, if separated, each strand of the molecule might act as a template on which a new complementary strand might be assembled, leading finally to two identical DNA molecules. Indeed, replication does take place in this fashion when cells divide, but the events leading up to the actual synthesis of complementary DNA strands are sufficiently complex. As depicted in the following drawing, the DNA of a cell is tightly packed into chromosomes. First, the DNA is wrapped around small proteins called histones These bead-like structures are then further organized and folded into chromatin aggregates that make up the chromosomes. An overall packing efficiency of 7,000 or more is thus achieved. Clearly a sequence of unfolding events must take place before the information encoded in the DNA can be used or replicated.

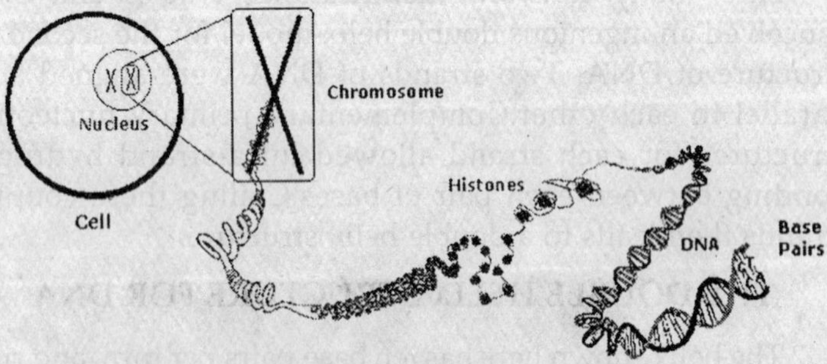


fig. double helix structure for DNA

Once the double stranded DNA is exposed, a group of enzymes act to accomplish its replication.

*Topoisomerase:* This enzyme initiates unwinding of the double helix by cutting one of the strands.

*Helicase:* This enzyme assists the unwinding. Note that many hydrogen bonds must be broken if the strands are to be separated..

*SSB:* A single-strand binding-protein stabilizes the separated strands, and prevents them from recombining, so that the polymerization chemistry can function on the individual strands.

*DNA Polymerase:* This family of enzymes link together nucleotide triphosphate monomers as they hydrogen bond to complementary bases. These enzymes also check for errors (roughly ten per billion) and make corrections.

*Ligase:* Small unattached DNA segments on a strand are united by this enzyme.

Polymerization of nucleotides takes place by the phosphorylation reaction .

Di- and triphosphate esters have anhydride-like structures

and are consequently reactive phosphorylating reagents, just as carboxylic anhydrides are acylating reagents. Since the pyrophosphate anion is a better leaving group than phosphate, triphosphates are more powerful phosphorylating agents than are diphosphates.The DNA polymerization process that builds the complementary strands in replication, could in principle take place in two ways. Alternatively, these assignments could be reversed. In practice, the former proves to be the best arrangement. Since triphosphates are very reactive, the lifetime of such derivatives in an aqueous environment is relatively short. However, such derivatives of the individual nucleosides are repeatedly synthesized by the cell for a variety of purposes, providing a steady supply of these reagents. In contrast, the growing DNA segment must maintain its functionality over the entire replication process, and can not afford to be changed by a spontaneous hydrolysis event. As a result, these chemical properties are best accomodated by a polymerization process that proceeds at the 3'-end of the growing strand by 5'-phosphorylation involving a nucleotide triphosphate.

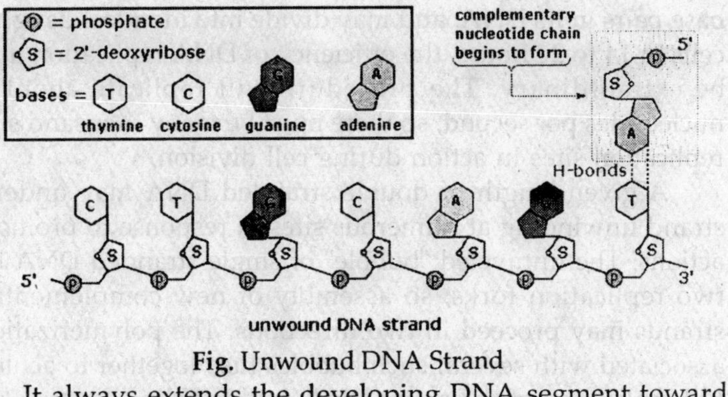


**Fig.** Unwound DNA Strand

It always extends the developing DNA segment toward the 3'-end. when a nucleotide triphosphate attaches to the free 3'-hydroxyl group of the strand, a new 3'-hydroxyl is generated. There is sometimes confusion on this point, because the original DNA strand that serves as a template is read from the 3'-end toward the 5'-end, and authors may not be completely clear as to which terminology is used.

Because of the directional demand of the polymerization,

one of the DNA strands is easily replicated in a continuous fashion, whereas the other strand can only be replicated in short segmental pieces.

This is illustrated in the following diagram. Separation of a portion of the double helix takes place at a site called the replication fork. As replication of the separate strands occurs, the replication fork moves away, unwinding additional lengths of DNA. Since the fork in the diagram is moving toward the 5'-end of the red-coloured strand, replication of this strand may take place in a continuous fashion (building the new green strand in a 5' to 3' direction).

This continuously formed new strand is called the leading strand. In contrast, the replication fork moves toward the 3'-end of the original green strand, preventing continuous polymerization of a complementary new red strand. Short segments of complementary DNA, called Okazaki fragments, are produced, and these are linked together later by the enzyme ligase. This new DNA strand is called the lagging strand. When you consider that a human cell has roughly 109 base pairs in its DNA, and may divide into identical daughter cells in 14 to 24 hours, the efficiency of DNA replication must be extraordinary. The procedure will replicate about 50 nucleotides per second, so there must be many thousand such replication sites in action during cell division.

A given length of double stranded DNA may undergo strand unwinding at numerous sites in response to promoter actions. The unraveled "bubble" of single stranded DNA has two replication forks, so assembly of new complementary strands may proceed in two directions. The polymerizations associated with several such bubbles fuse together to achieve full replication of the entire DNA double helix

## REPAIR OF DNA DAMAGE AND REPLICATION ERRORS

One of the benefits of the double stranded DNA structure is that it lends itself to repair, when structural damage or replication errors occur. Several kinds of chemical change may cause damage to DNA:

- Spontaneous hydrolysis of a nucleoside removes the heterocyclic base component.
- Spontaneous hydrolysis of cytosine changes it to a uracil.
- Various toxic metabolites may oxidize or methylate heterocyclic base components.
- Ultraviolet light may dimerize adjacent cytosine or thymine bases.

All these transformations disrupt base pairing at the site of the change, and this produces a structural deformation in the double helix.. Inspection-repair enzymes detect such deformations, and use the undamaged nucleotide at that site as a template for replacing the damaged unit. These repairs reduce errors in DNA structure from about one in ten million to one per trillion.

## RNA AND PROTEIN SYNTHESIS

The genetic information stored in DNA molecules is used as a blueprint for making proteins. Why proteins? Because these macromolecules have diverse primary, secondary and tertiary structures that equip them to carry out the numerous functions necessary to maintain a living organism. As noted in the protein, these functions include:

- Structural integrity (hair, horn, eye lenses etc.).
- Molecular recognition and signaling (antibodies and hormones).
- Catalysis of reactions (enzymes)..
- Molecular transport (hemoglobin transports oxygen).
- Movement (pumps and motors).

The critical importance of proteins in life processes is demonstrated by numerous genetic diseases, in which small modifications in primary structure produce debilitating and often disasterous consequences. Such genetic diseases incluse Tay-Sachs, phenylketonuria (PKU), sickel cell anemia, achondroplasia, and Parkinson disease. The unavoidable conclusion is that proteins are of central importance in living cells, and that proteins must therefore be continuously prepared with high structural fidelity by appropriate cellular chemistry. Early geneticists identifed genes as hereditary units that determined the appearance and / or function of an organism (i.e. its phenotype).

We now define genes as sequences of DNA that occupy specific locations on a chromosome. The original proposal that each gene controlled the formation of a single enzyme has since been modified as: one gene = one polypeptide. The intriguing question of how the information encoded in DNA is converted to the actual construction of a specific polypeptide has been the subject of numerous studies, which have created the modern field of Molecular Biology.

## THE CENTRAL DOGMA AND TRANSCRIPTION

Francis Crick proposed that information flows from DNA to RNA in a process called transcription, and is then used to synthesize polypeptides by a process called translation. Transcription takes place in a manner similar to DNA replication. A characteristic sequence of nucleotides marks the beginning of a gene on the DNA strand, and this region binds to a promoter protein that initiates RNA synthesis.

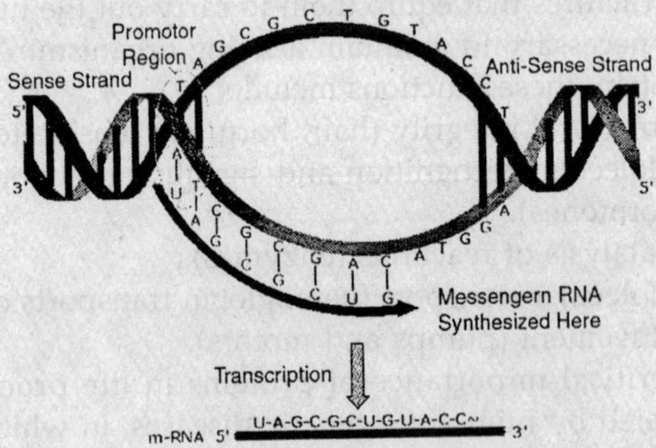


**Fig.** Central Dogma and Transcription

The double stranded structure unwinds at the promoter site. and one of the strands serves as a template for RNA formation, as depicted in the following diagram. The RNA molecule thus formed is single stranded, and serves to carry information from DNA to the protein synthesis machinery called ribosomes. These RNA molecules are therefore called messenger-RNA (mRNA). An important distinction must be made here. One of the DNA strands in the double helix holds the genetic information used for protein synthesis. This is called the sense strand, or information strand The complementary strand that binds to the sense strand is called the anti-sense strand (coloured green) and it serves as a template for generating a mRNA molecule that delivers a copy of the sense strand information to a ribosome.

The promoter protein binds to a specific nucleotide sequence that identifies the sense strand, relative to the anti-sense strand. RNA synthesis is then initiated in the 3' direction, as nucleotide triphosphates bind to complementary bases on the template strand, and are joined by phosphate diester linkages. A characteristic "stop sequence" of nucleotides terminates the RNA synthesis. The messenger molecule is released into the cytoplasm to find a ribosome, and the DNA then rewinds to its double helix structure.

In eucaryotic cells the initially transcribed m-RNA molecule is usually modified and shortened by an "editing" process that removes irrelevent material. The DNA of such organisms is often thousands of times larger and more complex than that composing the single chromosome of a procaryotic bacterial cell. This difference is due in part to repetitive nucleotide sequences (ca. 25 percent iu the human genome). Furthermore, over 95 percent of human DNA is found in intervening sequences that separate genes and parts of genes.

The informational DNA segments that make up genes are called exons, and the noncoding segments are called introns. Before the mRNA molecule leaves the nucleus, the nonsense bases that make up the introns are cut out, and the informationally useful exons are joined together in a step known as RNA splicing. In this fashion shorter mRNA molecules carrying the blueprint for a specific protein are sent on their way to the ribosome factories.

The Central Dogma of molecular biology, which at first was formulated as a simple linear progression of information from DNA to RNA to Protein, is summarized in the following illustration. The replication process on the left consists of passing information from a parent DNA molecule to daughter molecules. The middle transcription process copies this information to a mRNA molecule. Finally, this information is used by the chemical machinery of the ribosome to make polypeptides.

As more has been learned about these relationships, the central dogma has been refined to the representation displayed on the right. The relatively simple RNA-protein structures of a virus may induce RNA replication or a reverse transcription from RNA to DNA (magenta arrow), but these are exceptional, albeit significant transfers.

The AIDS virus, for example, functions by the action of reverse transcriptase enzymes accompanying its RNA. Direct translation of DNA information into protein synthesis (orange arrow) has not yet been observed in a living organism. Finally, proteins appear to be an informational dead end, and do not provide a structural blueprint for either RNA or DNA.

## TRANSLATION

Translation is a more complex process than transcription. This would, of course, be expected. After all, the coded messages produced by the German Enigma machine could be copied easily, but required a considerable decoding effort before they could be read with understanding. In a similar sense, DNA replication is simply a complementary base pairing exercise, but the translation of the four letter (bases) alphabet code of RNA to the twenty letter (amino acids) alphabet of protein literature is far from trivial.

Clearly, there could not be a direct one-to-one correlation of bases to amino acids, so the nucleotide letters must form short words or codons that define specific amino acids. Many questions pertaining to this genetic code were posed in the late 1950's:

- How many RNA nucleotide bases designate a specific amino acid?

  If separate groups of nucleotides, called codons, serve this purpose, at least three are needed. There are

43 = 64 different nucleotide triplets, compared with 42 = 16 possible pairs.

- Are the codons linked separately or do they overlap? Sequentially joined triplet codons will result in a nucleotide chain three times longer than the protein. If overlapping codons are used then fewer total nucleotides would be required.
- If triplet segments of mRNA designate specific amino acids in the protein, how are the codons identified? For the sequence ~CUAGGU~ are the codons CUA & GGU or ~C, UAG & GU~ or ~CU, AGG & U~?
- Are all the codon words the same size?

In Morse code the most widely used letters are shorter than less common letters. Perhaps nature employs a similar scheme.

Physicists and mathematicians, as well as chemists and microbiologists all contributed to unravelling the genetic code. Although earlier proposals assumed efficient relationships that correlated the nucleotide codons uniquely with the twenty fundamental amino acids, it is now apparent that there is considerable redundancy in the code as it now operates. Furthermore, the code consists exclusively of non-overlapping triplet codons.

Clever experiments provided some of the earliest breaks in deciphering the genetic code. Marshall Nirenberg found that RNA from many different organisms could initiate specific protein synthesis when combined with broken E.coli cells (the enzymes remain active). A synthetic polyuridine RNA induced synthesis of poly-phenylalanine, so the UUU codon designated phenylalanine. Likewise an alternating ~CACA~ RNA led to synthesis of a ~His-Thr-His-Thr~ polypeptide.

The RNA alphabet, and an equivalent DNA codon table would have all the U nucleotides replaced by T. Methionine and tryptophan are uniquely represented by a single codon. At the other extreme, leucine is represented by eight codons. The average redundancy for the twenty amino acids is about three. Also, there are three stop codons that terminate polypeptide synthesis. The translation process is fundamentally straightforward. The mRNA strand bearing the transcribed code for synthesis of a protein interacts with relatively small RNA molecules (about 70-nucleotides) to which individual amino acids have been attached by an ester bond at the 3'-end.

These transfer RNA's (tRNA) have distinctive three-dimensional structures cosisting of loops of single-stranded RNA connected by double stranded segments.

This cloverleaf secondary structure is further wrapped into an "L-shaped" assembly, having the amino acid at the end of one arm, and a characteristic anti-codon region at the other end.

The anti-codon consists of a nucleotide triplet that is the complement of the amino acid's codon(s). Models of two such tRNA molecules are shown to the right.

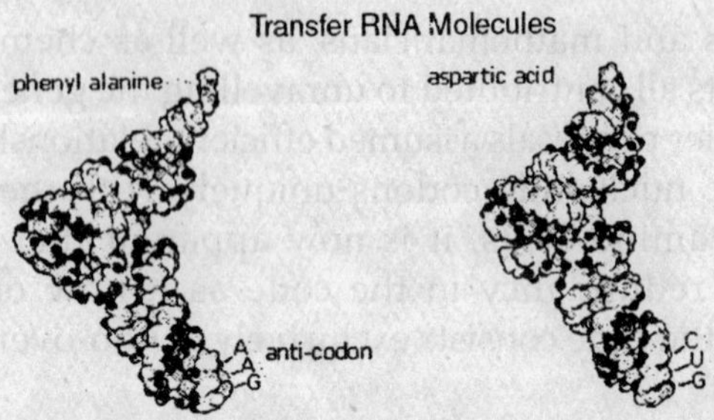


**Fig.** tRNA Molecules

A cell's protein synthesis takes place in organelles called ribosomes.

Ribosomes are complex structures made up of two distinct and separable subunits (one about twice the size of the other). Each subunit is composed of one or two RNA molecules associated with 20 to 40 small proteins.

The ribosome accepts a mRNA molecule, binding initially to a characteristic nucleotide sequence at the 5'-end. This unique binding assures that polypeptide synthesis starts at the e codon. A tRNA molecule with the appropriate ant-codon then attaches at the starting point and this is followed by a series of adjacent tRNA attachments, peptide bond formation and shifts of the ribosome along the mRNA chain to expose new codons to the ribosomal chemistry.

## ANALYSIS OF STRUCTURAL SIMILARITIES AND DIFFERENCES BETWEEN DNA AND RNA

We know that living organisms have the ability to reproduce and to pass many of their characteristics on to their offspring. From this we may infer that all organisms have genetic substances and an associated chemistry that enable inheritance to occur. It is instructive to consider the essential requirements such genetic materials must fullfill.

*Information:* Biologically useful information, especially instructions for protein sysnthesis, must be incorporated in the material.

*Stability:* The inherited information must be stable (unchanged) over the lifetime of the organism if accurate copies are to be conveyed to the offspring. Infrequent changes may take place.

*Reproduction:* A method of faithfully replicating the information encoded in the material, and transmitting this copy to the offspring must exist.

*Mutability:* Despite the inherent stability, the material must be capable of incorporating stable structural change, and passing this change on to succeeding generations.

Since this genetic substance has been identified as the nucleic acids DNA and RNA, it is instructive to examine the manner in which these polymers satisfy the above requirements.

## INFORMATION STORAGE

The complexity of life suggests that even simple organisms will require very large inheritance libraries. Although the four nucleotides that make up of DNA might appear to be too simple for this task, the enormous size of the polymer and the permutations of the monomers within the chain meet the challenge easily. DNA has four letters in its alphabet so the number of words that can be formed increase exponentially with the number of letters per word. Thus, there are 42 or 16 two letter words, and 43 or 64 three letter words. Assuring the stability of information encoded by the DNA alphabet presents a serious challenge.

If the letters of this alphabet are to be strung together in a specific way on the polymer chain, chemical reactions for attaching (and removing) them must be available. Simple carboxylic ester or amide links might appear suitable for this purpose step-growth polymerization, but these are used in lipids and polypeptides, so a separate enzymatic machinery would be needed to keep the information processing operations apart from other molecular transformations.

The overall stability of such covalent links presents a more serious problem. Under physiological conditions (aqueous, pH near 7.4 & 27 to 37°C) esters are slowly hydrolyzed. Amides are more stable, but even a hydrolytic cleavage of one bond per hour would be devastating to a polymer having tens of thousands to millions such links. Furthermore, short difuctional linking groups, such as carbonates, oxylates and malonates show enhanced reactivity, and their parent acids are unstable or toxic.

**Table: Hydrolysis at 35°C and pH 7**

| *Ester* | *Rate of Hydrolysis* | *Relative Rate* |
|---|---|---|
| Ethyl Acetate $CH_3C0_2C_2H_5$ | $1.0*10^{-2}$ | $5*10^{6}$ |
| Trimethyl Phosphate $(CH_30)_3PO$ | $3.4*10^{-4}$ | $2*10^{5}$ |
| Dimethyl Phosphate $(CH_3O)2PO_2(-)$ | $2.0*10^{-9}$ | 1.0 |

Phosphate is an ubiquitous inorganic nutrient. Mono, di and triesters of the corresponding acid (phosphoric acid) are all known. Because of their acidity (pKa H" 2), the mono and diesters are negatively charged at physiological pH, rendering them less susceptible to nucleophilic attack. Clearly, a polymer in which monomer units are joined by negatively charged diphosphate ester links should be substantially more stable than one composed of carboxylate ester bonds. The negative charge found on all biological phosphate derivatives serves other purposes as well.

- The diphosphate ester links that join the nucleotides

units of DNA are formed by phosphorylation reactions involving nucleotide triphosphate reagents. These reagents are the phosphoric acid analogs of carboxylic acid anhydrides, a functional group that would not survive the aqueous environment of a cell.

The high density of negative charge on the triphosphate function not only solubilizes the organic moiety to which it is attached, but also reduces the rate at which it is hydrolyzed.

- Living cells must conserve and employ their chemical reagents within a volume defined and enclosed by a membrane barrier. These lipid bilayer membranes have hydrophobic interiors, which resist the passage of ions. Indeed, special trans-membrane structures called ion channels exist so that controlled ion transport across a membrane may take place.

  Small neutral organic molecules, such as adenosine, cytidine and guanosine, may pass through lipid membranes, albeit at a reduced rate, but their mono, di and triphosphate derivatives are more tightly sequestered in the cell.

## WHY IS 2'-DEOXYRIBOSE THE SUGAR MOIETY IN DNA?

Common perhydroxylated sugars, such as glucose and ribose, are formed in nature as products of the reductive condensation of carbon dioxide we call photosynthesis. The formation of deoxysugars requires additional biological reduction steps, so it is reasonable to speculate why DNA makes use of the less common 2'-deoxyribose, when ribose itself serves well for RNA.

At least two problems associated with the extra hydroxyl group in ribose may be noted. First, the additional bulk and hydrogen bonding character of the 2'-OH interfere with a uniform double helix structure, preventing the efficient packing of such a molecule in the chromosome. Second, RNA undergoes spontaneous hydrolytic cleavage about one hundred times faster than DNA.

This is believed due to intramolecular attack of the 2'-hydroxyl function on the neighbouring phosphate diester, yielding a 2',3'-cyclic phosphate.

If stability over the lifetime of an organism is an essential characteristic of a gene, then nature's selection of 2'-deoxyribose for DNA makes sense. The following diagram illustrates the intramolecular cleavage reaction in a strand of RNA.

Structural stability is not a serious challenge for RNA. The transcripted information carried by mRNA must be secure for only a few hours, as it is transported to a ribosome. Once in the ribosome it is surrounded by structural and enzymatic segments that immediately incorporate its codons for protein synthesis. The tRNA molecules that carry amino acids to the ribosome are similarly short lived, and are in fact continuously recycled by the cellular chemistry.

## THE THYMINE VS. URACIL ISSUE

Structural formulas for the three pyrimidine bases, cytosine, thymine and uracil are shown on the right. The carbon atoms that are part of these compounds may be categorized as follows. All of these compounds are apparently put together from a three-carbon malonate-like precursor and a single high oxidation state carbon species.

Such biosynthetic intermediates are well established. Thymine is unique in having an additional carbon, the green methyl group. Biosynthesis of this compound must involve additional steps, thus adding constructional complexity to the DNA molecules in which it replaces uracil.

The reason for the substitution of thymine for uracil in DNA may be associated with the repair mechanisms by which the cell corrects damage to its DNA. One source of error in the code is the slow hydrolysis of heterocyclic enamines, such as cytosine and guanine, to their corresponding lactams. This changes the structure of the base, and disrupts base pairing in a manner that can be identified and then repaired. However, the hydrolysis product from cytosine is uracil, and this mismatched species must somehow be distiguished from the uracil-like base that belongs in the DNA. The extra methyl group serves this role nicely.

## THE VITAMINS

Vitamin, any of the organic (carbon-containing) compounds that the body requires in small amounts to maintain health and function properly. Children additionally need vitamins to grow. The body gets most of its vitamins from the foods we eat.

A healthy diet with plenty of fruits and vegetables should provide nearly all of the vitamins a person needs. Scientists have classified 13 compounds as vitamins. They have given most of these vitamins letter or letter plus number names, such as A, B12, and D. Most vitamins are produced by plants. Some, such as vitamin D, are produced only by animals. A few vitamins are made by the body itself. For example, bacteria in the digestive tract help produce vitamin K, and the skin uses sunlight to produce vitamin D.

Vitamins are also manufactured for sale as supplements for people who need additional vitamins to meet their body's requirements. For example, doctors often prescribe vitamin supplements for pregnant and nursing women to provide the additional nutrients needed by a rapidly growing fetus or infant. Older adults may not meet their vitamin requirements through food because the body's ability to absorb vitamins is impaired with age. Expressed in milligrams (mg) or international units (IU) for adults and children of normal health, these recommendations provide useful guidelines for daily intake. Such guidelines are useful not only for professionals in nutrition but also for the growing number of families and individuals who eat irregular meals and rely on prepared foods, most of which are now required to carry nutritional labeling.

## VITAMINS ARE VITAL

The word vitamine, later shortened to vitamin, was coined by Polish-American chemist Casimir Funk in the early 20th century. Funk was searching for the then-unknown substance in foods that prevents such diseases as beriberi, rickets, and scurvy. After experiments on pigeons, Funk guessed that an amine-a compound containing nitrogen and hydrogen-in foods was responsible. He called this substance a vital (necessary for life) amine, or vitamine. Scientists later discovered that not all vitamins were amines, however.

Vitamins help the body carry out essential biochemical processes. Vitamins generally combine with proteins to create enzymes that promote chemical reactions. These enzymes play an important role in metabolism-the reactions that break down the fats, carbohydrates, and proteins in food so that the body can use them for energy and cell repair.

The enzymes also promote reactions involved in the formation of bone, hormones, blood cells, nervous-system chemicals, and genetic material. Without vitamins, many of these reactions would slow down or cease. The different vitamins are not chemically related, and most differ in their actions in the body.

Deficiency of particular vitamins can lead to various diseases. Too little vitamin C, for example, can cause rickets, a disease in which the bones fail to develop properly. Bowlegs or "knock knees" are signs of rickets. Rickets and the other vitamin-deficiency diseases that interested Casimir Funk have largely disappeared in the developed world as the result of fortified foods and improved nutrition. But these diseases still occur in the developing world, especially where malnutrition is common.

Interest in vitamins grew during the 1990s, especially in those vitamins that act as antioxidants-vitamins A, C, and E. Antioxidants neutralize molecules known free radicals that cause cell damage. Free radicals have been linked to a number of disorders, including Alzheimer's disease, arteriosclerosis, cancer, diabetes, and Parkinson disease. Researchers theorized that increased doses of antioxidants might also prevent and even cure such diseases. These claims have so far not panned out, however, as scientific studies have failed to demonstrate the preventive abilities of antioxidants. Research on antioxidants and other vitamins continues.

## WATER-SOLUBLE VITAMINS

Water-soluble vitamins include eight well-recognized members of the vitamin B complex: thiamine, riboflavin, pyridoxine, pantothenic acid, niacin, biotin, folic acid and vitamin B12; the water-soluble essential nutritional factors: choline, inositol, ascorbic acid; and vitamins with less-defined activity for fish: p-aminobenzoic acid, lipoic acid and citrin.

**Table: Water-Soluble Vitamin Test Diet H-440 1/**

| *Complete test diet (g)* | | *Vitamin mix Mineral mix (mg) (mg)* | | | |
|---|---|---|---|---|---|
| Vitamin-free casein | 38 | Thiamine HCl | 5 | USP XII No. 2 | plus |
| Gelatin | 12 | Riboflavin | 20 | AlCl3 | 15 |
| Corn oil | 6 | Pyridoxine HCl | 5 | ZnSO4 | 300 |
| Cod liver oil | 3 | Choline chloride | 500 | CuCl | 10 |
| White dextrin | 28 | Nicotinic acid | 75 | MnSO4 | 80 |
| α-Cellulose mixture 2/ | 9 | Calcium | 50 | KI | 15 |
| | | pantothenate | | | |
| α-Cellulose | 8 | Inositol | 200 | CaCl2 | 100 |
| Vitamins | | Biotin | 0.5 | per 100 g of salt mixture | |
| | | Folic acid | 1.5 | | |
| Mineral mix | 4 | L-Ascorbic acid | 100 | | |
| Water | 200 | Vitamin B12 4/ | 0.01 | | |
| Total diet as fed | 300 | Menadione (K) | 4 | | |
| | | -Tocopherol acetate (E) 3/ | 40 | | |

1. *Diet preparation:* Dissolve gelatin in cold water. Heat with stirring on water bath to 80°C. Remove from heat. Add with stirring - dextrin, casein, minerals, oils, and vitamins as temperature decreases. Mix well to 40°C. Pour into containers; move to refrigerator to harden. Remove from trays and store in sealed containers in refrigerator until used. Consistency of diet adjusted by amount of water in final mix and length and strength of beating.

2. *Delete two parts:* Cellulose and add two parts CMC for preliminary feeding.
3. *Dissolve:* Tocopherol in oil mix.
4. *Add vitamin:* B12 in water during final mixing.

The first eight, though required only in small amounts in the diet, play major roles in growth, physiology and metabolism. Choline, inositol and ascorbic acid are required in appreciable quantities in the diet and sometimes are not referred to as vitamins but as major dietary nutrients. A typical vitamin test diet for fish control over the water-soluble vitamins

## THIAMINE

Experimental avian polyneuritis, a condition resembling beriberi, was produced by Eijkman in Java in 1886 and the anti-beriberi factor was crystallized and named "vitamine". The term "thiamine" was introduced when the chemical nature of the factor was established. Thiamine was isolated from rice polishings in 1926 and synthesized in 1936.

Thiamine hydrochloride is a water-soluble, colourless, monoclinic, crystalline compound. It is comparatively stable to dry heat but is rapidly broken down in neutral or alkaline solutions and is split by sulphites into constituent pyrimidine and thiazole moieties. It has a characteristic yeast-like odour. The pyrimide ring is relatively stable, but the thiazole ring is easily opened by hydrolysis.

Several derivatives are stable to heat and appear to be more completely soluble in weak alkaline solutions than thiamine itself and still show biological activity in animals. These derivatives include thiamine propyl disulphide, benzoylthiamine disulphide, dibenzoylthiamine, and benzoylthiamine monophosphate. Both thiamine hydrochloride and thiamine mononitrate have been successfully used as the active vitamin in test diets for fish nutrition studies.

### Positive Functions

Thiamine functions in all cells as the coenzyme

cocarboxylase, thiamine pyrophosphate, which participates in the oxidative decarboxylation of pyruvic acid to acetate for entry into the tricarboxylic acid (TCA) cycle.

## THIAMINE PYROPHOSPHATE

Thiamine pyrophosphate is also a coenzyme of the transketolase system by which direct oxidation of glucose occurs in the cytoplasm of cells via the pentose phosphate pathway. Erythrocyte (RBC) levels of metabolites of this system have been used to indicate thiamine status in experimental animals, including salmon and trout.

Thiamine is essential for good appetite, normal digestion, growth and fertility. It is needed for normal functioning of the nervous tissue and the requirement is determined by the caloric density of the diet.

### Deficiency Syndrome

Deficiency signs in salmonids include impaired carbohydrate metabolism, nervous disorders, poor appetite, poor growth, and increased sensitivity to shock. A trunk-winding symptom in eels has been reported, together with haemorrhage at the base of the fins.

Skin congestion and subcutaneous haemorrhage occurs in carp fed thiamine-deficient diets. Typical symptoms observed in salmonids, carp and catfish . Thiamine deficiency has also been reported in marine flatfish started on clam neck diets stored long enough for thiaminase present to hydrolyze the thiamine in the ration. Typical nervous paralysis occurred with rapid mortality from physical shock.

**Table: Vitamin Deficiency Syndromes**

| *Vitamin* | *Symptoms in salmon, trout, carp, catfish* |
|---|---|
| Thiamine | Poor appetite, muscle atrophy, convulsions, instability and loss of equilibrium, oedema, poor growth |
| Riboflavin | Corneal vascularization, cloudy lens, haemorrhagic eyes - photophobia, dim vision, incoordination, abnormal pigmentation of iris, striated constrictions of abdominal wall, |

| | dark colouration, poor appetite, anaemia, poor growth |
|---|---|
| Pyridoxine | Nervous disorders, epileptiform fits, hyperirritability, ataxia, anaemia, loss of appetite, oedema of peritoneal cavity, colourless serous fluid, rapid postmortem rigor mortis, rapid and gasping breathing, flexing of opercles |
| Pantothenic acid | Clubbed gills, prostration, loss of appetite, necrosis and scarring, cellular atrophy, gill exudate, sluggishness, poor growth |
| Inositol | Poor growth, distended stomach, increased gastric emptying time, skin lesions |
| Biotin | Loss of appetite, lesions in colon, colouration, muscle atrophy, spastic convulsions, fragmentation of erythrocytes, skin lesions, poor growth |
| Folic acid | Poor growth, lethargy, fragility of caudal fin, dark colouration, macrocytic anaemia |
| Choline | Poor growth, poor food conversion, haemorrhagic kidney and intestine |
| Nicotinic acid | Loss of appetite, lesions in colon, jerky or difficult motion, weakness, oedema of stomach and colon, muscle spasms while resting, poor growth |

| *Vitamin* | *Symptoms in salmon, trout, carp, catfish* |
|---|---|
| Vitamin B12 | Poor appetite, low haemoglobin, fragmention of erythrocytes, macrocytic anaemia |
| Ascorbic acid | Scoliosis, lordosis, impaired collagen formation, altered cartilage, eye lesions, haemorrhagic skin, liver, kidney, intestine, and muscle |
| p-Aminobenzoic acid | No abnormal indication in growthappetite, mortality |

## Requirements

In determining dietary thiamine requirements of fish, some considerations must be placed upon the composition of dietary ingredients in the ration. Fat content of the diet may affect not only caloric intake but also the thiamine requirement because cocarboxylase participates in the oxidation of that through a-ketoglutarate. Therefore, fish on a high fat diet and low thiamine intake might take longer to develop deficiencies and will give an erroneous requirement.

## Sources and Protection

Common sources for thiamine are dried peas, beans, cereal bran, and dried yeast. Fresh glandular tissue is also a good source for thiamine and other members of the vitamin B water-soluble complex, but is seldom used in modern commercial fish diets.

Thiamine can be easily lost by holding diet ingredients too long in storage or by preparing the diet under slightly alkaline conditions or in the presence of sulphide. Wet or frozen diets pose a problem because moisture content increases the chance of enzymatic (thiaminase) hydrolysis and subsequent destruction of thiamine. Obviously, wet or moist diet, preparations containing any fresh fish or shellfish tissue must be used immediately.

## Antimetabolites and Inactivation

Acetylcholine is an antagonist to thiamine and pyrithiamine. Oxythiamine and normal butylthiamine are specific antimetabolites. Several thiaminases occur which destroy thiamine. These rupture the thiazole ring at the sulphur bond making the residue inactive. Freshwater fish tissues have high thiaminase activity as do tissues from clams, shrimp and mussels. Thiaminases have also been found in beans and mustard seed and in several micro-organisms.

Thiaminase activity is low in most saltwater fish tissues, however, and the enzyme is inactivated by heating or prolonged pasteurization. Thiamine present in fresh Torula yeast is relatively unavailable to fish, but the yeast becomes an excellent thiamine source after rupturing the cells by steam treatment or by dehydration.

## Clinical Assessment

Clinical assessment of thiamine status have been made by measuring erythrocyte transketolase activity. Typical saturated levels for thiamine activity in sea salmon range from 15 to 20 g of thiamine/g of wet liver tissue. Fingerling chinook or coho salmon reared in 10 or 15 C water on test diets containing the listed amount or more of thiamine hydrochloride: assayed at 8-10 g of thiamine/g of wet liver tissue. These liver storage levels and normal erythrocyte transketolase activity in the absence of any deficiency sign will indicate an adequate thiamine intake for that fish population.

## RIBOFLAVIN

Following the recognition that the original water-soluble B consisted of more than one factor, a new growth-promoting factor consisting of yellow-green pigments was isolated in 1879. This new factor was designated G by some and 83 by others. The second water-soluble vitamin discovered was given its specific chemical name riboflavin. Lactoflavin, hepatoflavin, and ovoflavin were also shown to be identical with the pure riboflavin. Riboflavin occurs in the free form only in the eye, whey and urine.

**Table: Vitamin Requirements for Growth 1**

| *Vitamin (mg/kg dry diet)* | *Rainbow trout* | *Brook trout* | *Brown trout* | *Atlantic salmon* | *Chinook salmon* | *Coho salmon* |
|---|---|---|---|---|---|---|
| Thiamine | 10-12 | 10-12 | 10-12 | 10-15 | 10-15 | 10-15 |
| Riboflavin | 20-30 | 20-30 | 20-30 | 5-10 | 20-25 | 20-25 |
| Pyridoxine | 10-15 | 10-15 | 10-15 | 10-15 | 15-20 | 15-20 |
| Pantothenate | 40-50 | 40-50 | 40-50 | * | 40-50 | 40-50 |
| Niacin | 120-150 | 120-150 | 120-150 | * | 150-200 | 150-200 |
| Folacin | 6-10 | 6-10 | 6-10 | 5-10 | 6-10 | 6-10 |
| Cyanocobalamin | * | * | * | * | 0.015-0.02 | 0.015-0.02 |
| myo-Inositol | 200-300 | * | * | * | 300-400 | 300-400 |
| Choline | * | * | * | * | 600-800 | 600-800 |

| | | | | | | |
|---|---|---|---|---|---|---|
| Biotin | 1-1.5 | 1-1.5 | 1.5-2 | - | 1-1.5 | 1-1.5 |
| Ascorbate | 100-150 | * | * | * | 100-150 | 50-80 |
| Vitamin A | 2 000-2 500 I.U. | * | * | - | * | * |
| Vitamin E 2/ | * | * | * | - | 40-50 | * |
| Vitamin K | * | * | * | - | * | * |

1. Fish fed at reference temperature with diets at about protein requirement
2. Requirement directly affected by amount and type of unsaturated fat fed

* Denotes a requirement, the level of which has not been established

**Table: (Concluded) Vitamin Requirements for Growth 1**

| Vitamin (mg/kg dry diet) | Carp | Channel catfish | Eel | Sea bream | Turbot | Yellowtail |
|---|---|---|---|---|---|---|
| Thiamine | 2-3 | 1-3 | 2-5 | * | 2-4 | * |
| Riboflavin | 7-10 | * | * | * | * | - |
| Pyridoxine | 5-10 | * | * | 2-5 | * | * |
| Pantothenate | 30-40 | 25-30 | * | * | * | * |
| Niacin | 30-50 | * | - | * | - | - |
| Folacin | - | * | * | - | * | - |
| Cyanocobalamin | - | * | - | * | - | - |
| myo-Inositol | 200-300 | * | - | 300-500 | - | - |
| Choline | 500-600 | * | - | * | * | - |
| Biotin | 1-15 | * | * | - | * | * |
| *Vitamin (mg/kg dry diet)* | *Carp* | *Channel catfish* | *Eel* | *Sea bream* | *Turbot* | *Yellowtail* |
| Ascorbate | 30-50 | 30-50 | - | * | - | * |
| Vitamin A | 1 000-2 000 I.U. | * | - | - | - | * |
| Vitamin E 2/ | 80-100 | * | - | - | - | * |
| Vitamin K | * | * | - | - | - | - |

1. Fish fed at reference temperature with diets at about protein requirement
2. Requirement directly affected by amount and type of unsaturated fat fed

* Denotes a requirement, the level of which has not been established

## RIBOFLAVIN

Riboflavin is a yellow-brown crystalline pigment. The vitamin is very slightly soluble in water but soluble in alkali. It is insoluble in most organic solvents. Riboflavin is stable to oxidizing agents in strong mineral acids and in neutral aqueous solution. It is also stable to dry heat but is irreversibly decomposed on irradiation with ultraviolet rays or visible light, breaking down to lumiflavin. Riboflavin phosphate is the chemically active group of Warburg's 'yellow' enzyme.

### POSITIVE FUNCTIONS

Riboflavin functions in the tissues in the form of flavin adenine dinucleotide (FAD) or as flavin mononucleotide (FMN). The flavo-proteins function as enzymes of tissue respiration and are involved in hydrogen transport to catalyze the oxidation of reduced pyridine nucleo-tides (NADH and NADPH). Thus, they function as coenzymes for many oxidases and redunctases such as cytochrome e reductase, D- and L-amino acid oxidases, xanthine and aldehyde oxidase, succinic dehydrogenase, glucose oxidase and fumaric dehydrogenase. Riboflavin is also involved with pyridoxine in the conversion of tryptophan to nicotinic acid and is most important in the respiration of poorly vascularized tissues such as the cornea of the eye. Riboflavin is involved in the retinal pigment during light adaptation and lack of it causes impaired vision and photophobia in experimental animals, including fish.

### DEFICIENCY SYNDROME

Tissue riboflavin is exhausted in young salmonids after 10 to 12 weeks on riboflavin-deficient diets in 10-15 °C water systems. Poor appetite and poor diet efficiency are the first signs, followed by photophobia, mono or bilateral cataract's, corneal vascularization, eye haemorrhage, incoordination, and general anaemia. Dark pigmentation, coupled with striated constrictions of abdominal wall in salmon have been noted. Skin atrophy has been reported for some fish species and abnormal pigmentation of both skin and iris has been noted. Replacement of riboflavin in the diet reduced the symptoms except when cataracts have developed. This irreversible condition will continue in monolateral cataracts throughout the life of the fish, whereas bilateral cataracts largely result in eventual starvation and death of the afflicted animal. The first specific signs have consistently appeared in and about the eyes of salmonids, carps and catfish.

## REQUIREMENTS

The requirements of fish for riboflavin under experimental conditions in 10-15°C water supplies Values for the trout are slightly lower than those reported for salmon. The requirements may vary depending upon the balance of other dietary ingredients, caloric density and environmental conditions under which the fish is raised. The values established under these standard test conditions should serve as tentative-levels which will satisfy biological demands for normal growth, health, and physiological function. Most of these studies have been made on very young fish, often initially feeding fry, with the logical assumption that their vitamin needs would be more than that of larger fish having advanced metabolic enzyme systems, capable of synthesizing at least some of these vitamins to partially meet the requirements.

## SOURCES AND PROTECTION

Riboflavin is widely distributed in plants and in animal glandular tissues. Milk, liver, kidney, heart, yeast, germinated grains, peanuts, soybeans and eggs are rich sources. Keeping feeds from sunlight or intense artificial light is necessary to minimize loss of the vitamin by conversion to lumiflavin.

## ANTIMTABOLITES AND INACTIVATION

When the ribose group in the molecule is replaced by other groups, analogues are formed which either have reduced activity or become antimetabolites. Galactoflavin is an antagonist to riboflavin and inhibits growth of rats when the diet contains this compound. Flavin monosulphate inhibits D-amino acid oxidase and appears to act as a competitor and inhibitor for growth of Lactobacillus casei. The hydroxyethyl analogue is an antagonist for riboflavin function in both rats and bacteria and also shows antifungal activity.

## CLINICAL ASSESSMENT

Liver tissue of actively feeding sea salmon contains 6-8 g of riboflavin/g of wet tissue. In a freshwater environment, young fish fed test diets in 15°C water systems showed liver storage of 3.5-4.0 ωg/g. Riboflavin content of blood plasma does not change significantly in riboflavin deficiency in experimental animals.

However, the erythrocyte riboflavin content has been reported at around 10 g/100 ml blood for man on low riboflavin intake and approximately twice that level on high intake. Urinary excretion of riboflavin has been used clinically to indicate dietary status, with 200ωg/24 it suggesting adequate intake and less than 100ωg indicating low intake. Excretion of 50ωg or less daily is strong indication of extended dietary deficiency. Several studies show that excretion of less than 200ωg of riboflavin/g of creatinine would be indicative of deficiency.

## PYRIDOXINE

A new factor which could cure dermatitis in rats was reported by Györgi in 1935 and was named vitamin B6. The name pyridoxine is also attributed to Györgi. The active material was isolated in 1939 and pyridoxine was subsequently synthesized. First quantitative requirements for fish were described in 1944. Pyridoxine deficiency has since been reported in trout and in salmon.

### Pyrodoxine Hydrochloride

Compounds which have vitamin B6 activity include: pyridoxine, pyridoxal and pyridoxamine. Pyridoxine hydrochloride is readily soluble in water and is, heat stable in either acid or alkaline solution. It acts as a coenzyme in a number of enzyme systems. Pyridoxine is sensitive to ultraviolet light in neutral or alkaline solutions. Pyridoxamine and pyridoxal in dilute solutions are labile compounds which are rapidly destroyed on exposure to air, heat, or light. Therefore, most vitamin supplementation is in the form of pyridoxine hydrochloride and analysis for pyridoxine activity by microbiological assay of diet ingredients probably measures pyridoxal phosphate and other intermediates as well.

## Positive Functions

Pyridoxal phosphate is the coenzyme, codecarboxylase, involved in the decarboxylation of amino acids. It is also the co-factor of the 22 different transaminases present in animal tissues. As codecarboxylase, pyridoxal phosphate has been shown to participate in the decarboxylation of 5-hydroxytryptophan to produce 5-hydroxytryptamine or serotonin. As a co-factor of the enzyme desulphydrase converts cysteine to pyruvic acid.

Pyridoxal phosphate is the co-factor for the synthesis of -amino-levulinate, the latter a precursor of haeme. Many neuro hormones require pyridoxal phosphate as a coenzyme in their synthesis. Pyridoxine is also involved in fat metabolism, especially of the essential fatty acids. It is involved in the synthesis of messenger RNA which determines amino acids sequence in polypeptide synthesis. Pyridoxine plays a most important role in protein metabolism and, as a result, carnivorous fish have stringent requirements for the vitamin in the diet and stores are rapidly exhausted.

## Deficiency Syndrome

Signs of pyridoxine deficiency in fish Since salmonids, ictalurids, and very young cyprinids are carnivorous, with protein requirements for young animals between 40-50 percent of the ration, pyridoxine stores are rapidly exhausted when fish are held on pyridoxine-deficient rations. Acute deficiency signs occur in salmon after 14-21 days on a high protein diet of pyridoxine and the entire population dies within 28 days in 12-15 C water. Deficiency is followed by general nervous disorders, and alteration in control of melanophore contraction. Postmortem rigor mortis occurs very rapidly. Rapid and gasping breathing with flexing of the opercules is a common observation and oedema in the peritoneal cavity with colourless serous fluid occurs in fish on some experimental treatments. Salmon, trout, carp and yellowtail exhibit premortem rigor a few hours before death. Recovery at late stage deficiency is very unlikely unless fish are injected with pyridoxal phosphate. Handling the animals generally induces more damage than vitamin administration will correct. Recovery among fish still feeding is equally rapid and dramatic upon administration of pyridoxine hydrochloride in the diet. Deficiency signs disappear within a day or two. Erythrocyte and plasma transaminase activities reflect the deficiency state.

### Requirements

Pyridoxine requirements for trout, salmon, carp, and yellowtail Supplementing commercial fish diets with pyridoxine is desirable because of the important role the vitamin plays in protein metabolism.

### Sources and Protection

Good sources of pyridoxine are yeast, whole cereals, egg yolk, liver and glandular tissues. Pyridoxine compounds in phosphorylated form present in agricultural products are fairly stable but are sensitive to ultraviolet radiation. Some pyridoxal phosphate will be lost on exposure to air. Free forms or pyridoxal and pyridoxamine are rapidly destroyed by air, light and heat when moist.

### Antimetabolites and Inactivation

Antagonists may compete for reaction sites of the apoenzyme or may react with pyridoxal phosphate to form inactive compounds. Deoxypyridoxine is a potent pyridoxine antagonist because of competition for apoenzyme sites but is a useful agent to accelerate pyridoxine deficiency in experimental animals. This same compound inhibits tyrosine decarboxylase. Methoxypyridoxine is another antagonist. Oxopyrimidine (2-methyl-4-amino-5-hydroxymethyl-pyrimidine) produces liver damage and inhibits glutamate decarboxylase.

## Clinical Assessment

Plasma and erythrocyte transaminase activity in most instances reflects pyridoxine status of fish although high tryptophan load in the diet tends to increase the pyridoxine requirement.

Liver storage measured by microbiological assay showed 5-6 V-g of pyridoxine activity/g in fresh sea salmon liver; whereas, fingerling salmon fed a 50 percent protein diet in freshwater had 2-3 mg/g of wet tissue. Assay of diet for pyridoxine by microbiological methods gives a truer representation of total pyridoxine activity including intermediates.

## Pantothenic Acid

A chick dermatitis was cured by Elvehjem and Köehn in 1935 using a factor containing -alanine. The term pantothenic acid was given earlier by Williams to a yeast growth factor which was later recognized as the anti-dermatitis factor. Pantothenic acid was synthesized by Stiller in 1940. Phillips observed clubbed gills in trout fed pantothenic acid-deficient diets in 1945. Rucker detected the same condition in salmon fed low pantothenic acid diets. Pantothenic acid may be considered as dihydroxydimethylbutyric acid bonded to -alanine. The free acid is a yellow, viscous oil and therefore the compound generally used in fish diet preparation is the calcium salt. This salt is a white crystalline powder readily soluble in water, mild acid, and is almost insoluble in organic solvents. It is stable to oxidizing and reducing agents and to autoclaving, but is labile to dry heat, hot alkali, or hot acid. Pantothenol has almost as much activity as pantothenic acid for growth of chicks. Pantothenic acid acetate, benzoate, and diphosphate esters are biologically active for animals but not for lactic acid bacteria. The optical isomer L-pantothenic acid, appears physiologically inert. Some organisms may utilize a portion of the molecule. Bacteria appeared to require only the dihydroxydimethyl-butyric acid, and some yeasts utilize only g-alanine. Animals, however, need the entire pantothenic acid molecule or its reduced alcohol form.

## Positive Functions

Pantothenic acid is part of acetyl coenzyme A which occurs in many enzymatic processes involving 2-carbon compounds. It has been shown to be required by all animal species studied and by many micro-organisms.

The acetyl coenzyme A system is involved in the acetylation of aromatic amines and choline; condensation reactions for synthesis of acetate, fatty acids, and citrate; the oxidation of pyruvate and acetaldehyde; and is essential for the development of the central nervous system. The 2-carbon fragment called 'active acetate', or acetyl coenzyme A, is an essential intermediate in metabolism. It is involved in acylation of acetate, succinate, benzoate, propionate and butyrate. Pantothenic acid is involved in adrenal function and for the production of cholesterol. Coenzyme A is also involved in many other steps of intermediate metabolism of carbohydrates, fats and proteins.

## Deficiency Syndromes

Deficiency signs for pantothenic acid Under standard test conditions with deficient diet fed in 10-15°C water systems, salmon and trout exhaust pantothenic acid stores within 8-12 weeks. Fish stop feeding and close examination of gill filament show proliferation of epithelial surface plus swelling and clubbing together of the filaments and lamellae. The opercules become distended and the surface of the gills is often covered with an exudate. Fish become prostrate or sluggish. Necrosis, scarring and cellular atrophy of the tender gill elements occur and anaemia develops after prolonged deficiency. The same type of symptom has been observed in salmon, trout, eel, carp and catfish. After replacement of pantothenic acid in the diet, recovery is rapid for those fish still feeding and gross deficiency symptoms disappear after about four weeks on the recovery diet although evidence of necrosis and scarring remains.

## Sources and Protection

Good sources for pantothenic acid are cereal bran, yeast,

liver, kidney, heart, spleen and lung. Fish flesh is a relatively rich source, although the content is only about 20 percent of pantothenic acid found in animal glandular tissue. Royal jelly probably contains the greatest amount as it contains over 500ωg of pantothenol/g dry weight. The calcium or sodium salt of pantothenic acid is relatively stable and can be incorporated into either moist or dry fish diets. Some loss is incurred during autoclaving and excessive heat should therefore be minimized during diet preparations. Since the free acid is labile to heat and also to acid and alkali, some loss can be expected during moist diet preparation or during storage. Certain cereal brans may have pantothenic acid bound in a form unavailable to fish because of the low digestibility coefficient and should not be relied upon as the sole pantothenic acid source in the diet.

## Antimetabolites and Inactivation

Since pantothenic acid affects the respiration of many types of cells, compounds like 6-mercaptopurine, 2,6-diaminopurine, and 8-azaguanine, which inhibit growth of tumors are antagonistic toward pantothenic acid. Pantoyltaurine is an antimetabolite of pantothenic acid and has been used to accelerate deficiency syndromes in experimental animals. Methyl- -pantothenic acid was reported to interfere with the formation of acetyl coenzyme A and accelerates deficiency syndromes in animals.

This compound inhibits sulfanilamide acetylation in pigeon liver homogenates but did not prevent citric acid formation. Pantothenic acid can be used to overcome these inhibitory effects, including the reversal of mitosis blockage by 7-mecaptopurine in animals. High levels of calcium pantothenate in the diet also has a transient effect on the migratory urge of salmon.

## Clinical Assessment

Assay of pantothenic acid content in the diet may be misleading unless care is exercised in proper hydrolysis of the raw materials being assayed since pantothenic acid is only slowly liberated by normal hydrolytic procedures. Complete hydrolysis with enzyme preparations will liberate all of pantothenic acid from biologically active material of glandular tissues, fish flesh, yeast, and bran.

Salmon feeding actively in the oceans showed liver pantothenic acid content of 18-20 /g of fresh tissue. Young chinook and coho salmon fingerlings reared in fresh water at 12-15°C showed maximum liver storage at about 14-16ωg of pantothenic acid/g of fresh liver tissue. Load tests and acetylation reactions of sulfanilamide have been used in man and other experimental animals but have not yet been extended for assessment of pantothenic acid status in fish.

## NIACIN

Nicotinic acid was synthesized in 1873 but was left on the shelf as an organic compound unrelated to the severe pellagra afflictions occurring throughout the world at that time. Sixty years later the compound was shown to be present in coenzymes I and II and two years thereafter Elvehjem cured 'black tongue' in dogs with the vitamin.

### Nicotinic Acid

Niacin is the preferred nomenclature. Nicotinic acid amide or niacinamide is the common form in which the vitamin is physiologically active. Niacin is a white, crystalline solid, soluble in water and alcohol. It is stable in the dry state and may be autoclaved for short periods without destruction. It is also stable to heat in mineral acids and alkali. Niacin is both a carboxylic acid and an amine and forms quaternary ammonium compounds because of its basic nature. Acidic characteristics include salt formation with alkali. Niacin can be esterified easily, then converted to amides. Niacinamide is a crystalline powder soluble in water and ethanol and the dry material is stable up to about 60°C. In aqueous solutions it is stable for a short period when autoclaved. It is the form in which the vitamin is normally found in niacinamide adenine dinucleotide (NAD) and in niacinamide adenine dinucleotide phosphate (NADP).

## POSITIVE FUNCTIONS

The major function of niacin in NAD and NADP is hydrogen transport in intermediary metabolism. Most of these enzyme systems function by alternating between the oxidized and reduced state of the coenzymes NAD-NADH and NADP-NADPH.

Oxidation-reduction reactions may be anaerobic as when pyruvate acts as the hydrogen acceptor and lactate is formed, or the reactions may be coupled to electron transport systems with oxygen as the ultimate hydrogen acceptor; such aerobic reactions occur in respiration. Both NAD and NADP are involved in the synthesis of high energy phosphate bonds which furnish energy for certain steps in glycolysis, in pyruvate metabolism, amino acid and protein metabolism, and in photosynthesis.

An interrelationship between thiamine and niacin exists since both vitamins are coenzymes in intermediary metabolism whereby food is oxidized to furnish energy for physiological functions, to maintain homeostasis, and for body temperature maintenance in homeotherms.

## DEFICIENCY SYNDROME

Stores of niacin are more slowly exhausted under experimental conditions than are some of the other vitamins resulting in less defined and more slowly developing symptoms. Niacin deficiencies in fish were experimentally induced in the late forties and early fifties by using basal diets which had a low niacin content. Loss of appetite and poor food conversion were the first signs noted.

Then the fish turned dark and went off feed, followed by the appearance of lesions in the colon, erratic motion, oedema of the stomach and colon and muscle spasms while fish were apparently resting. A predisposition to sunburn in fish confined in the open in shallow ponds or raceways Common carp showed a congestion of the skin with subcutaneous hemorrhages. Common symptoms of niacin deficiency in most fish studied were muscular weakness and spasms, coupled with poor growth and poor food conversion.

### Requirements

In homeotherms on a balanced test ration the niacin requirement is generally estimated to be about ten times that of the thiamine requirement. These rations generally contain considerable carbohydrate material to furnish energy to maintain body temperature. In fish the requirement appears to be twenty to thirty times that of the thiamine needs determined for the same test conditions and test rations.

This difference may be due to low carbohydrate content of young fish diets and the higher protein content of these rations. Conversion of tryptophan to niacin occurs in the mammalian liver and possibly also in the liver of fish. This conversion may account for the slow development of the niacin-deficiency syndrome in fish. However, after 10-14 weeks on diets devoid of niacin, deficiency symptoms did occur in several species of fish. The symptoms were reduced by replacement of niacin in the ration even when high protein diets containing an excess of tryptophan were fed. Too much niacin inhibits growth.

### Sources and Protection

Niacin is found in most animal and plant tissues. Rich sources are yeast, liver, kidney, heart, legumes, and green vegetables. Wheat contains more niacin than corn and the vitamin is also found in milk and egg products. The vitamin is very stable since it is generally found in coenzyme form in raw materials. Niacin added to the diet as a supplement remains relatively unaltered during diet manufacture, processing, and storage.

### Antimetabolites and Inactivation

Pyridine-3-sulphonic acid and 3-acetylpyridine are compounds structurally related to niacin and are antimetabolites for this vitamin in animals and in micro-organisms. Additional niacin can overcome the anti-metabolic effect.

Deficiency symptoms in rats may be induced by 6-amino niacinamide. The symptoms are reversed by addition of ten times more niacinamide than the antimetabolite. Thioacetamide has been reported to be a niacin antagonist in fish.

### Clinical Assessment

Salmon feeding actively in the oceans showed liver niacin content of 70-80ωg of niacin/g of fresh tissue. About half of this amount is present in fingerling salmon raised in fresh water at 12-15°C and fed test rations containing 40-50 percent protein and niacin supplements 500-750 mg/kg dry diet. Urinary metabolities of niacin have been measured in other animals on standard niacin load in test rations containing a standard tryptophan load. The technique is well developed to measure the N1-methyl derivative in mammalian urine. These data have not been reported for fish but metabolism chambers are available for collecting branchial and urinary wastes from large fish intubated with different diet material.

## BIOTIN

Certain foods, notably liver and kidney, were found to be protective against a form of dermatitis resulting from the consumption of egg white by rats, in the early thirties. It was soon learned that a specific component of egg albumin, avidin, rendered dietary biotin unavailable, hence producing the symptoms. Biotin was variously called coenzyme R and 'vitamin H'. It was isolated by Du Vigneaud in 1941 and synthesized by workers of Merck and Company in 1943. Biotin was once thought to be part of factor H for fish. Blue slime patch disease due to biotin deficiency was reported in trout.

### Biotin

Biotin is a monocarboxylic acid slightly soluble in water and alcohol and insoluble in organic solvents. Salts of the acid are soluble in water. Aqueous solutions or the dry material are stable at 100 C and to light.

The vitamin is destroyed by acids and alkalis and by oxidizing agents such as peroxides or permanganate. Biocytin is a bound form of biotin isolated from yeast, plant, and animal tissues. Other bound forms of the vitamin can generally be liberated by peptic digestion. Oxybiotin has partial vitamin activity but oxybiotin sulphonic acid and other analogues are antimetabolites inhibiting the growth of bacteria.

Avidin, a protein found in raw egg white, binds biotin and makes it unavailable to fish and other animals. Heating to denature the protein makes the bound biotin available again to the fish.

Biocytin or -biotinyl lysine (the epsilon amino group of lysine and the carboxyl of biotin being combined in a peptide bond) is hydrolyzed by the enzyme biotinase making the protein-bound biotin available.

## POSITIVE FUNCTIONS

Biotin is required in several specific carboxylation and decarboxylation reactions, including the carboxylation of pyruvic acid to form oxaloacetic acid. It is part of the coenzyme of several carboxylating enzymes fixing $CO_2$ such as propionyl coenzyme A involved in the conversion of propionic acid to succinic acid in methylmalonyl coenzyme A. Biotin is also involved in the conversion of acetyl $CO_2$ to malonyl coenzyme A in the formation of long chain fatty acids. It has possible involvement in citrulline synthesis and may have effects on purine and pyrimidine synthesis. It is involved in the conversion of unsaturated fatty acids to the stable cis form in the synthesis of biologically active fatty acids.

## DEFICIENCY SYNDROME

Some signs of biotin deficiency in salmonids are skin disorders, muscle atrophy, lesions in the colon, loss of appetite, and spastic convulsions. Haematology discloses fragmentation of erythrocytes. Poor growth is a common symptom and has been reported for salmonids, common carp, goldfish (Carassius auratus) and eel. Blue slime patch disease in brook trout deficient in biotin appears typical for this species. Fish reared in 10 -15 C water exhaust biotin stores in 8-12 weeks and the first signs are anorexia, poor food conversion, and general listlessness before the more acute deficiency symptoms become detectable.

### Sources and Protection

Rich sources of biotin are liver, kidney, yeast, milk products, and egg yolks. Nut meats contain good supplies of biotin. The diet should be protected from strong oxidizing agents or conditions which promote oxidation of ingredients. Raw egg white should not be incorporated into moist fish diets. Cooking will inactivate the avidin.

### Antimetabolites and Inactivation

Many biotin homologues with different side chain lengths inhibit the growth of bacteria. Oxybiotin, in which sulphur is replaced by oxygen, has about the same biological activity as the natural biotin. On the other hand, oxybiotin sulphonic acid inhibits biotin activity.

### Clinical Assessment

Measurement or urinary excretion of biotin in animals does not prove a good clinical method since biotin is synthesized by several organisms in the gut. In fish, this technique needs to be explored and may well be valuable as a clinical tool since the gastrointestinal tract of many freshwater fish contains only small quantities of bacteria. Biotin is one of the most expensive vitamins to add to fish rations. Salmon feeding actively in the oceans had liver biotin concentrations of 10-12 mg of wet liver tissue. The concentrations in the liver of young salmon fingerlings in fresh water fed test diets containing an excess of biotin were between 6-8 mg of the vitamin/g of tissue. Fish with these levels of biotin in the liver should probably be in sound biotin nutritional status.

## FOLIC ACID (FOLACIN)

A factor effective in curing metaloblastic anaemia in monkeys was found in yeast and liver concentrates and was designated vitamin M in 1935. In 1939 an anti-anaemic factor was found in liver and called vitamin BC. These were later shown to be the same substance with folic acid as the active ingredient. Folic acid was synthesized in 1946 and was soon used in fish diets for preventing purified diet anaemia.

## Folic Acid

Folic acid crystallizes into yellow spear-shaped leaflets which are soluble in water and dilute alcohol. It can be precipitated with heavy metal salts. It is stable to heat in neutral or alkaline solution, but unstable in acid solution. It deteriorates when exposed to sunlight, or during prolonged storage. Several analogues have biological activity including pteroic acid, rhizopterin, folinic acid, xanthopterin and several formyltetrahydropteroyl-glutamic acid derivatives. These have closely allied ring structures and many have been isolated as derivatives in various animals or microbiological preparations. One simple form, xanthopterin, present in the pigments of insects, is shown below and is of special interest because of early work with this compound as the anti-anaemic factor H for fish.

## Positive Functions

Folic acid is required for normal blood cell formation and is involved as a coenzyme in one-carbon transfer mechanisms. In the presence of ascorbic acid, folic acid is transformed into the active (5-formyl-5, 6,7,8) tetrahydrofolic acid. Folic acid is involved in many one-carbon metabolism systems such as serine and glycine interconversion, methionine-homocysteine synthesis, histidine synthesis, and pyrimidine synthesis.

Several coenzyme forms of the active vitamin have been isolated. Folic acid is involved in the conversion of megaloblastic bone marrow to normoblastic type. It has a role in blood glucose regulation and improves cell membrane function and hatchability of eggs.

## Deficiency Syndrome

Macrocytic normochromic anaemia occurs in several experimental animals, including fish fed diets devoid of the folic acid. Increasing numbers of senile cells are observed as the deficiency progresses until only a few old and degenerating cells are found in the blood of deficient fish. Anterior kidney imprints disclose only adult cells and no preforms present. Other signs observed have been poor growth, anorexia, general anaemia, lethargy, fragile fins, dark skin pigmentation, and infarction of spleen.

## SOURCES AND PROTECTION

Yeast, green vegetables, liver, kidney, glandular tissue, fish tissue, and fish viscera are good sources of folic acid. Insects contain xanthopterin which has folic acid activity. At one time the yellow pigment of xanthopterin was identified as the fish anti-anaemic factor H, but subsequent experiments showed only partial activity and that folic acid itself was a much more potent antimacrocytic anaemia factor. Insects may contribute significantly to the folic acid requirements of wild fish, but in scientific fish husbandry artificial diets are more reliable sources. Activity is lost during extended storage and when material is exposed to sunlight. Therefore, dry feeds should be carefully protected during manufacture and moist diet rations should be carefully preserved. Both types of fish diets should be fed soon after manufacture to assure minimal loss of folic acid activity.

## ANTIMETABOLITES AND INACTIVATION

One antagonist of folacin is 4-aminopteroylglutamic acid or aminopterin. This material, when incorporated in the diet of guinea pigs and rats, induces anaemia and leucopenia and has been used to treat leukemia in man. Amethopterin (4-amino-N10-methylpteroylglutamic acid) can also be used to induce deficiency by inhibiting purine, pyrimidine, and nucleic acid protection; viz., inhibition of nucleic acid synthesis results in macroytic anaemia.

## CLINICAL ASSESSMENT

Haematology is used as a simple clinical tool to assess haemopoiesis in fish. Anterior kidney imprints easily disclose normal distribution of immature cells and preforms undergoing reticulosis. Salmon actively feeding in the ocean and young salmon fingerlings raised in fresh water and fed diets rich in folacin show liver storage of 3-4ωg of folic acid/g of wet tissue. Microbiological assay is preferred for assessment of total folic acid in dietary raw materials because the total biological activity it measures includes all the various coenzyme forms and folic acid analogues.

Assessment of the dietary intake of folic acid is important for intensive cold water fish husbandry. In pond culture, aquatic and terrestrial insects, algae, etc., may also be available and folic acid in the supplementary feed may not be as critical. Since folic acid is labile in storage, excess amounts are generally added to manufactured feed in anticipation of storage losses.

However, prudent fish husbandry dictates rapid use of manufactured rations with minimum storage. Routine periodic haematology of fish assures proper nutritional status for maximum production and sound health. The author has noted in several series of experiments that when fish diseases occur through inadvertent contamination of the water supply, those groups of fish partially or completely deficient in folic acid were among the first to show acute disease symptoms. Therefore, folic acid must also play an important role in resistance to disease.

## VITAMIN B12

The antipernicious anaemia factor found to be contained in liver was isolated and crystallized in the mid forties. This substance, named vitamin B12 by its discoverers, was later to be recognized as essential for growth of chicken fed diets entirely of plant origin and was designated animal protein factor (APF). When anaemic salmon were injected with crystalline B12 in combination with folic acid and xanthopterin positive haemopoiesis occurred within a few days, and the salmon showed rapid recovery from the anaemia. Vitamin B12 or cynacobalamin has the following approximate chemical structure.

### Cyanocobalamin

The molecule has a planar group and a nucleotide group lying nearly at right angles to one another. This cobalt-containing vitamin has a net charge of one at the central cobalt atom to which is attached a replaceable cyano group. Vitamin B12 is stable to mild heat in neutral solution, but is rapidly destroyed by heating in dilute acid or alkali. Crude concentrates are more unstable and rapidly lose activity. The compound is similar to the porphyrins in its spatial configuration with a central cobalt atom linked to four reduced pyrrole rings in the haeme series. Replacing the cyanide ion with a variety of anions produce derivatives which have comparable biological activities; viz., hydroxocobalamin, nitritocobalamin, chlorocobalamin and sulphatocobalamin.

**Positive Functions**

Cyanocobalamin is involved with folic acid in haemopoiesis. It is required by many micro-organisms and is a growth factor for many animals. The animal protein factor present in fish and animal by-products was not recognized until crystalline vitamin B12 was injected into anaemic chinook salmon fingerlings in 1949 and positive haemopoiesis was observed. A coenzyme incorporating vitamin B12 is involved in the reversible isomerization of methyl-malonyl coenzyme A to succinyl coenzyme A and in the isomerization of methylaspartate to glutomate. Cyanocobalamin is involved in the coenzyme for the methylation of homocystine to form methionine. It is also involved in several other one-carbon reactions and in the synthesis of labile methyl compounds. One vitamin B12 containing coenzyme acts in methylation of the purine ring during thymine synthesis. Vitamin B12 is also involved in cholesterol metabolism, in purine and pyrimidine biosynthesis, and in the metabolism of glycols.

**Deficiency Syndrome**

Deficiency signs in young pigs, chicks, and rats show abnormal blood elements, poor growth, and pernicious anaemia. An intrinsic factor is necessary for good absorption of the vitamin from the gut. This factor is a low molecular weight mucoprotein which normally occurs in gastric juice, and especially in hog gut mucosa. Pernicious anaemia in chinook and coho salmon is characterized by fragmented, erythrocytes with many aberrant forms present. Haemoglobin levels are inconsistent and erythrocyte counts have a range extending from frank anaemia to a near normal blood pattern. Cyanocobalamin stores in fish tissues are slowly exhausted and only after 12-16 weeks on test do the symptoms appear in deficient salmon populations. Poor appetite, poor growth, poor food conversion, and some dark pigmentation can be observed before frank anaemia is detected.

### Sources and Protection

Rich sources of vitamin B12 are found in fish meal, fish viscera, liver, kidney, glandular tissues, and slaughter house wastes. Since vitamin B12 is labile on storage, and in mild acid solution is easily destroyed by heating, care must be exercised in diet preparation containing flesh or meat scraps.

### Antimetabolites and Inactivation.

The vitamin B12 coenzymes are very unstable under light, which rapidly decomposes the coenzymes. Photo-sensitivity is increased in dilute acid solutions.

### Clinical Assessment

Generalized anaemia with fragmentation of erythrocytes and extremely variable haemoglobin levels and erythrocyte counts indicate possible B12 deficiency. Prompt response in individual fish is obtained by injecting B12 alone or in combination with folic acid in the ratio of 1 part vitamin B12 to 100 parts folic acid. Careful interpretation of haemotological data will enable one to distinguish one form of anaemia from the other.

## ASCORBIC ACID

The isolation of ascorbic acid was due to Szent-Györgyi of Hungary and C.G. King of the U.S.A. Vitamin C synthesis was accomplished in 1933 after the chemical structure of ascorbic acid was established by British and Swiss workers. McCay and Tunison reported scoliosis in brook trout fed formalin-preserved meat in 1934 and McLaren observe haemorrhages in trout fed rations low in ascorbic acid. It was not until the sixties that a critical need for L-ascorbic acid by trout and salmon was demonstrated. L-ascorbic acid is a white, odourless, crystalline compound, soluble in water but insoluble in fat solvents.

It is readily oxidized to dehydroascorbic acid, the less biologically potent form. Ascorbic acid is very stable in acid solution because of the preservation of the lactone ring, but in alkaline solution hydrolysis occurs rapidly and vitamin activity is lost. It is very heat labile and prone to atmospheric oxidation, especially in the presence of copper, iron, or several other metallic catalysts. The reduced form is the most biologically active but several derivatives or salts are obtainable which have varying degrees of ascorbate activity.

## Positive Functions

L-ascorbic acid acts as a biological reducing agent in hydrogen transport. It is involved in many enzyme systems for hydroxylation; i.e., hydroxylation of tryptophan, tyrosine, or proline. It is involved in the detoxification of aromatic drugs and also acts in the production of adrenal steroids. Ascorbic acid is necessary for the formation of hydroxy proline which is a constituent of collagen, a component of intercellular material in bones and soft tissues. Ascorbic acid plays a synergistic role with vitamin E as intracellular antioxidants and free radical traps. The conversion of folic acid to folinic acid requires vitamin C. Ascorbic acid is involved in the formation of chondroitin sulphate and intercellular ground substance. Labelled ascorbic acid fed to fish previously deficient in the vitamin was shown to be rapidly mobilized and fixed in areas of rapid collagen synthesis and became concentrated in the thick collagen of the skin and in cartilagenous bones, as well as in the glands of the anterior kidney. Ascorbic acid is also involved in erythrocyte maturation.

## Deficiency Syndrome

Scurvy with impaired collagen formation, perifollicular haemorrhages, loose teeth and poor bone formation, anaemia, and oedema have been reported in other animals. Deficiency signs in fish are generally related to impaired collagen formation.

Fish soon show hyperplasia of jaw and snout. The same symptoms have been observed in trout, salmon, yellowtail, carp, guppies and char. Histologically, hypertrophy of the adrenal tissues and haemorrhage at the bases of fins have been observed in coho salmon. Deficiency signs cease to develop and new growth becomes normal upon replacement of ascorbic acid in the ration. Anaemia eventually develops in extremely deficient fish and scoliosis and lordosis do not repair but are walled off by new growth around the afflicted areas of the spine when ascorbic acid is once again added to the ration.

## Requirements

In the rainbow trout, reasonable blood and anterior kidney storage levels were obtained with an intake of about 100 mg of vitamin C/kg of dry ration in 10°, 12°, or 15° C water systems. When wound repair experiments were initiated, however, or when fish were exposed to other stress then the requirements doubled or tripled.

When severe abdominal or intramuscular wounds were inflicted, young fish needed at least 500 mg of active ascorbate for tissue repair comparable with control fish receiving 1 g or more of ascorbate in the diet/kg of dry diet. Coho salmon appear to need about half of these requirements for adequate tissue levels and for maximum severe wound repair rates.

The requirement for ascorbic acid is related to stress, growth rate and size of the animal, as well as to the other nutrients present in the diet. A compromise value of about 200 mg of ascorbic acid/kg diet for trout and salmon raised in freshwater systems between 10-15°C would ensure reasonable tissue storage levels and furnish some excess for mild stress conditions and for ascorbic acid loss from the diet through oxidation. Large common carp can synthesize some ascorbate and the requirement for this species may be dependent on fish size and the environment in which they are reared.

**Table: Growth and Tissue Ascorbate**

| Vitamin C diet treatment | Trout | | | Salmon | | |
|---|---|---|---|---|---|---|
| | *Average weight at 24 weeks* | *Ascorbate concentrate* [1] | | *Average weight at 24 weeks* | *Ascorbate concentrate* [1] | |
| | | *blood* | *kidney* | | *blood* | *kidney* |
| *mg/100 g* | *g* | *g/g* | *g/g* | *g* | *g/g* | *s/s* |
| 0 | 2.4 | [2] | [2] | 5.0 | 22.3±2.2 | 89 |
| 5 | 9.6 | 34.4±2.9 | 125 | 6.0 | 30.5±1.2 | 132 |
| 10 | 10.6 | 34.6±1.3 | 137 | 5.7 | 35.8±1.6 | 265 |
| 20 | 10.1 | 38.8±3.3 | 132 | 6.1 | 34.2±2.3 | 183 |
| 40 | 10.2 | 46.8±6.2 | 162 | 6.3 | 33.7±2.0 | 225 |
| 100 | 10.8 | 51.0±4.6 | 247 | 6.0 | 37.8±2.3 | 321 |

1. Average of five samples for blood (± S.D.) and two for head kidney tissue
2. No fish available for assay

## Sources and Protection

Ascorbic acid is widely distributed in nature with citrus fruits, cabbage, liver, and kidney tissue good sources for the vitamin. Fresh insects and fish tissues contain reasonable amounts of the vitamin. Synthetic ascorbic acid is also readily available. Fish food should be protected from oxidizing agents and kept sealed or frozen until used to prevent loss of the vitamin.

## Antimetabolites and Inactivation

D-ascorbic acid, the optical isomer of the active form, has no vitamin activity and competes for sites of several enzyme reactions mediated by L-ascorbic acid. 6-deoxy-L-ascorbic acid, dehydroascorbic acid and L-glucoascorbic acid have very little activity.

## Clinical Assessment

Ascorbic acid status for experimental animals is normally attempted by tissue ascorbate analysis. Most of the assays previously used measure total ascorbate and not biologically active L-ascorbic acid. Consequently these are fraught with errors and misconception of true vitamin C status. In fish tissues, blood and liver do not adequately reflect the ascorbic acid intake and status, whereas assay of the anterior kidney which contains adrenal tissue provides a fairly representative picture of tissue storage of the vitamin.

Stress rapidly reduces the ascorbic acid content of this tissue with concurrent production of adrenal steroids. Conversely, dietary repletion is reflected by up to four or five-fold storage levels from the deficient state.

Examination of fragile support cartilage in the gill filaments under low magnification will detect early hypovitaminosis before clinically acute symptoms become noticeable.

However, the best tissue for routine clinical analysis to assess vitamin C status in trout and salmon appears to be the anterior kidney with samples selected from the junction of the two forward wings. Tissues are blotted free of blood with filter paper, and then assayed for total ascorbate by one of the improved quick methods to determine total ascorbate.

**Inositol**

Muscle 'sugar' was discovered by Scherer in 1850 and was characterized as inositol in 1887.

Inositol was shown to be an alopecia (a form of hairlessness) preventing factor for mice in 1940. Poor growth and poor food passage in inositol-deficient fish was observed in salmon and carp.

**Inositol**

Seven optically inactive and two optically active isomers of hexahydroxycyclohexane can exist. One of the optically active forms, myo-inositol, is a white crystalline powder soluble in water and insoluble in alcohol and ether. The material can be synthesized, but is easily isolated from biological material in free or combined forms. The mixed calcium-magnesium salt of the hexophosphate is phytin. Isomers have little biological activity but will compete in chemical reactions. Inositol is a highly stable compound.

## POSITIVE FUNCTIONS

Myo-inositol is a structural component in living tissues. It has lipotropic action by preventing accumulation of cholesterol in one type of fatty liver disease and is involved with choline in maintaining normal lipid/metabolism. It is a growth-promoting substance for micro-organisms and prevents an alopecia in mice. It is a reserve carbohydrate in muscle as well as a major component of phosphoglycerides in animal tissues.

## DEFICIENCY SYNDROME

Poor growth, increased gastric emptying time, oedema, dark colour, and distended stomachs are symptoms observed in salmon, trout, carp and catfish held for long periods on inositol-deficient test rations. A 'spectacle-eye' condition for rats has not been observed under the experimental conditions used in fish studies. The major deficiency sign is inefficiency in digestion and food utilization and concomitant poor growth leading to a population of fish with distended abdomens.

## SOURCES AND PROTECTION

Myo-inositol occurs ubiquitously in large amounts wherever biological tissue is found. Wheat germ, dried peas, and beans are rich sources. Brain, heart, and glandular tissues are very good sources of biologically active inositol. Citrus fruit pulp and dried yeast also contain inositol. The compound is stable.

## ANTIMETABOLITES AND INACTIVATION

Seven optically inactive and one optically active but biologically inactive stereo isomers occur. Since inositol is synthesized in the biologically active form by many microorganisms in the gut, only large quantities of chemically synthesized inactive isomers added to diets would interfere with the metabolism of inositol for growth and normal physiological function.

Biologically inactive stereoisomers of myo-inositol do not compete for critical sites in metabolism. Methyl derivatives and mono-, di- and triphosphoric acid esters occur naturally. Salts of the hexaphosphate or phytin make the bound inositol practically unavailable to the animal.

## CLINICAL ASSESSMENT

Assessment in fish has been based on lack of deficiency signs coupled with the most efficient food conversion. Salmon feeding actively in oceans show 1-1.5 mg of inositol/g of fresh liver tissue and young fingerlings raised in freshwater at 10-15°C had 600-700ωg/g of liver tissue. An alternate, better assessment may be based on a standard muscle section or whole carcass analysis for free or for bound inositol. Projection of inositol intake from normal fish diet ingredients should indicate an excess of this particular vitamin.

## CHOLINE

Methylation as a basic metabolic process was postulated by Hoffmeister. Methyl transfer was shown in vivo by Thompson and the interrelationships between choline, methionine, and homo-cystine were shown by du Vigneaud in 1939-42. Trout fed low choline rations developed haemorrhagic kidneys and salmon showed an aversion to food in choline-deficient diet.

### Choline

Choline is a very strong organic base and forms many derivatives widely distributed in animal and vegetable tissue. One derivative, acetylcholine, is involved in the transmission of nerve impulses across synapses. Choline is very hygroscopic, very soluble in water, and is stable to heat in acid, but decomposes in alkaline solutions.

### Positive Functions

Choline acts as a methyl donor in trans-methylation reactions. It is a lipotropic and antihaemorrhagic factor preventing the development of fatty livers. It is involved in the synthesis of phospholipids and in fat transport. Acetylcholine transmits the excitory state across the ganglionic synapses and neuromuscular junctions. Choline is essential for growth and good food conversion in fish.

### Deficiency Syndrome

Deficiency signs include poor growth and poor food conversion accompanied by impaired fat metabolism. Haemorrhagic kidneys and intestines have been reported in trout and increased gastric emptying time has been observed in salmon.

### Sources and Protection

Rich sources of choline are wheat germ, beans, brain, and heart tissue. Choline hydrochloride, the commercially available form, may inactivate -tocopherol and vitamin K when in direct contact with these vitamins and care should be exercised in selecting properly protected (gelatin coated) fat soluble vitamins in fish diet preparation.

### Clinical Assessment

Choline status of fish can be estimated from the assay of choline content of the dietary ingredients with the absence of deficiency signs. Maximum liver storage may not be the best criteria to determine choline nutritional status but has been used to assess the tentative requirement listed for the two species of salmon.

## p-AMINOBENZOIC ACID

Para-aminobenzoic acid is a white crystalline powder which is water soluble and heat and light stable in aqueous and mild alkaline solution. It is a growth promoting vitamin for micro-organisms which require the vitamin for folic acid synthesis. Large intake has been shown to counteract the antimetabolite effect of sulphonamides in bacterial culture. No positive function or deficiency signs have been observed in fish.

## LIPOIC ACID

Lipoic acid is both fat soluble and water soluble. Its active form is the amide derivative known as lipoamide. Lipoic acid functions as a coenzyme in a -ketoacid decarboxylation. It was discovered independently in several laboratories during the period 1945-50 and shown to be an essential component of multienzyme, the pyruvate dehydrogenase complex. Pyruvate is converted to 'active acetaldehyde' which in turn is picked up by lipoamide. Subsequent oxidation of the aldehyde results in the reduction of lipoamide to a disulphydryl form. The multi-enzyme unit also includes thiamine pyrophosphate, coenzyme A, and flavin adenine dinucleotide. Glandular tissues are good sources of lipoic acid. No requirements have been determined for fish.

## FAT-SOLUBLE VITAMINS

Fat-soluble vitamins A, D, E, and K differ from the water-soluble vitamins in their accumulative action. Little evidence has been recorded for hypervitaminosis with the water-soluble vitamins since these compounds are rapidly metabolized and excreted when intake exceeds liver or tissue storage capacity, but hypervitaminosis is a common occurrence in fish and other animals when large quantities of any one of the fat-soluble vitamins are ingested. Toxicity symptoms involving vitamins A and D are indistinguishable from deficiency symptoms for the same vitamins. On the other hand, symptoms of excess vitamins E and K intake are more discrete. Fish rations may often be enriched with fish oils to increase caloric density of the ration resulting in excessive intake of the fat-soluble vitamins.

### VITAMIN A

A fat-soluble factor that promoted growth in rats by Hopkins and by Osborne and Mendel at the beginning of this century. McCollum and Simmonds cured xerophthalmia, a disease of the eye, with this material. The chemical structure of vitamin A and its relationship to -carotene was shown by Von Euler in 1928. Active vitamin A was synthesized in the mid-thirties. The relationship of the vitamin A alcohols to naturally occurring -carotene Retinene, the aldehyde form of vitamin A, has been isolated from the retina of dark adapted eyes and is involved in vision in dim light. Retinoic acid, which is the oxidized form of vitamin A alcohol, has been shown to have some vitamin A activity. Vitamin A1 is found in saltwater fish, whereas vitamin A is more abundant in freshwater fish. Interconversion of the two forms in living fish tissue has been demonstrated. Fish oils contain vitamin A as free alcohols or esters. Vitamin A alcohol occurs as a light coloured viscous oil which is heat labile and subject to air oxidation. Beta-carotene occurs as an orange, crystalline compound which is more stable to heat and oxidation. Vitamin A is water insoluble but is soluble in fat and organic solvent.

## POSITIVE FUNCTIONS

Vitamin A is essential in maintaining epithelial cells, preventing atrophy and keratinization of epithelial tissues. It is combined with a protein in visual purple and is important in night vision. Vitamin A also prevents xerophthalmia in rats and young children. Vitamin A promotes growth of new cells and aids in maintaining resistance to infection.

## THE SYNDROME OF DEFICIENCY OR EXCESS

Hypovitaminosis A is characterized by poor growth, poor vision, keratinization of epithelial tissue, xerophthalmia, night blindness, haemorrhage in the anterior chamber of the eye, haemorrhage of the base of the fins, and abnormal bone formation. Nerve degeneration has been reported in pigs, chickens, rats, rabbits, and ducks, but only occasionally observed in fish after long periods of deficiency.

Hypervitaminosis A fish and in other animals and involved enlargement of liver and spleen, abnormal growth, skin lesions, epithelial keratinization, hyperplasia. of head cartilage, and abnormal bone formation resulting in ankylosis and fusion of vertebrae. Hypervitaminosis A is reflected in very high liver oil vitamin A content and elevated serum alkaline phosphatase. Removal of excess vitamin A from the diet promotes rapid recovery.

## SOURCES AND PROTECTION

Cod liver oil is the best known source of vitamin A

although black sea bass, swordfish, or ling cod oils contain much more of the vitamin. Whale liver oil contains kitol which has little or no biological activity until heated above 200° C when one molecule of biologically active vitamin A is generated per molecule of whale kitol. This biologically inactive kitol may be deposited in the whale to avoid hypervitaminosis A during excessive vitamin A intake. The possibility of hypervitaminosis A occurs when tuna, shark, or ling cod viscera are used in preparation of moist fish diets. Synthetic vitamin A preparations, such as vitamin A palmitate, are available and are often used to supplement rations low in fish meal, fish viscera, or carotenes. Some fish species seem able to utilize b-carotene as a vitamin A source, whereas others are unable to split the g-carotene molecule and vitamin A must be added to the diet.

## CLINICAL ASSESSMENT

Vitamin A status can best be assessed by the absence of deficiency signs and by assay of liver oil for vitamin A content. Assay for the vitamin in blood or plasma has not been found to be useful.

## VITAMIN D

Rickets was induced with test diets by Hopkins, and Mellanby cured the disease in dogs by adding cod liver oil to the ration. Ultraviolet light exposure was shown to have an anti-rachitic effect, and provitamin D activity had been ascribed to ergosterol. Crystalline vitamin D was isolated by Angus and activated 7-dehydrocholesterol was isolated by Windaus. Vitamin $D_2$ or ergocalciferol, is one of several biologically active forms of vitamin D. Vitamin $D_3$ or activated 7-dehydrocholesterol has the chemical formula C27H44O and contains a more simplified, unsaturated 8-carbon side chain. Vitamin $D_3$, also known as cholecalciferol, is formed in most animal tissue by the rupture of one of the ring bonds of 7-dehydrocholesterol by ultraviolet radiation. Cholecalciferol is a white, crystalline compound soluble in fat and organic solvents and is stable to heat and oxidation in mild alkali or acid solution.

## POSITIVE FUNCTIONS

Vitamin D is essential for maintaining calcium and inorganic phosphate homeostasis. It is involved in alkaline phosphatase activity, promotes intestinal absorption of calcium, and influences the action of parathyroid hormone on bone. Fish may sequester calcium from water through the gill membrane; thus the major function of vitamin D for other animals may not be necessary to satisfy calcium requirements for fish.

## SYNDROME OF DEFICIENCY OR EXCESS

Rickets and abnormal bone formation for animals has not been observed in fish fed low vitamin D diets. However little work has been done under carefully controlled conditions with young growing fish and only alteration in alkaline phosphatase activity has been reported on different vitamin D diet intake. Hypervitaminosis D, however, has been reported. Brook trout fed large doses of vitamin D showed impaired growth, lethargy, and dark colouration. High intake of vitamin D mobilizes phosphorus and calcium from the bone and tissues and may result in fragile bones, poor growth, and poor appetite related to the nausea in man afflicted with hypervitaminosis D. This area needs to be explored because of the potential for hypervitaminosis D in fish fed diets containing various fish viscera which might contain large amounts of the vitamin D. Tuna liver oil may contain, for example, 100 to 1 000 times as much active vitamin D as cod liver oil.

## SOURCES AND PROTECTION

Vitamin D requirements of many animals can be met by ultraviolet irradiation of the skin where cholesterol derivatives are converted to calciferol. Since the vitamin is fat soluble and accumulates in lipid stores, fish liver oil is a rich source of the material. Content varies tremendously in liver oil, however, with values of about 25 I.U./g present in shark liver oil and over 200 000 I.U./g in albacore tuna liver oil. Cod liver oil contains from 100-500 I.U./g and animal liver contains some vitamin D. One international unit (I.U.) is equal to 0.025 mg of crystalline vitamin D.

## CLINICAL ASSESSMENT

Absorption maxima in the ultraviolet region can be used to detect provitamins D in the non-saponifiable fraction of oils. Concentrated preparations of vitamin D can be assayed by the Carr-Price antimony trichloride reaction when assay is necessary to determine biologically active materials in liver oil of fish on different treatments. The chick assay may not apply to fish liver storage levels because the biologically active form for fish has not been determined. Therefore, clinical assessment for hypervitaminosis A must rely on crude methods for determination of vitamin D in liver oil samples by the Carr-Price reaction or by measuring absorption in the UV spectrum.

## VITAMIN E

The existence of an antisterility vitamin was postulated by Evans and Bishop. This factor was named 'vitamin E' by Sure. The active tocopherol was isolated, characterized, and synthesized by Karrer. Vitamin E is composed of a class of compounds known as tocopherols. One of the most important tocopherols is -tocopherol.

Eight naturally occurring tocopherol derivatives have been isolated and all belong to the D series. Synthetic -tocopherol is a racemic DL- -tocoperhol mixture. The derivatives of tocol or of tocotrienol are named, alpha, beta, gamma, delta, epsilon, eta, zeta1 and zeta2-tocopherol. The pure tocopherols are fat-soluble oils which are capable of esterification to form crystalline compounds. The tocopherols are stable to heat and acids in the absence of oxygen, but are rapidly oxidized in the presence of nascent oxygen, peroxides, or other oxidizing agents.

The tocopherols are sensitive to ultraviolet light and are excellent antioxidants in the free form. The esters are more stable and are commonly used as dietary supplements-anticipating hydrolysis in the gut and absorption of the free alcohol to act as an active intra- and intercellular antioxidant. Ethyl derivatives on the aromatic ring are also active. Oxidation products of a-tocopherol can be reduced with hydrosulphite to -tocopherylhydroquinone or, in the presence of ascorbic acid, to -tocopherol.

**Table: The Tocopherols**

| | |
|---|---|
| Tocopherol (alpha) | 5,7,8-trimethyltocol |
| Tocopherol (beta) | 5,8-dimethyltocol |
| Tocopherol (gamma) | 7,8-dimethyltocol |
| Tocopherol (zeta2) | 5,7-dimethyltocol |
| Tocopherol (eta) | 7-methyltocol |
| Tocopherol (delta) | 8-methyltocol |
| Tocopherol (epsilon) | 5,8-dimethyltocotrienol |
| Tocopherol (zeta1) | 5,7,8-trimethyltocotrienol |

## POSITIVE FUNCTIONS

The tocopherols act as extra and intracellular antioxidants to maintain homeostasis of labile metabolites in the cell and tissue plasma. As physiological antioxidants, these usually protect oxidizable vitamins and labile unsaturated fatty acids. Vitamin E prevents encephalomalacia in chicks, erythrocyte haemolysis in several animals and steatitis in mink, pigs, and farm animals. The tocopherols also prevent exudative diathesis in chicks, white muscle disease in lambs and calves, and dietary liver necrosis in rats. They act with selenium and with vitamin C for normal reproductive activity and are involved in prevention of nutritional muscular dystrophy in the chick, the yellowtail and carp. The tocopherols act as free radical traps to stop the chain reaction during peroxide formation and stabilize unsaturated carbon bonds of polyunsaturated fatty acids and other long-chain labile compounds.

Vitamin E is involved in the maintenance of normal blood capillary permeability and the integrity of heart muscle. It was first shown to be involved in prevention of sterility and foetal resorption in rats. It may likewise affect embryo membrane permeability and hatch-ability of fish eggs.

## SYNDROME OF DEFICIENCY OR EXCESS

One of the first signs is erythrocyte fragility closely followed by anaemia, ascites, xerophthalmia, poor growth, poor food conversion, epicarditis, and ceroid deposits in spleen and liver. Muscle dystrophy and xerophthalmia have been in yellowtail and carp. Non-specific forms of degenerative conditions have been described in several species of fish fed large quantities of polyunsaturated fatty acids with inadequate tocopherol in the ration. Hypervitaminosis E results in poor growth, toxic liver reaction, and death.

## SOURCES AND PROTECTION

Vegetable oils are rich sources of tocopherols. Synthetic - tocopherol in esterfied acetate or phosphate form is commonly used as a diet supplement. These esters are much more stable than the free form which is rapidly lost by air oxidation or in the presence of labile compounds like the polyunsaturated fish oils. Addition of antioxidants such as BHT (butylated hydroxyanisole) and BHT (butylated hydroxytoluene) protects fats and other labile compounds in the ration from oxidation, but these antioxidants have no vitamin E activity.

## CLINICAL ASSESSMENT

The erythrocyte fragility test indicates the physiological state of fish. Absence of histologically detectable ceroid in liver and spleen from representative samples in the population is a good clue on the presence of adequate amounts of physiological antioxidants in the fish. A barbituric acid test for oxidation of components has not been applied to fish tissues as a clinical tool for assessment of nutrition state except when liver oils become saturated with very labile polyunsaturated fatty acids such as when feeding squid or saury oils.

These assays are good indicators of the state of oxidation of the finished ration, but the absence of deficiency signs and normal erythrocyte fragility are better clinical tests. Analysis for tocopherol is difficult and time-consuming and only applicable under critical research experiment situation.

## VITAMIN K

The name vitamin K (for 'koagulation') was proposed by Dam. Dam isolated the vitamin from alfalfa and from fish meal in 1939. It was synthesized later that year. The K vitamins $K_2$ contain 6,7, or 9 isoprene units in the side chain which varies from 30-45 carbon atoms. Many isosteres of vitamin K have been identified in animal tissues, plant tissues, and micro-organisms. Although fairly stable compounds, they are destroyed upon oxidation and exposure to ultraviolet radition. Menadione is very reactive and is subject in aqueous media to chemical interaction.

## POSITIVE FUNCTIONS

Vitamin K is involved in the hepatic synthesis of blood clotting proteins - prothrombin and proconvertin. Substituted forms of vitamin K are strongly bacteriostatic and may serve as an alternate defence mechanism for bacterial infections. Vitamin K is structurally related to flavo compounds which function as electron carriers in the electron transport system. The primary role of vitamin K is to maintain a fast normal blood clotting rate which is so important to fish living in a water environment.

## SYNDROME OF DEFICIENCY OR EXCESS

Prothrombin time in salmon fed diets devoid of vitamin K was increased three to five times and, during prolonged deficiency states, anaemia and haemorrhagic areas appeared in the gills, eyes, and vascular tissues. Increased blood clotting time has also been reported for other fish reared on diets with low vitamin K content. Interrelationships with other vitamins have not been documented in fish where the primary deficiency signs are slow blood clotting and haemorrhage, severe anaemia, or death in wounded fish. Haemorrhagic areas often appear in fragile tissues such as the gills.

## SOURCES AND PROTECTION

Vitamin K sources are green, leafy vegetables. Alfalfa leaves are among the best sources of vitamin K. Low levels are found in soybeans and animal liver. Synthetic menadione is also available. Although vitamin K in ground alfalfa is fairly stable, synthetic material should be protected from exposure to ultraviolet light and to excessive oxidizing or reducing conditions. The use of rapidly cured dehydrated alfalfa is essential to minimize formation of the physiological antagonist, dicumarol.

The diet should be kept dry, prepared with minimum exposure to air oxidation, and fed as soon as practicable after manufacture to minimize vitamin K loss.

## ANTIMETABOLITES AND INACTIVATION

Dicumarol was isolated from spoiled sweet clover hay and shown to prevent the normal function of vitamin K in maintenance of normal blood clotting times. Early work was with cattle and hogs but was soon extended to rats, other experimental animals, and man.

Dicumarol is an anticoagulant and has been used to prevent thrombosis in animals and man. It is not an antimetabolite competing for vitamin K sites but plays another role in preventing normal blood clotting. Vitamin K counteracts the dicumarol effect quantitatively. Another anticoagulant, warfarin, is a common rat poison which has five to ten times the anticoagulant activity of dicumarol. The effects of warfarin can also be reversed by administration of vitamin K.

## CLINICAL ASSESSMENT

Vitamin K status of fish is determined by measuring blood clotting time. Chemical methods for determining menadione content of active metabolite after storage are available but have not been applied to measure vitamin K activity in the liver or tissues of young, growing fish.

# Index